The Reception of Darwinism in the Iberian World

BOSTON STUDIES IN THE PHILOSOPHY OF SCIENCE

VOLUME 221

THE RECEPTION OF DARWINISM IN THE IBERIAN WORLD

Spain, Spanish America and Brazil

Edited by

THOMAS F. GLICK

Boston University, Boston, U.S.A.

MIGUEL ANGEL PUIG-SAMPER

Centro de Estudios Históricos, Madrid, Spain

and

ROSAURA RUIZ

Universidad Nacional Autónoma de México

SPRINGER SCIENCE+BUSINESS, MEDIA, B.V.

A C.I.P. Catalogue record for this book is available from the Library of Congress.

ISBN 978-94-010-3885-0 ISBN 978-94-010-0602-6 (eBook)
DOI 10.1007/978-94-010-0602-6

Printed on acid-free paper

Spanish Edition: El Darwinisimo en España e Iberoamérica,
©1999, Universidad Nacional Autónoma de México,
Consejo Superior de Investigaciones Científicas y Ediciones Doce Calles.

Roberto Moreno de los Arcos
1943-1996

In memoriam

TABLE OF CONTENTS

PART THREE: THEORETICAL PERSPECTIVES

PREFACE

I

Twenty-five years ago, at the Conference on the Comparative Reception of Darwinism held at the University of Texas in 1972, only two countries of the Iberian world—Spain and Mexico—were represented.[1] At the time, it was apparent that the topic had attracted interest only as regarded the "mainstream" science countries of Western Europe, plus the United States. The Eurocentric bias of professional history of science was a fact. The sea change that subsequently occurred in the historiography of science makes 1972 appear something like the antediluvian era. Still, we would like to think that that meeting was prescient in looking beyond the mainstream science countries—as then perceived—in order to test the variation that ideas undergo as they pass from center to periphery.

One thing that the comparative study of the reception of ideas makes abundantly clear, however, is the weakness of the center/periphery dichotomy from the perspective of the diffusion of scientific ideas. Catholics in mainstream countries, for example, did not handle evolution much better than did their correligionaries on the fringes. Conversely, Darwinians in Latin America were frequently better placed to advance Darwin's ideas in a social and political sense than were their fellow evolutionists on the Continent.

The Texas meeting was also a marker in the comparative reception of scientific ideas, Darwinism aside. Although, by 1972, scientific institutions had been studied comparatively, there was no antecedent for the comparative history of scientific ideas. Since that time, there have been a few such efforts, but they have been sporadic, rather than sustained. One might inquire what accounts for the drag or reticence of the field to involve itself in such efforts. Eurocentrism, as we have indicated, has been part of the problem. But another part has been an inability to identify the variables in play. Here we think that the papers collected here have a positive and creative role to play, because (with the exception of Brazil) the linguistic and associated cultural variables are common to all the cases considered, directing the reader's attention to social and political variables.

[1] Thomas F. Glick, ed., *The Comparative Reception of Darwinism* (Austin, University of Texas Press, 1974; 2nd ed., University of Chicago Press, 1988).

Thomas F. Glick, Rosaura Ruiz, and Miguel Ángel Puig-Samper (eds.), The Reception of Darwinism in the Iberian World, ix-xii.

Iberian Darwin studies were slow in getting started and in some countries reception studies suffered from a mistaken view of the nature of positivism there. In the countries of Latin America, positivism was the great definer of progressive, modernizing political programs. But the notion took hold—because of the overweening influence of João Cruz Costa in Brazil and Leopoldo Zea in Mexico, among others—that positivism of the Comtian variety precluded the reception of Darwinism, which only made gains in the those countries under the sway of Spencer. The problem was first, that those authors had little interest in science and did not regard it as falling under the purview of "philosophy" and, second, they failed to recognize that in those countries with a high Comtian tradition—Brazil and Mexico—the Comtian phase was followed by an equally—if not more—tenacious Spencerian phase, bringing with it—in both instances—the ineluctable introduction of the Darwinian paradigm—first in social thought, then in biological.

II

The papers which follow are a selection of those presented at the "Colloquium on the Reception of Darwinism in Iberoamerica: A Comparative Analysis," held in Cancún, Mexico, November 10-14, 1997, under the directorship of Rosaura Ruiz and the sponsorship of the National Autonomous University of Mexico.

We begin with a series of essays on the reception of Darwin in Latin America. The first three are written by the pioneers of this endeavor. Marcelo Montserrat, whose insistent pursuit of the Darwinian story in Argentina influenced us all, follows the impact of Darwinism in Argentine political debates of the late nineteenth century. Here, as in other Iberoamerican countries, Darwinism played an important role in the creation of ideology. Both Darwinism and Spencerism were used in Argentina to justify the extermination of natives. Thomas F. Glick began his career studying Darwinism in Spain. Here he follows the Uruguayan case and describes a two-phase reception: the first limited, surprisingly, to cattlemen only; the second, the standard debate between positivists and believers. He concludes that centralization of power is the crucial variable in the reception of Darwin in Latin countries. Pedro M. Pruna suggests that the reception of Darwinism in Cuba, although it follows the common Latin American pattern in general, also displays some distinctive features due to its having been a Spanish Colony until 1898. Inasmuch as Darwinism was opposed to the official ideology (that of the Church), its acceptance could be taken for political dissidence, with all the risks that that implied. The fourth *histórico,* Roberto Moreno, whose study of Darwinism in Mexico and anthology of texts has been a beacon for us all, is no longer with us. This volume is dedicated to his memory.

Now we proceed into new territory. Heloisa Bertol Domingues and Magali Romero Sá analyze the way Darwin's ideas were received by Brazilian scientists. In Brazil, as in Mexico, the first discussions on evolution focused on the human species. In Brazil, however, views on human evolution were highly colored by theories of racial and cultural differentiation. It is interesting to see how the great evolutionary biologist Fritz Müller, based in Brazil, and the Emperor Pedro II, a correspondent of Quatrefages, shaped the Brazilian debate from opposing conceptual perspectives. Marcos Cueto, among other topics, includes an analysis of the reception of Darwinism in Peru during the second half of last century and the relationship between Natural History and Darwinism. Peruvian naturalists continued an older tradition of basing their perspective on an explicit distinction between Peruvian and European science. Such a focus was not related to nationalism per se, but to what Cueto calls "self-perception." For example, a cornerstone of Peruvian Biology was the impact of altitude upon adaptation. (In Mexico, too, there was a concern over the study of the effect of altitude in the development of living beings. Alfonso L. Herrera argued that species whose habitats were at high altitudes were inferior to related species living near sea level.)

Neither did Spanish naturalists, members of the Spanish Society of Natural History, polemicize on Darwin, according to Francisco Pelayo, even though many had explicit positions. Some, like Juan Vilanova tried to harmonize Darwin's ideas with Biblical statements. Others like Salvador Calderón narrowed the problem to a narrow set of issues: either one must accept some kind of theory of extinction or else recognize an evolutionary process. To Calderón, the integration of such contradictory points of view is impossible. Pelayo shows how paleontological data played a major role in shaping the responses of both supporters and opponents of evolution. Susana Pinar examines Spanish botanists of the 19th century and their reaction to evolution. In this period, botanists discussed such concepts as adaptation to the environment, species, natural selection, and the succession of plants, all. in the context of the new systematics. Miguel Angel Puig Samper examines debates on Darwinism in Spanish anthropology—contemporaneous with those in other disciplines—beginning in the 1870s. The Spanish Anthropology Society was an early forum of evolutionary discussion. However no full-fledged confrontation or debate emerged there, in spite of the presence of anti-Darwinians like Joaquín Hysern and Juan Vilanova and Darwinians like Francisco M. Tubino.

The second set of essays concerns the extension of "Darwinian" ideas into the social realms of Social Darwinism and Eugenics. Laura Suárez analyses the attempts to improve races in Mexico. Following Galton, some Mexican intellectuals tried to legitimize race and class segregation in order to promote the progress of society. Unlike in Argentina, where Eugenics was used applied to social problems arising from immigration, in Mexico it was used against Indians and in favor of miscegenation. Armando García González analyzes the relationship between Darwinism, Mendelism and Eugenics in Cuba during the first half of this century. Darwin's ideas were known in Cuba at least from 1868 when the Count of Pozos Dulces lectured on the subject at the Havana Academy of Medicine. Cuban eugenics was colored by the nearly simultaneous introduction of the ideas of Galton and Mendel in an atmosphere already prepared by substantial support for Darwin. Ricardo Campos Marín and Rafael Huertas analyze the development of Degeneration Theory in Spain, in particular, its implications for public health and to a certain extent social policy. They describe debates over the influence of heredity and environment in the development of conditions such as insanity or alcoholism. They also comment on the changes induced in the concept of degeneration by the introduction of Darwinian or evolutionist concepts generally. Alvaro Girón Sierra shows how Spanish Anarchists adopted Darwinism in order to suit their own ideology. The principal ideologues, Kropotkin and Bakunin, wanted to limit the action of the struggle for existence to the remote past, because of the incompatibility of the concept of competition with the Anarchists' notion of "Moral Economics." The influence of Nietzche further complicated this polemic. Marta Irurozqui claims that social Darwinism could have been used in Bolivia to justify the domination of white people. However, Chile's victory in the Pacific War (1879-1883) obliged an ideological shift: national unity required the inclusion of "Cholos." Once the war was over, attention reverted to the traditional Latin American scapegoats, the Indians.

The third and final section explores two very different analytical approaches to Darwinism. In the first, Thomas F. Glick and Mark Henderson present an analytical model for assessing the fate of scientific ideas when they cross cultural boundaries. In this model, receptions can be "thetic," "antithetic," "corrective," "extensional" or some combination of the four modes. Besides its heuristic value, the model suggests that there is a pervasive psychological dimension to reception and that the intentions of those receive ideas weigh as heavily in their reception as does their truth content, however assessed. Turning then to biological theory, Rosaura Ruiz and Francisco J. Ayala present a scheme of the basic aspects of Darwinism introduced in different countries. Their proposal is based upon a historical and epistemological analysis of the construction of Natural Selection Theory and its most important components.

Thomas F. Glick
Rosaura Ruiz
Miguel Angel Puig-Samper

April 2001

MARCELO MONTSERRAT

THE EVOLUTIONIST MENTALITY IN ARGENTINA: AN IDEOLOGY OF PROGRESS

> When a story is true, a story that exposes mysteries not yet revealed in fiction, our need to examine it closely becomes more compelling. Moreover, the discovery of a crack in the hard skin of reality fascinates all of us.
>
> Adolfo Bioy Casares

It is not the aim of this essay to offer the reader a long list of ideas, facts, institutions, or scientific processes, which are more likely to be found elsewhere, in more painstaking chronological order.[1] Instead, the purpose of this article is to look back at a body of scientific ideas that appeared in the 1880s and which assisted in the birth and development of an evolutionist mentality at the very core of the Argentine intellectual and political elite, within the general frame of a more extensive ideology: thus, the idea of Progress.

FIRST SYMPTOMS

Although it is true that the reception of specifically Darwinian ideas—not forgetting that evolutionism here had a rather strong Spencerian flavor—became a controversial issue during the decade 1870-1880, it is advisable to go back a few years to put into perspective some of the significant events that took place some time before then. In

[1] The reader may refer to José Babini, *Historia de la ciencia en la Argentina* (Buenos Aires, Solar, 1986), preliminary study by Marcelo Montserrat. Here, I would like to thank to Editorial Sudamericana, Buenos Aires, for the partial use of the material published in Gustavo Ferrari and Ezequiel Gallo, eds., *La Argentina del Ochenta al Centenario* (Buenos Aires, 1980), pp. 785-818. My thanks also to Patricia Tubby, secretary of the Humanities Department at San Andrés University, Argentina, for her technical contribution, to the university itself for its support, and last, but not least, to Natacha Delgado for the English translation, and to Sue Michaelson for her assistance in editing the English version.

Thomas F. Glick, Rosaura Ruiz, and Miguel Ángel Puig-Samper (eds.), The Reception of Darwinism in the Iberian World, 1-27.

1862, a violent book, initially published on the pages of the newspaper *La Tribuna* by José Manuel Estrada—twenty years old at the time—came out under the title, *El génesis de nuestra raza.*[2] The writings were aimed at Gustavo Minelli. A careful reading of this murky text allows us to infer that Darwin has not yet entered the intellectual circle in Buenos Aires, a phenomenon highly understandable, considering that it was only three years after the publication of *The Origin of Species.*

One of the first to read the book in Argentina was a young man, about the same age as Estrada himself, named William Henry Hudson. At the "Los Veinticinco Ombúes," a small ranch located in the old Quilmes district where the Hudson family had originally settled and to which they returned in 1856 after financial reverses in Chascomús, William Henry stopped reading Gibbon's *Decline and Fall* to concentrate on the book by Darwin that his brother Daniel had recently brought from England. Hudson later wrote a few lines about the impression the reading left on him:

> Despite my determination to put the question off my mind, or subconscious mind, like a dog with a bone which refuses it to drop in defiance of its master's commands, went on revolving it... What explanation was possible but that of community of descent? How incredible it appeared that this had not been seen years ago—yes, even before it was discovered that the world was round and was one of a system of planets revolving around the sun... All this starry knowledge was of little or no importance compared to that of our relationship with all the infinitely various forms of life that share the earth with us. Yet it was not till the second half of the nineteenth century that this great, almost self-evident truth had won a hearing in the world!... And again, insensibly and inevitably, the new doctrine has led to modifications of the old religious ideas and eventually to a new and simplified philosophy of life. A good enough one so far as this life is concerned, but unhappily it takes no account of another, a second and lasting life without change of personality." [3]

---

[2] *El génesis de nuestra raza / Refutación de una lección del Dr. D. Gustavo Minelli sobre /la misma materia /por José Manuel Estrada* (Buenos Aires, La Bolsa, 1862; orignally published in *La Tribuna*). Alejandro Korn records this debate, but mentions a "certain doctor Gustavo Nivelli" (for Minelli), in his *Obras Completas* (Buenos Aires, Claridad, 1949), pp. 194-195. For further information about Minelli, see Marcelo Montserrat, "La influencia italiana en la actividad científica argentina del siglo XIX," in Francis Korn, ed., *Los italianos en la Argentina* (Buenos Aires, Fundación Giovanni Agnelli, 1983), pp. 105-123. There is an Italian translation in *Euroamericani* (Torino, Fondazione Giovanni Agnelli, 1987), II , 141-169.

[3] William H. Hudson, *Far Away and Long Ago*, reprint ed. (London, Eland, 1991), pp 329-330 (original ed., 1918). See the excellent biography by Alicia Jurado, *Vida y obra de W. H. Hudson* (Buenos Aires, Fondo Nacional de las Artes, 1971), pp. 51-52. I disagree with the author on one point: Hudson could not have been only fifteen or sixteen years old, but eighteen at least, when he read *The Origin.*, inasmuch as Hudson was born in 1841 and Darwin's book appeared in 1859. In the same book (p. 58), there is an account of an unusual exchange between Darwin and Hudson on the Pampas Woodpecker--unusual because Darwin did not usually get involved in public debates--in the pages of the *Proceedings* of the Zoological Society in London in 1870. See the note by Roy Bartholomew, "El joven Hudson y el viejo Darwin", in *Revista de la Universidad Nacional del Centro* (Tandil), 1 (1977), 113-125, even though the consequences of the episode seem to be far-fetched. Hudson's position on Darwinism appears even more problematic in his posthumous work, *A Hind in Richmond Park*, the result of his vague skepticism on science and progress. For a recent view of Hudson, see David Elliston Allen, *The Naturalist in Britain (A Social History)* (Harmondsworth, Penguin, 1978), pp. 228-229. I have dealt with Hudson in "La recepción literaria de la ciencia en la

But, besides Hudson's spiritual qualms, that partly seemed to echo Darwin's own, let us return to Estrada. His heated allegations were directed against two of Minelli's judgments: the denying of the human race as an unity and the casting of doubt on mankind as God's direct creation; the latter, a more philosophical, rational and dogmatic issue while the first, a rather scientific matter.[4]

To judge by the sources that Estrada cites we can observe the status in the Argentine cultural milieu of many of the components of the evolution controversy to come. The name of Darwin is not heard yet, but instead there are the catastrophist theses of Cuvier and the acquiescent efforts of Biblical geology. The young Estrada was feverishly excited in favor of the Diluvian theory and the charges against uniformitarian geology: "Contrary to those arguments, it has been stated that such phenomena are provoked by currently existing causes, and everything produced by convulsive action can be referred to regularly observed earthquakes. Such is an opinion most scientists have rejected and it is only followed by the Hutton and Lyell School in England, where this science has been mainly studied."[5]

Estrada purports to find threatening traces of Lamarck in Minelli's ideas, and so he asks with raging fury: "Do you believe in racial progress, do you believe in Man before Adam?... Hence, you do not believe in the existence of the soul, you believe in a brutish mortal without fate; in Man without self-awareness and without a notion of absolute justice, you believe in Virey and Lamarck, you believe in Proudhon and Lucretius? Well then, you do not believe in God. You are an Atheist."[6] Estrada's harsh tone—probably fueled by his readings of Cardinal Wiseman—is combined with a pugnacious accusation against materialist atheism, whose subversive nature is emphasized by the mention of Proudhon.

The intensity of the transformist debate can be felt in the paragraph dedicated to "all of you, who wish to reduce man to a descendant from a split-tail porpoise, or from a sick monkey with a long nose."[7] Along with tedious digressions on human language, in which names such as Humboldt, Schlegel, Herder and of course, *Naturphilosophie* come to the surface, there is also some room left for Buckland's geological findings, that supposedly bear a meaning they actually do not have, and the universal creation "that spins in invariable harmony and solemn security" as it had for six thousand years."[8] This literary pamphlet anticipates the violence of the replies to positivism that

---

Argentina: el caso darwiniano," *Redes*, 2 (1995), 99-117, and in *Usos de la memoria (Razón, ideología e imaginación históricas)* (Buenos Aires, Sudamericana/Universidad de San Andrés, 1996), pp 150-169. See also my *Ciencia, historia y sociedad en la Argentina del siglo XIX* (Buenos Aires, CEAL, 1993).

[4] Estrada, Génesis de nuestra raza, p. iii.

[5] *Ibid.*, pp. 58-59.

[6] Ibid., p 49. On Darwinian scientists and Natural Theology, see Charles Coulston Gillispie, *Genesis and Geology (A Study in the Relations of Scientific Thought, Natural Theology and Social Opinion in Great Britain, 1790-1850)*, 2nd ed. (Cambridge, Harvard University Press, 1969).

[7] Estrada, *Génesis de nuestra raza*, pp. 36-7.

[8] The lack of temporal depth attributed to the universe in classical thought is remarkable. Estrada, it is true, did not reach the distressing accurateness of the calculation made by archbishop Ussher and Dr. John

were to come ten years later and which again ignited the impassioned spirit of Estrada. In August 1880, at a lecture delivered at the local Catholic Club which we shall examine later in greater detail, the young tribune once again attacked transformism, which by then had assumed a fully Darwinian guise.

Within scientific institutions, too, some restlessness was starting to show. Juan Ramorino, professor in the Natural Sciences Department at Buenos Aires University, expressed his personal view to the University Rector, José María Gutiérrez, on the occasion of a new syllabus for the Introductory Course in Natural History. Ramorino, who was of Italian background, writes: "the third part of the program is devoted to Botany. The study of single plants is preceded by several considerations on organic elements in general, and by additional issues such as spontaneous generation, species transformation, and the circulation of matter, which have become objects of discussion in most current debates...without coming to any definitive solution as yet."[9] Ten years had passed since the publishing of *The Origin...* and Argentina had begun to echo some of the European disputes.

But already in 1870, there was a strong champion of the anti-Darwinian cause in Buenos Aires: Carlos Germán Conrado Burmeister, who had been appointed director of the Buenos Aires Museum in 1862, thanks to both Mitre's invitation and Sarmiento's suggestion. This Prussian savant—who liked to confine himself to the building on the corner of Alsina and Perú streets—had acquired a certain notoriety after publishing his book *Historia de la Creación* in 1843, some time before Humboldt published *Kosmos*. His work became rapidly known in Europe and Burmeister, in each succeeding edition, continued adding the data necessary to keep it fresh and updated. In one such entry he announced his great disdain of Darwinian theories, an opinion he would only revise in 1889, at the age of 84, not long before his death.[10]

In the 1870 French edition (published after the eighth German edition) a hint of scathing criticism appears. The *Historia* starts with a resolute reaffirmation of Vulcanism as the hypothesis that best explained basic global transformations, admitting only very slightly a Neptunian position. It is worthwhile to point out his acceptance of Vulcanism, then a rigid position opposing any kind of transformism. On Vulcanism he writes that "if there ever existed a hypothesis unable to be empirically proven before the eyes, but supported by strong facts, it was Vulcanism ... All the phenomena of the

---

Lightfoot of Cambridge University, who held that the world had been created at 9 o'clock in the morning on Sunday, October 23, 4004 BC. With good reason Michel Foucault asserted that "there is not and cannot be any hint of evolutionism or transformism in classical thought, since time is never considered the beginning of the development of living creatures. It was only conceived as a revolution in the space in which they live;" *Las palabras y las cosas* (Mexico, Siglo XXI, 1968), p. 151. See also S. Toulmin and J. Goodfield, *El descubrimiento del tiempo* (Buenos Aires, Paidós, 1968), ch. 8 and 9, and François Jacob, *La lógica de lo viviente (Una historia de la herencia)* (Barcelona, Laia, 1973), ch. 3: *El tiempo*.

9 See Horacio Camacho, *Las ciencias naturales en la Universidad de Buenos Aires (Estudio Histórico)*, (Buenos Aires, Eudeba, 1971), pp. 134-36 (note by Ramorino to the Rector, May 24, 1869).

10 On Burmeister, see Max Birabén, *Germán Burmeister: Su vida, su obra*, (Buenos Aires, Ediciones Culturales Argentinas, 1968).

Earth's surface work in its favor and confirm it admirably..." By contrast, he says of Darwinism that "these two opinions (gradual creation and transformism) are, as far as I am concerned, equally acceptable, since neither one of them has one single shred of evidence taken from historical data. In fact, they have only dogmatic and hypothetical value. Recently, the latter hypothesis has been revised and further developed by Darwin, and his attempt was warmly welcomed. However, we must admit, that we are unable to give excessive credit to Darwin's arguments and followers, because it would be useless for the empirical sciences to think up hypothetical concepts about this matter and subsequently find themselves lost in complicated debates concerning its probability, with no end in sight."[11]

Inasmuch as the two ideas in question presented such different approaches, Darwin's sharp warning in the final passage of *The Origin...* should not be forgotten: "Although I am fully convinced of the truth of the views given in this volume... I by no means expect to convince experienced naturalists whose minds are stocked with a multitude of facts all viewed during a long course of years, from a point of view directly opposite to mine."[12]

It is possible that the fear of a paradigm shift—to adopt the terminology used by T. S. Kuhn—influenced Burmeister and the *normal science* of which he was a worthy representative. This becomes absolutely clear in his refusal to accept an animal origin for Man, saying that he was "only behaving as an accurate naturalist. Man and the ape," Burmeister writes revealingly, "are different from each other, not only zoologically but also psychologically, and since we cannot abandon the principle of the invariability of specific traits without causing a whole revolution in scientific zoology, rather we have every reason to believe that such differences have always existed and that they will continue to do so in the future."[13] In sum, "True science should not be devoted to such outrageous ideas because, lacking any positive and scientific evidence, they will always be considered nothing more than mere hypotheses."[14]

A FANTASTIC OVERTURE

The new decade brought some important changes. Between 1870 and 1873, a number of foreign scientists hired by the national administration at Burmeister's suggestion, arrived in Argentina and were appointed to the School of Physics and Mathematics at Córdoba University and its Academy of Exact Sciences (which became the National

[11] Hermann Burmeister, *Histoire de la Création (Exposé scientifique des phases de dévelopment du globe terrestre et des ses habitants)* (Paris, F. Savy, 1870), pp. 6, 176.
[12] Charles Darwin, *The Origin of Species*, (New York, Collier, 1963), p. 499.
[13] Burmeister, *Histoire de la Création*, pp. 642-643. See T. S. Kuhn, *The Structure of Scientific Revolutions*, 2nd ed. (Chicago, University of Chicago Press, 1970), pp. 151, 171-173.
[14] *Anales del Museo Público de Buenos Aires*, I: 20, cited by Alberto Palcos, "Reseña histórica del pensamiento científico (1862-1930)," in *Historia Argentina Contemporánea* (Buenos Aires, El Ateneo, 1966), II, 2nd Section, p. 28.

Sciences Academy in 1871). It was precisely in Córdoba that, in the same year, the National Observatory was inaugurated and Benjamin Gould was appointed as director. He had arrived in Argentina with some colleagues in September 1870 and immediately set to work. The Argentine Scientific Society was created in July 1872. One year later, Eduardo Ladislao Holmberg, born in 1852, began to work for the creation of an Argentine Academy of Arts and Literature which lasted a decade, gathering together intellectuals "whose objective of nationalizing literature and art. . . ran against manners and tastes based on foreign education, as embodied by members of the Literary Scientific Circle.[15]

Holmberg himself warned against the increasing interest in Natural Science around town. Early on he emphasized the fact that "it was a common saying, not only among students but also all over the country, that Zoology was the province of butchers, Botany of green grocers, and Mineralogy of miners, or occasionally of marble workers." However, in the first pages of his novel, *Dos partidos en lucha*, published in 1875, Holmberg asks himself: "To which library can we go to today, without finding that half of the books offered display a close relationship to the sciences?" He goes on to mention scientific journals like the Bulletin of the Academy of Sciences of Córdoba, the Argentine Scientific Annals, Agricultural Annals, Entomological Annals, all of which joined the Annals of the Public Museum, the only journal in existence at the beginning of the decade. If we browse through some of these publications, we can find Claude Bernard, Lyell, and Agassiz, along with Flammarion, Verne and Mayne Reid.

In 1869, while Holmberg was beginning his preparatory studies at the University, another young man, only slightly older than he, Florentino Ameghino, set out for Mercedes as an assistant in an elementary school. From there, he embarked on a long series of explorations, with Ramorino's support, and which sent paleontological specimens to the Museum of Natural History in Milan. Some years later, in 1873, Francisco Pascasio Moreno, a cousin of Holmberg, embarked on expeditions, first to Carmen de Patagones, and later to the mouth of the Santa Cruz River. Encouraged by Burmeister, he published a detailed description on Patagonia in the *Revue d'Anthropologie*, whose editor was Paul Broca. In 1872 Holmberg, also twenty-years-old at the time, traveled to Río Negro, sponsored by the Argentine Scientific Society. Thus, was history weaving a fine web around three great naturalists: Ameghino, Moreno and Holmberg.[16]

---

[15] Martín García Mérou, *Recuerdos literarios* (Buenos Aires, Eudeba, 1973), p. 244 (original ed., 1891). Alfred Ebélot, the French engineer who directed the works of Alsina's "ditch" and accompanied Roca to the desert, was a penetrating witness to the cultural transition of the 70s and 80s in a remarkable short play, *La Pampa* (Buenos Aires, Eudeba, 1961), p. 107: "Thus came *Europeanization.* Being Argentinean seemed an insult." The original French edition is from 1889 and was translated into Spanish by the author himself the following year.

[16] Among numerous biographies of these figures, see Fernando Márquez Miranda, *Ameghino (Una vida heroica)* (Buenos Aires, Nova, 1951); Carlos A. Bertomeu, *El Perito Moreno, centinela de la Patagonia* (Buenos Aires, El Ateneo, 1949); Luis Holmberg, *Holmberg, el último enciclopedista* (Buenos Aires, private edition, 1952). See also, Antonio Pagés Larraya's "Estudio Preliminar" to Holmberg, *Cuentos fantásticos*,

In 1874 Buenos Aires University was reorganized, as result of which the School of Physics and Natural Sciences was opened in 1875. In this timid but discernible atmosphere of intellectual renewal, Holmberg embarked on the study of Medicine, graduating in 1880. He was a fellow student of José María Ramos Mejía and belonged to the same generation which in 1882 graduated from the Law School Nicolás Matienzo, Juan Agustín García, Rodolfo Rivarola, Luis M. Drago, and Ernesto Quesada. In 1874, he not only married Magdalena Jorge Acosta and published his first scientific paper—about the Arachnidae—in the *Agricultural Annals*, but he also had time to translate Dickens' *Pickwick Papers* and to work on his novel, *Dos partidos en lucha.*

It is not surprising that Darwinism came knocking at the door of a new Republic which was hungry for new ideas, but an unusual aspect of it rested on the fact that the first public profession of a Darwinian credo came out in a science fiction novel written by a twenty-two-year-old medical student.[17]

This is unmistakably a first literary endeavor. The reader notices that Holmberg is very well informed, that he is also quite subtle in the use of irony and even imagination applied to social criticism, while often unabashed in his exposition. The brief introduction, dated December 1874, anticipates the plot through a well-known fictional device: the real author is supposed to be someone called Ladislao Kaillitz—a slightly altered version of the original Kannitz of the Holmberg family—a Darwinian who hands the manuscript to the narrator, just before departing for Europe. After a quote by "our dear friend and poet Rafael Obligado" the action begins. How does the plot feed the fourteen chapters of *Dos partidos en lucha?* Holmberg takes advantage of local political conflicts surrounding the city and the nation—the presidential election on April 12th, in which Mitrism, in Buenos Aires, overcame the Avellaneda-Acosta line, which claimed victory in almost every province of the country; the Electoral College election that resulted in 146 votes for Avellaneda against 79 for Mitre, and the Mitrist Rebellion that was finally put down by December of that year—spinning around them a web both of cultural and political misunderstandings. The preparations for such a secular coup, which was defeated by the railways, the telegraph, and a few Remingtons (according to a brief report presented by the American minister Thomas Osborn to his government), along with local political meetings and a ferocious fight that flared up among local newspapers, all appear in the opening pages of the novel, after a description of a short journey to Río Negro, meant as a kind of homage to the Darwinian periplus performed four decades before.

---

still the best analysis of Holmberg's literary work. I have studied the personality and influence of Holmberg in "Holmberg y el Darwinismo en la Argentina" in *Criterio*, 47 (1974), 591-598.

[17] The full title is *Dos partidos en lucha / fantasía científica / publicada / por / Eduardo L. Holmberg / Imprenta El Arjentino / Buenos Aires / calle Piedad n. 134. / 1875* (148 pages). Paul Broca's article *Los Akkas, raza pigmea del Africa Central*, is reproduced on pp. 140-148. Citations included in the text are to this edition.

Holmberg begins the novel rhetorically, regretting his ignorance of the English naturalist: "However I, who had just passed my last exam at the University, had no idea who Darwin was (p. 3)." He is not long in starting a series of famous attacks on Burmeister: "a savant, perhaps too wise, as will be understood by those who are well informed (p. 7)."

Even paying due to Burmeister's scientific achievements, we must still discern a certain Prussian stubbornness in him concerning new ideas. In 1862, the French engineer Adolfo Sourdeaux, former captain of the naval forces that had blockaded Buenos Aires, had to bear the harshness of the *Burmeisterian dictum* with regard to the artesian wells that he introduced in our country. In Burmeister's opinion, such semi-surging waters did not exist, and it was completely "insane to look for them." Yet the Frenchman did not surrender and kept on digging until he belligerently announced in an article of *La Tribuna*, dated March 1862, that "in spite of the fatal forecasting of certain savants, a group formed by eunuch scientists who are totally useless and disturb the work of others; in spite of such a verdict decided by those infallible judges, comfortably seated on their cozy sofas, who all know, claim all and declare all. . . I must announce that artesian wells do exist in this country. Furthermore, at a depth of 92 yards, we have found in Barracas water similar to that found in La Piedad, the latter having been an object of doubt and scorn from such gentlemen." This odd Prussian-French dispute grew more fierce, because Burmeister, presiding over a new scientific board, then declared the water not potable. The local administration appointed a new commission of doctors and chemists which, at last, determined that the artesian wells were perfectly healthy. Burmeister was not exactly in a good position at the end of the conflict.

In 1873, Ameghino did not do much better than Sourdeaux, when he asked Burmeister to concern himself with some human fossils found in one his previous excursions to Mercedes. The savant excused himself with disdain: "I am not inspired with confidence on such findings; I do not believe in them; and even if they were what you have told me, they absolutely lack any importance and do not appear interesting to me." Shortly after, Burmeister charges him once more, saying: "Self-taught people like him are well known to be arrogant; the life of a school teacher in a very small town, where truly wise men are needed, increases the quality of unauthorized wisdom obtained by teachers in such places, where they are often surrounded by people who do not know any better." Years later, when Ameghino, in spite of these unhappy incidents, decided to honor him by naming a fossil mammal—*Orocanthus Burmeister*—after him. Burmeister furiously rejected such a courtesy.[18]

But let us return to Holmberg's fictional work. *Dos partidos* features three distinctive characters: Francisco P. Paleolítez, Juan Estaca, and Pascasio Griffitz. Each

[18] The artesian well saga is told with unbeatable charm by Noel H. Sbarra, *Historia de las aguadas y el molino*, 2nd ed. (Buenos Aires, Eudeba, 1973), pp 115-121. On the relationship between Burmeister and Ameghino, see Márquez Miranda, *Ameghino*, pp. 189-193.

of them secretly bears some relation to the author's allies and enemies.[19] Griffitz is a complete Darwinian, who keeps in the cellars of his large and secret residence zoological and botanical collections classified according to a scheme of gradual perfection. "I am going to tell the Truth," he declares. "I serve a scientific doctrine: Darwinism. Sooner or later it will become a political doctrine, and this is why I require a certain amount of secrecy in my conduct (p. 45)." This singular hybrid of Vernian hero and Buenos Aires plotter is one of the best characters in the novel.

Last but not least, there is a debate between Darwinism and "Rabianism." Rabián is the local anti-transformist leader. At the first public scientific meeting specially called to elucidate "whether we have descended from apes or if we should believe, as some pretend, that we are the result of spontaneous generation in different epochs, particular to each species (p. 52)," Paleolítez speaks on behalf of a doctrine "sacred to some since it does not reject the Mosaic narrative which says that we come from dirty clay and which is far more noble than to descend from monkeys" (p. 54). Then, a mysterious "unknown" character makes an appearance. He bears a suspicious resemblance to Ameghino and says that "before arguing as anthropologists, we implicitly recognize that we are geologists" (p. 59) and we can also count on a "physician of moral diseases," a very credible allusion to José María Ramos Mejía, friend of the author since adolescence.

Meanwhile, Holmberg praises Verne and Mayne Reid—published as serial stories in the most important newspapers—Flammarion, Figuier and, in chapter 8 (which Holmberg himself grades "as slightly heavy"), he expresses some criticism of local Patricians, accusing them of being frivolous and Tartuffe-like. Suddenly, the setting is changed: from a "Rabianist" salon, it jumps to Regents Street, in the heart of London. From there, we go to the Zoological Garden where Charly (Darwin) and Dick "Old Bones" (Richard Owen) are carrying out the dissection of an "anthropomorphic ape" that turns out to be an *Akka,* a Central African Pygmy discovered by Doctor Livingstone.[20] After the Argentineans assert that "not only do we not owe anything to Englishmen, but also never want to" (p. 90), they invite Darwin to a second scientific meeting in Buenos Aires, and Queen Victoria puts at his disposition the fastest ship available. The *Hound*—a nautical mirror of the *Beagle* under Fitz Roy's command—brings him to Montevideo in less than two weeks.

In Buenos Aires, spirits are high: one of the few who manages to keep calm is the Darwinian leader Pascasio Griffitz, who says: "If the Rabianists triumph, propaganda for the *status quo* will cast the shadow of its foolish rigidity everywhere. Science will

---

[19] It is quite hard to find the key to the cryptogram. It is clear that Holmberg had fun when matching up names: Francisco P. suggests Moreno; Pascasio Griffitz reflects Moreno's matronymic, Kannitz; and Juan Estaca may conceal Ramorino. Remember that Moreno was not originally an evolutionist, but professed the ideas of his mentor Burmeister.

[20] The literary convergence is still paradoxical here since Owen, a great anatomist, was also a passionate opponent of Darwin, so much so that T. H. Huxley included him among those who cultivated "the mistaken zeal of the Bibliolaters." See Ronald Millar, *The Piltdown Men* (New York, Ballantine, 1974), ch. 4.

not progress and if it does, it will be in a negative way, in my view. But, if we, the followers of Darwin, happen to triumph...no one will doubt that social norms must be totally changed. So either a philosophical revolution of incalculable transcendence will break out, or the most extreme indifference will envelope us, to the point of not taking any scientific doctrine into account in the development of society" (p. 105).

On August, 28, 1874 --"the year that more powder was burnt in the Argentine Republic than ever before" (p. 110)--Darwin arrives in Buenos Aires and at ten o' clock, is warmly welcomed by President Sarmiento: "I have the great honor to welcome this illustrious English reformer..." (p. 112). After an introduction to vice-president Alsina, the English scientist greets Mitre, telling him "I cherish and admire you, yet I do not understand you" (p. 113). Finally, he congratulates the president-elect, Avellaneda.

The second meeting is held in the Teatro Colón, Congress and the Cathedral having been previously rejected, because "How can we discuss at an apostolic Roman Catholic temple a doctrine that so openly attacks—according to some people—our religious beliefs?" (p. 116). The old Teatro Colón which was then located next to Plaza de Mayo, is the fancied setting, where the truth or falsity about evolutionism will be decided. Decorated for the occasion, a large banner is unfurled up-front, waving the motto ***Struggle for life*** over the barely tasteful painted figures of three great monkeys clumsily fighting over a giant carrot, while the crowd is seated. The National Anthem and *God Save The Queen*—"the moderate anthem of Albión" (p. 126)--are played along with *Die Wacht am Rhein*, (a reference to Burmeister), after which the debate begins.[21] Darwin makes a preliminary presentation, but is interrupted by Paleolítez, Estaca and even by an explorer who comes in, holding an Akka in his hands. Shortly after, the discussion focuses on the origin of life. Estaca stubbornly supports the most rigid concept of creationism, before the bewildered eyes of the evolutionists, and Griffitz presents a hypothesis based on the spontaneous generation of a universal germ called *Protobia*. The debate will be fired up by Paleolítez expressing, once again, some anti-transformist anatomical observations, to the alternate bafflement or applause of the audience, "minus those who had fallen asleep" (p. 135). And so the decision is made to perform surgery on the *Akka*, in order to ascertain its real nature.

Meanwhile, Darwin steps forward and states that "everything is linked, or if you prefer the Linnean aphorism, *Natura non facit saltum.* Even in the tiniest details he discerns the admirable gradation among beings (p. 135)." Griffitz supports the Englishman with a general Spencerian presentation of evolution, based on the old belief that human society followed a progressive course from East to West. Evolution has not stopped since then, and "if it is true that during many years it has been chained to Europe, it has nevertheless been heard in America where the presence of a new Emperor

---

21 This fictitious episode evokes a real one told by Ismael Bucich Escobar (Martín Correa). At the end of August 1870, the singer Carlota Patti, sister of the famous Adelina, performed at the Teatro Colón with pianist Teodoro Ritter and violinist Pablo Sarasate. The audience insisted on Patti singing *La Marseillèse* (the Franco-Prussian war had just started), but she refused to do so and diplomatically sang the Argentine national anthem instead. See *Visiones de la Gran Aldea*, 2nd ser. (1870-1) (Buenos Aires, 1933), p. 71.

of the World is felt" (p. 136). But the impetuous Griffitz goes a step further: humanity should yield to evolutionary law and to life—"whose minister is Death"—only to perish in a great geological cataclysm, laying the groundwork for "a great metamorphosis of life forms" (p. 137). From the ashes of humanity, a new being will be born, "in whom human form will be slightly modified, although its ultra human intelligence will reach the highest developmental stage." That creature will present an utterly wicked synthesis of "all evils that preceded our own species, mankind" (p. 137). With this Wagnerian omen, Griffitz concludes the presentation. Then, the *Akka* is brought onto the stage where, after receiving some chloroform, surgery is performed on its fifth costal space. This is an *experimentum crucis* suggested by Darwin to observe the *Akka*'s heart functioning, in order to fully demonstrate that it is a race "halfway between monkey and man" (p. 142). The experiment concludes with the pained scream of Paleolítez: "Gentlemen. . .we have been defeated, the ideas of Darwin have triumphed" (p. 138). So ends this Darwinian fantasy, with a nod to the generous spirit of the novel's putative author, Ladislao Kaillitz.[22] And in case any reader might doubt the strictly Darwinian *bona fides* of the young Holmberg, a footnote on page 139 is initialed by "E. L. H., *Darwinist.*"

AN OPEN DEBATE

In the middle of that decade of the 1870s, new groups that would play leading roles in the 1880s entered the debate. Evolutionism—whether in a discreet Darwinian version or a radical Spencerian stance[23]--became the key element of the mental gear of our local version of positivism, which it pervaded with militant biological progressivism both in style and content.

Let me concentrate on two principal, and antagonistic, figures of the 1880s—Miguel Cané and Eduardo Wilde—as they approached our subject in the 70s. In August 1874,

[22] Psychopathology, phrenology and spiritism are always present in Holberg's narratives. Spiritism was popular in Buenos Aires from 1870. The names of Wallace, co-discover of natural selection, and Crookes-- both enthusiasts of spiritism-- appear to be have connected with the activities of the Constancia lodge. See "*La casa endiablada*", in E. L. Holmberg, *Cuentos fantásticos* (Buenos Aires, Hachette, 1957), p. 320. Spiritism, when construed as "a disguised materialism," turns out to have been complementary to the "positive spirit;" see Jacques Barzun, *Darwin, Marx, Wagner (Critique of a Heritage)* (New York, Doubleday Anchor, 1958), p. 105 n. 7. The Catalan chemist, Miguel Piuggarí, then lecturer at Buenos Aires University, also reflected this linkage, though he opposed both. See R. F. Recoder, "Químicos de antaño," INTI No. 24 (Buenos Aires, 1973).

[23] It is often diffcult to identify the precise origin of the ideas of any one particular thinker. As Etienne Gilson pointed out in *De Aristóteles a Darwin (y vuelta)* (Pamplona, EUNSA, 1976), p 149: "Beyond any doubt Spencer was the one who, at the beginning of the movement, coined the key way of thinking in the years 1850-1910 from the notion of evolution. The merging of Darwinism and Spencerism was almost instantaneous, despite the absence of good will on the part of both authors." It is true that Darwin's opinion of Spencer was not a good one, as Nora Barlow confirms in her edition of *The Autobiography of Charles Darwin (1809-1882)* (London, Collins, 1958), which includes omissions from the first edition prepared and expurgated by Darwin's son, Francis.

while Holmberg was writing science fiction, Eduardo Wilde, then a young physician, poking fun at electoral politics and at the electorate's ability to absorb political slogans, explained: "Men have a striking resemblance to apes, a truth proved long before Darwin demonstrated their linkage to such animals."[24] Some months later, Miguel Cané, 24 years old, published one of the few reviews of *Dos partidos en lucha.* A calm and accurate appraiser of the talent of his colleague, Cané was amazed that Holmberg "had enough courage to publish a book in Buenos Aires, which is something like reciting one of Petrarch's verses at the local stock market." Although he admits "that the general plan is, as the French like to put it, absolutely *manqué*," he stresses several good points presented in the book and states, in a romantic vein, that we "would sacrifice our dignity by accepting Darwin's disgusting theory about the transformation of species, if the resurrection of this sensitive phenomenon turns out really to be true." Ten years later, Cané asked a similar question—and got an answer similar to the one Disraeli gave in 1864: "Is man a monkey or an angel? I, sir, prefer him to be an angel. I reject with indignation and disgust those novel theories."[25]

In Argentina, meanwhile, persons very different from the English Tory leader shared similar feelings. In March 1875, the academic activities at the local Teachers School began. This school was created the previous year by a decree signed by Mariano Acosta, governor of Buenos Aires and his education minister, Amancio Alcorta. There, Eduardo L. Holmberg began his teaching career, and gradually, the neo-progressive zeal of the local double of T. H. Huxley had its effect: the names of Laplace and Darwin, "both familiar from the beginning of the course" and their "subversive doctrines" caused the young professor nothing but trouble. According to his biographer, there was one occasion when seventy-nine of Holmberg's students (all women), following the orders of an "outer voice," after his question about why "when snow or ice melts in the mountains, it flows through the plains instead of rising to the skies like geysers," jointly replied, "Because it is God's will," while only one of them attempted a reference to the law of gravity.[26] The following year, pressure from the Ministry of Public Instruction, which threatened to implicate Holmberg, fell upon the school principal, Emma Nicolay de Capprile, but Sarmiento, who was then both senator and General Director of Provincial Schools decided in favor of the intrepid Darwinian teacher, carrying the issue to the presidency itself.

In 1877 (the year in which the first edition of *The Origin of Species* appeared in Spanish[27]), another singular episode questioned the role of evolutionism in local

---

[24] Eduardo Wilde, *Tiempo perdido*, 4th ed. (Buenos Aires, W. M. Jackson, 1945), p 181 (original ed., 1878). The observation appears in a chapter with the curious title, *"El chocolate Perón es el mejor chocolate."* It is worth remembering that Darwin had already published *The Descent of Man* (1871) by that time.

[25] Miguel Cané, *Ensayos* (Buenos Aires, Sopena, n.d.; original ed., 1877), pp. 111-114. The essay is from 1875. This Victorian statement by Disraeli is taken from Geoffrey Bruun, *La Europa del siglo XIX* (México, F.C.E., 1964), p, 172.

[26] Luis Holmberg, *Dos partidos en lucha*, pp 119-120.

[27] Charles Darwin, *Origen de las especies por medio de la selección natural o por la conservación de las razas favorecidas en la lucha de la existencia*, tr. Enrique Godínez (Madrid, Biblioteca Perojo, 1877). On

intellectual circles. That July, some members of the Argentine Scientific Society, including Estanislao Zeballos, Valentín Balbín, and Miguel Puiggarí, moved to elect Darwin to honorary membership. The motion was accepted in August by an executive committee presided over by Guillermo White. Darwin's designation was promptly communicated to a society member in London, Walter F. Reid, who in a letter dated December 1877, said that he had given the diploma to the English naturalist at his home, in Down, Sussex, during one of his periodic health crises. Thus, Darwin became the third honorary member of the institution, after Guillermo Rawson (1874) and Germán Burmeister (1875). Sheer coincidence had established this bond between the three men, no doubt most irritating to our stubborn Prussian. But while the diploma was on its way to England, Carlos Berg—a Russian-born naturalist and employee of the Public Museum, thanks to Burmeister's invitation and faithful to his director's anti-Darwinism, made a proposal to amend the Society's statutes in order to make the election of honorary members more difficult. Berg's motion was rejected and at the end of November, he resigned from the Museum for personal reasons. It is not far fetched to imagine that this conflict was rooted in Darwin's election. It is interesting to note that the National Science Academy, based in Córdoba, decided to elect Darwin to honorary membership on August 3, 1878 (just two days before his election to the French Academy of Sciences) at a meeting presided over by Dr. Weyenbergh, along with the other voting members, Latzina, Hieronymus, Brackebusch, and Oscar and Adolfo Doering, with Tobías Garzón as secretary. Three months later, on September 21, 1878, president Avellaneda and his minister Bonifacio Lastra accepted the academic resolution and confirmed Darwin and Grisebach as honorary members. Darwin who, in a letter to Asa Gray, made an ironic reference to his election to the botany section of the French academy, did not hesitate to express his thanks for the honorary membership of the Córdoba academy by sending a copy of the sixth edition of *The Origin of Species*, together with an autographed photograph.[28]

With the new decade came an open and forthright period of debate. In August 1880, José Manuel Estrada returned to the limelight in a speech delivered at the Catholic Club about *Naturalismo y Educación*. His assault against evolution is forthright: "Natural history is enormous... I know of no one who will deny or doubt the rare talent of

---

evolutionism in Spain, see Diego Nuñez Ruiz, *La mentalidad positiva en España: desarrollo y crisis* (Madrid, Tucar, 1975), ch. 6; Thomas F. Glick, *Darwin en España* (Barcelona, Península, 1982), and Núñez, ed., *El darwinismo en España* (Madrid, Castalia, 1977). For an overview, see Thomas F. Glick, ed., *The Comparative Reception of Darwinism*, 2nd ed. (Chicago, University of Chicago Press, 1988).

[28] I owe this valuable information on the Sociedad Científica Argentina to Roberto F. Recoder, as well as to his mentors, Juan Nielsen and Cristóbal M. Hicken. As regards Darwin's election to the National Academy of Sciences, see Telasco García Castellanos, *Darwin (Homenaje en el centenario de la primera manifestación científica sobre el origen de las especies)* (Córdoba, National Academy of Sciences, 1958; *Miscelánea*, No. 36). Regarding Darwin's reaction to his French appointment, see Charles Darwin, *Autobiografía y cartas escogidas* (Selection by Francis Darwin) (Madrid, Alianza, 1977), II, pp. 420, 486. On the French milieu, see Yvette Conry, *L' Introduction du darwinisme en France au XIXe. siècle* (Paris, Vrin, 1974), ch. 1 and pp. 29-31, 434-435.

observation of these men like Darwin, so reckless in their theorizing, or who will reject its evil consequences, or disdain Haeckel's indecent trickery and the credulous and ridiculous pride of those who discover in a grain of lime phosphate the germ of a Genesian anchovy. . . Look at less boisterous scenes: nineteenth century society, shaped by naturalism, enhanced by physics and mathematics, ruled by a political economy under the rigid inspiration of Adam Smith, predecessor of MacLeod and of a long list of Sophists, over which tower those patriarchs of utilitarianism Bentham and Franklin, the great strategists of calculated virtue."[29] It is important to underline the thematic distance between the Estrada of the debate with Minelli in 1862, and this incarnation. During the eighteen years that had elapsed, the works of Darwin became widely known, Haeckel had begun his evolutionary apostolate and, from the other side of the ring, Christendom had heard the stentorian condemnation expressed in Proposition 80 of the *Syllabus of Errors.*

Almost two years later, on April 19th, 1882, Charles Robert Darwin passed away and on the 26th, he was solemnly buried at Westminster Abbey, not far from Newton's grave.[30] Exactly one month after his death, the Argentine Medical Circle, founded by José María Ramos Mejía, organized an homage at the National Theater of Buenos Aires. Sarmiento, still a very lively speaker at seventy-one, spoke first, followed by Holmberg.[31] Each represented a distinctive generation of Argentine intellectuals. Sarmiento, originally from the province of San Juan, began by analyzing the anti-transformist views of Burmeister and Agassiz and replies by Huxley and Ameghino. He notes that, while an exile in Chile, he had met the *Beagle* and its crew, but had not personally known Darwin. Then he ironically refers to the variation of species empirically performed by our local ranchers (Pereira, Duportal, Chas, Olivera, Kemmis and Lowry *et al.*), continuing with an analysis of the social influence of Darwinism, within the framework of a *sui generis* philosophy. His entire lecture is informed by the aura of harmony that the evolutionary theory has introduced into the comprehension of nature: "I, gentlemen, support the doctrine of the evolution"—Spencer *dixit*—"as a necessity for the soul, because I need to rest on a harmonious and beautiful principle, to finally silence the doubt, which is the soul's torment."[32] What Newton had accomplished by reducing the physical cosmos to a precise, though intangible, mechanism, Darwin has done to the organic world, placing it under a transforming and progressive law. Newton found order in space, Darwin in time. The idea of evolution,

---

[29] José Manuel Estrada, *Discursos* (Buenos Aires, Mundo Moderno, 1953), pp. 124, 126.

[30] Did Darwin die of Chagas' Disease contracted in Mendoza in March 1835? P. B. Medawar supports this hypothesis in *El arte de lo soluble* (Caracas, Monte Avila, 1969), pp. 83-94. Darwin himself says: "We slept in the village of Luxan, which is a small place surrounded by gardens, and forms the most southern cultivated district in the Province of Mendoza; it is five leagues south of the capital. At night, I experienced an attack (for it deserves no less a name) of the *Benchuca*, a species of *Reduvius*, the great black bug of the Pampas," *The Voyage of the Beagle* (New York, Collier, 1963), p. 333.

[31] Published as a book, *Carlos Roberto Darwin* (Buenos Aires, 1882). Citations in the text correspond to this edition.

[32] It may be read in *Obras*, A. Belín Sarmiento, ed. (Buenos Aires, 1899), XXII: 104-133.

extrapolated to the natural and human cosmos according to Spencerian formulas, becomes the universal key that provides a reason for all of reality. The following year, Sarmiento, in that "aged Facundo" that resulted in his unfinished book, *Conflicto y armonías de las razas en América*, felt the need to embellish his thoughts with the latest trends. In a letter to Francisco P. Moreno (included in the same book) Sarmiento writes, "You do a fine job tracing the evolutionist ideas of Spencer and I have so proclaimed when referring to social issues, but I leave you and Ameghino to proclaim Darwin's ideas, if you are willing to take them on too. I get along with Spencer splendidly since we both walk the same road."[33]

Holmberg's lecture that followed is a remarkable piece of work not only for the wealth of knowledge displayed but also for its polemical power. The tone is set in his introduction, with an attack on Burmeister and the Argentine National Science Academy. Shortly after, there is an homage to Claude Bernard—no admirer of Darwin—and the expression of a secular faith in the natural sciences conceived as agents for social and cultural renovation. This would be a shift of unpredictable consequences, because "the day that Darwin's doctrines are taught at Russian schools, the Emperors will have interposed their bodies against the bombs of nihilism" (p. 8). Wearing a quote from the Baron of Holbach like a badge—"Men cannot leave nature, not even in their thoughts"—Holmberg believes that "the doctrine of selection was not the result of one man, or one day; selection itself had adapted human intelligence to that mass of observations and facts accumulated during centuries" (p. 47). Holmberg, when expressing his views on the struggle for life—a key element of Darwin's theory—wanders down a path that leads him to an individualist approach to *social Darwinism.*[34] When posing the question of fairness of the fight against the Indian—just three years after Roca's successful Campaign in the Desert—he concludes that "we did away with the Indians, because Malthus' law is above individual opinions" (pp. 65-66). Although the subject merits more extensive treatment, it is sufficient here to note that Holmberg returned to this line of thought in his *Essential Botany* (1908), where he turns the struggle for existence into a social explanation, because "in all areas of human social life, between animals, between plants, struggle is unceasing and only the fittest wins," as in the case of Napoleon, an historical example of which he was particularly fond.[35]

---

[33] "Conflictos y armonías," in *ibid.*, XXXVII: 322-323.

[34] According to one of the trends described by Richard Hofstadter, *Social Darwinism in American Thought* (Boston, Beacon Press, 1955), ch. 1, 2, and 10. See also James Allen Rogers, "Darwinism and Social Darwinism," *Journal of the History of Ideas*, 33 (1972), 265-280, and Robert C. Bannister, *Social Darwinism (Science and Myth in Anglo-American Social Thought)* (Philadelphia, Temple University Press, 1979).

[35] I cite the text entitled *Evolución*, which is ch. 19 of *Botánica Elemental* (Buenos Aires, Sociedad Luz, 1915). Darwin was not fond of these historical applications of evolutionism. On Napoleon's case, see the letter to Charles Lyell (Jan. 24,1860) in Francis Darwin, ed., *Life and Letters of Charles Darwin*, 2 vols (London, John Murray, 1887), II: 262. The confusion between biological and social evolution is at the center of the relationship between Darwin and social Darwinism. Rogers, in the article cited above, thinks that "it

In the course of this lecture given before an audience of more than three thousand people, there is a reference to a presentation given at the Argentine Medical Circle by Pedro S. Alcacer, an anti-transformist medical student, on *La vida y el transformismo moderno*. This allusion is especially interesting because it presents the arguments used by local opponents of Darwin. "It is impossible to challenge Transformism with sentiment, nor with Le Play, Moigno, Burmeister, or Claude Bernard, with nothing except the free flow of understanding detached from all inhibiting influences," Holmberg concludes (p. 108).

The new ideas finally made their way to Congress itself. The debate over "Law 1420" became a kind of test of educational modernism. It went on throughout 1883-1884, and representatives of all views turned to science looking for support. If Goyena led off with a blazing critique against this modern scientific *hubris*, Achával Rodríguez preferred the shortcut of acquiescence, stating that "Genesis spoke of a truth of a natural order, albeit in a time when science was very young;" he adds that the modern wave theory of light explained scientifically "that cosmic matter which, according to Humboldt, might have been the first thing created by God, before anything else." Thus, "there is no opposition between the truths expressed by the Catholic religion and modern science, nor ever will be." [36]

The religion minister Eduardo Wilde replied, in a rather long speech, criticizing the Catholic viewpoint with this thesis statement, apparently irrefutable for him: "The law of progress has to be forcibly verified, and progress is in everything." No doubt he was unaware that another Wilde—Oscar—held "that progress is the ecstasy of imbeciles." Based on that central proposition, Wilde asserts "that today's science must be in contradiction, it has to be in contradiction, it can not be less than in contradiction with certain assumptions made by the Church" because, "Natural History and Biology. . .can be expected to contradict those beliefs, if they did not exist at the time of their revelation."[37]

Achával Rodríguez reentered the debate by proposing: "Let us see what is going on in Germany at this moment. . . On the discussion about Darwinism, Mr. Gessler—the Prussian minister of public instruction—used to say: 'I am not one of Darwin's followers.' To me it is a useless exercise to pit science against religion" (the latter point addressed to Wilde)--"although I do not know if he believes that Mr. Gessler's authority is less than his." With that, the representative from Córdoba concludes: "Well, Mr. President, the truth is that these ideas that started in Europe—despite the surprise some members of this House may experience because of the early onset of this reaction—are now being felt here."[38] The notion of a public reaction to Darwinism to which Achával

---

is the legacy of the Malthusian theory of human population" (p. 275); texts like Holmberg's lecture ought to be interpreted in this way.

[36] *Debate parlamentario sobre la ley 1420* (Buenos Aires, Raigal, 1956; estudio preliminar, selección, y notas de Gregorio Weinberg), pp. 311-312.

[37] *Ibid.*, pp. 206, 208.

[38] *Ibid.*, pp. 336-337.

Rodríguez alludes in support of his anti-secularist views was linked to the Haeckel-Virchow debate that erupted in Germany in 1877. Virchow—a brilliant biologist, a founder of cellular pathology, an anti-Bismarckian liberal politician who was on the barricades in Berlin in 1848, and who had also anticipated the *Kulturkampf* to come and founded the progressive *Fortschrittspartie*, had suggested a possible affinity between Darwinism and Socialism in order to counter the hyperbolic Haeckel, an accusation which in that place and time, in midst of Bismarck's campaign against Socialism was tantamount (following Erik Nordenskiöld) to an accusation of treason. In fact, shortly after this episode, the Prussian Minister of Education banned the teaching of Darwinism from all public schools, and "under the new education law, biology was completely excluded from the curricula for advanced classes, in order to protect students from new doctrines."[39]

At the beginning of the new decade, in an 1890 *causerie*, the refined intellect of Lucio V. Mansilla pondered evolutionism, among other things. In a review of a book by Jossuet in which modern science is accused of being anti-Christian, Mansilla displays a rare and balanced knowledge of the evolution debate and the personality of Darwin himself. He is discerning when he identifies the important issue of final causes as well as the debate with *natural theology*, asserting that "Darwin, seen in all these shadings, these outbursts of excitement that reflect his skills of argumentation, displays a much different image from the one presented by his dogmatic followers, those who want to make use of the *Origin of Species* to support the most absolute materialism." Mansilla goes on to say that not long before Darwin's death he had sent a letter to him from Rome, in which he gave a detailed account of the story of "two werewolves" abandoned as children by a local guerrilla from La Rioja. "Thus," he continues, "there is transformism and evolutionism, ascent and descent, progress and decadence. We should not feel ashamed of descending from any animal similar to a monkey, when there are some men in La Rioja who have turned out to be so brutal, and much less clever than monkeys themselves." In summary, Mansilla detects that "although rationalist views. . . everywhere strive towards unity and harmony, I would not hold my breath. . . Our civilized societies are not perfect, neither have they come to the last stage of civilization, if they ever will. Progress is indefinite and evolutionary.[40]

---

[39] See Erik Nordenskiöld, *Evolución histórica de las ciencias biológicas* (Madrid, Espasa Calpe, 1949), pp. 585 and following; and Yvette Conry, *Introduction du darwinisme en France*, pp. 26-28. Darwin rejected such biological-political arguments, whether from the right or left, although in some of his letters and in his *Descent of Man*, his position softened. On Virchow's reaction, see Francis Darwin, *Life and Letters*, p. 237, and the article by Rogers cited above, especially pp 269-274. Regarding the debate in Germany, see E. M. Radl, *Historia de las teorías biológicas* (Madrid, Revista de Occidente, 1931), II, 133-144. See Werner Sombart, *El burgués* (Buenos Aires, Oresme, 1953), ch. 16 and especially page 359 n. 246, where in the discussion on heredity, Lamarckism is deemed liberal or social-democrat, Weismannism, conservative.

[40] Lucio V. Mansilla, *Entre-nos (Causeries del jueves)* (Buenos Aires, Hachette, 1963), pp. 554-562. The *Causeries* were collected in five volumes between 1889 and 1890.

AN IDEOLOGY OF PROGRESS

Charles Morazé was dead right when he stated that "from 1870 onwards, in every corner of Europe, to have a scientific spirit, to be "positive," meant to subscribe to evolutionism."[41] Something similar took place in Argentina, on its way towards Europeanization. But, as in the Old Continent, evolutionism—besides its obvious contribution to biology—was enlisted as a scientific tool to legitimize a very powerful social ideology: the idea of Progress. J. B. Bury, in his classic book, has shown the amazing history of progressivism in the Western World. *The Origin of Species* ushered in the *third* stage of the fortunes of the idea of Progress, which became a general article of faith in the seventies and eighties of the last century, an element of the mental outlook of educated people."[42]

The intuitions of the Enlightenment required a more accurate scientific base. Spencer would elevate the notion of progress to the philosophical heights of an irresistible and universal law. The topic was no longer about inferring the perfectibility of man from a psychology that claimed the plasticity of the human nature at the hands of the legislator and the educator, similar to what went on in the eighteenth century.[43] Now, it was the nature of mankind itself which was unavoidably attached to the general laws of change, that would lead the species to final harmony through an inevitable unconscious adaptation process assisted by legislation and education. The progress of mankind was thus considered a necessary fact, a logical concomitant of cosmic evolution. Thus, the enthusiastic support for this kind of *secular religion* that spread from the top to the bottom of the social edifice, providing the inner certainty that Sarmiento invoked in his posthumous homage to Darwin "to finally silence doubt, which is the torment of the soul."[44]

Spencer's optimism was a wonderful spiritual prescription, axiomatic for the organizers of our political society. But the levels and intensities—conscious or unconscious—of this collective credo were many and diverse. From the exclusive political clubs of people who called themselves progressives, or from those shops proudly opened by immigrants under the name of *El Progreso* in a Patrician city that was struggling to become *bourgeoise,* to the inexpensive editions of Francisco Sempere (published in Valencia, under the sponsorship of Blasco Ibáñez) which would do so much for the propagation of Haeckelian evolutionism among the anarchist working class. From the local Zoological Garden, zealously directed by Holmberg from 1888 to 1903, winning popular affection and personal acknowledgment of his persona on the

[41] Charles Morazé, *El apogeo de la burguesía*, (Buenos Aires, Labor, 1965), p. 300.

[42] J. B. Bury, *The Idea of Progress* (New York, Dover, 1955), pp. 334, 335, 337.

[43] See Marcelo Montserrat, "La introducción de la ciencia moderna en Argentina: el caso Gould," *Criterio* 44 (1971), 726-729; "Sarmiento y los fundamentos de su política científica," in *Sur*, no. 341 (July-December 1977), 98-109; and "Sarmiento, propulsor de la ciencia," in *Ciencia e Investigación*, 42 (1988), 277-283.

[44] See Karl Löwith, *El sentido de la historia (Implicaciones teológicas de la filosofía de la historia)* (Madrid, Aguilar, 1956), especially ch. 4.

cover of *Caras y Caretas* magazine to the Buenos Aires production of the Spanish operetta *La Verbena de la Paloma* that proclaimed with a quaint Madrid accent that "nowadays, the sciences are progressing: a tremendous event (*Hoy las ciencias se adelantan, que es una barbaridad*)." From the Teachers' School at Paraná, a rising center for the new ideas of Jorge Stearns, Pedro Scalabrini, and Alfredo Ferreira, to the "German wizards from Cordoba" as Gould called professors Lorentz, Doering and others who took part in Roca's campaign in the desert. From the daring drive of Luis Jorge Fontana, one of Burmeister's collaborators and intrepid expeditionary of the Gran Chaco, to the taxonomic urges of the botanist Cristóbal M. Hicken, Holmberg's disciple and founder of the *Darwinion*, a botanical institute whose motto was *In Aggregatio Evolutio Maxima,* or the remarkable contributions by Miguel Lillo in Tucumán. From the southern lands explored by the unwearied *perito* Moreno or by Estanislao Zeballos, to the austral hemisphere described by Gould in his *Uranometría Argentina* (1879) and also in the research he carried out until his return to the United States in 1885. In all these stirring expressions and in many more are conveyed, beyond any words, the positive faith in evolutionary progress.

Not in vain did Alfredo Ebélot write in the *Revue de Deux Mondes* in 1876, that "a very particular feeling overtakes a Frenchman of this century—a century which is, if I may say so, rather critical, rational, and slightly obnoxious—when he happens to find himself in the company of some real savages," because along with the horror, the curiosity and the sympathy felt, there is also a sense of standing before a potential transformation, since "one of the characteristics of modern science, perhaps its worthiest one, is to refer everything having to do with matter and life to the laws of an ascending evolution."[45]

It was not because of a mere academic urge that Nicolás Avellaneda, shortly after completing his term as President of the Nation, responded favorably to Luis Jorge Fontana's request and gave his support to a grandiose volume commemorating Argentine culture titled *El Gran Chaco*, whose introduction reads: "All these works changed our intellectual development. They are worthy because of their direct utility and because, under their impulse, the propagation of the *scientific spirit* has begun. As that spirit expands like a gas, it brings enlightenment to our rational minds, strength to our convictions, and prosperity to our people. . . There is no such thing as a cohesive civilization without a scientific spirit. . . In the physical order, there is no phenomenon which is not bound by law; and social life too is composed of causes and effects. Thus nothing that exists in the present can avoid being explained by the past. . . . The scientific spirit has suppressed *arbitrary* understanding of the universe."[46]

Therefore, evolutionary progress ideologically articulated within an intensely biological matrix became the central feature of Argentine positivism. Riding the wave

---

[45] Alfred Ebélot, *Relatos de la frontera* (Buenos Aires, Solar/Hachette, 1968), p. 25. The article was originally published in the *Revue des Deux Mondes*, May 1, 1876.

[46] Luis Jorge Fontana, *El Gran Chaco* (Buenos Aires, Solar/Hachette, 1977), "Introducción," pp. 39-40. The original edition is from 1881.

of evolutionary biology, the "conquering bourgeoisie" of the 1880s found, through this surrogate of Providence, an ideology legitimized by modern science. That is precisely the reason why almost all our positivism seems "not like a philosophical concept emanating from a conclusion of physics or mathematics but as a hypostasis from biological data," according to Ricaurte Soler.[47]

So it is not strange that the ideological and scientific *pathos* of the beginning of the new century was captured by the paleontologist Florentino Ameghino. From his *Filogenia,* completed at the end of 1882 and published two years later thanks to Estanislao Zeballos, to *Mi Credo,* a lecture given at the Argentine Scientific Society in August 1906, as he was invested with honorary membership, Ameghino attempted an original and all-embracing construction with some biological ideas of his own. He defines the cosmos "as the junction of four infinite elements: immutable "infinite space" occupied by "infinite matter" in "infinite motion" through the succession of "infinite time." In that cosmos, the transformation and evolution of matter is bound by progressive, evolutionary laws, always in the direction of greater density, along with increasing complexity and diversification. Matter is also limited, moreover, to a "regressive, radiant evolution," tending towards increasing rarity, acquiring greater homogeneity and simplicity, in a sort of dialectical game of attraction and repulsion that finally reaches a universal balance. That is how he explains the infinite variety of aspects that matter presents and also how "the organism develops in an inverted, that is, radiant, form, and in a space of time infinitely short, all the concentrating motion carried out by the phenomena that have gone before us, from basidium to our begetters. Thus, ontogeny recapitulates phylogeny."

This cosmogony in which "the plurality of worlds is a very evident fact" concludes with a formidable anthropological prediction: "I firmly believe that Man will have the occasion to delay almost indefinitely the production of those fatal phenomena believed to come at a certain time in life," because "the length of a life span is not a payable order with a fixed rate but an open money market account which we can cancel—the later, the better. I do not think death need always be an inevitable and fatal consequence of life. . . "With his knowledge, man could . . .master evolution, give it some direction and frankly place himself on the path towards immortality." In this scheme, where "the notion of God vanishes before a much more real, positive, and greater conception of eternity of *infinite matter in infinite motion* within infinite space," is the summation of Ameghino's cosmological credo.[48]

---

[47] Ricaurte Soler, *El positivismo argentino* (Buenos Aires, Paidós, 1968), p. 66. See also Hugo E. Biagini, ed., *El movimiento positivista argentino* (Buenos Aires, Belgrano, 1985), especially pp. 210-222, and the superb work by Eduardo Zimmermann, *Los liberales reformistas. La cuestión social en la Argentina (1890-1916)* (Buenos Aires, Sudamericana / Universidad de San Andrés, 1994).

[48] Cited by Florentino Ameghino, *Conceptos fundamentales*, 4th ed. (Buenos Aires, W. M. Jackson, 1945), pp. 195-197, 213, 217, 219, 220, and 227. The way Ameghino constructs certain concepts on antagonistic and complementary "evolution-dissolution" antinomies is almost mimetically Spencerian.

This strong biological direction was especially influential in the analysis of historical and social phenomena. At the same time as José Ramos Mejía began to write a "collective psychology" under the title *Las multitudes argentinas* (in whose first chapter, *Biología de la multitud,* the social and political properties of multitudes are likened to the mechanics of organic bodies),[49] José Nicolás Matienzo—slightly influenced by Spencer—said in a very interesting book, that "political science should be positive and experimental, like the biological sciences, or else they risk degenerating into purely verbal exercises."[50]

Carlos Octavio Bunge did not hesitate to assert that "My concept of Law and State rely upon the biological ideas of adaptation, inheritance, and natural selection or the struggle for existence. Therefore, I may well use for my legal theory a modern type of research, inspired in the natural sciences and in tune with the positive trends of our time." Nevertheless, Bunge believed that the exaggeration of the principle of natural selection "has led to antisocial and fallacious doctrines such as Nietzsche's amoralism," because "what is normally understood by 'social Darwinism' or 'Nietzscheanism' is nothing more than a literary fantasy in which the most important facts of people's lives are underestimated." This is so because one specific trait in the struggle for existence among humans, the state of belonging to different communities, from the family up to the state, has achieved greater importance than the more individualistic struggle for existence among animals. This sociology or "anthropo-sociology that gets its specific theory from the evolutionary or genealogical theory" drives Bunge to emphasize ethnic differences, to praise Theodore Roosevelt's politics, and to reject "the neo-humanist principles" of the eighteenth century. On the road of the Enlightenment, "we have been given shots by the historical school against the smallpox of the philosophical school, and by the biology school against the chickenpox of romantic philanthropy." Bunge became a spokesman for Argentine imperialism: "I have never seen a people as white, as unique. . . . This young nation, so beautifully naïve, yet so beautifully generous, sees its black people and Indians as brothers...This nation is called, in maps and through history, the Argentine Republic. But just wait to see its teeth grow after it is weaned from the breasts which feed it. Just wait![51] Strangely enough, one of the executors of

[49] José María Ramos Mejía, *Las multitudes argentinas (Estudio de psicología colectiva)* (Buenos Aires, Belgrano, 1977), ch. 1, pp. 28, 37. The original edition is from 1899.

[50] José Nicolás Matienzo, *Le Gouvernment Représentatif Fédéral dans la République Argentine* (Paris, Hachette, 1912). The original Argentine edition--less complete than the French one--is from 1910.

[51] Carlos Octavio Bunge, *El Derecho (Ensayo de una teoría científica de la ética, especialmente en su fase jurídica),* 3rd ed. (Buenos Aires, Valerio Abeledo, 1907), pp 270-276, 299-304. The work was translated into French with the suggestive title, *Le Droit, c'est la Force*, by Emile Desplanque. According to the author, the French edition was done in such a way that if it be considered "a translator's abuse," as Ricaurte Soler asserted (*El positivismo argentino*, p. 182), the blame must be shared. See *Le Droit...* (Paris, Schleicher Frères), "Préface du traducteur," p. v. Moreover, the raw legal realism suggested by the title is confirmed by Bunge's definition of law: "This concept, derived from the above-stated biological bases, may be summed up in this way: law is the political expression of legal rules imposed more or less consciously by the dominating classes in order to preserve an economic status in their favor" (p. 304 of the Argentine edition).

American *manifest destiny* that Bunge admires so much was Theodore Roosevelt, who visited the southern Argentine region under the expert guidance of the *perito* Moreno in 1912.

The radical biologism of Carlos Octavio Bunge became even more intense in the ethical statements of Augusto Bunge, who did not hesitate to say: "In the protozoon that reacts against a less nutritious substance, a legal sanction and a moral criterion exist in a latent state. . . That is why I have said that that life is both law and ethics."[52]

Surprisingly however, Juan Bautista Justo, expressing a Socialist approach to history in his *Teoría y práctica de la historia*, also has a clearly articulated notion of progressivism, viewed in the light of biological evolution. "We walk through History without stopping. Mankind is always tending towards growth and transformation," writes Justo. This declaration of faith in continuous progress is only comprehensible within a biological framework: "From the moment that man has become intelligent enough to consider himself an animal, he must seek biology at the root of his history. . . The laws of life are the most general laws of History."[53]

But the pretension—both immoderate and frustrated—of reaching a thorough synthesis, a "scientific philosophy that clings to the results based on experience and which includes the current positivist biologism" would be attempted by José Ingenieros. His idea, condensed, that "the unity of reality (monism) unceasingly changes (evolutionism) because of natural causes" is only interesting to us for its sociological and political consequences, even when inconsistent.[54] When the whimsical Ingenieros considered the social evolution of Argentina at mid-century, he had first to explain the foundations of his thought. Humanity, because "it is a living species, is bound by biological laws. Because it lives in social clusters it falls under the rule of sociological laws. Because of its ability to transform and use the natural powers existing within its environment, it evolves according to economic laws which are a special part of the preceding group." But, when Ingenieros goes from theory to application, the results are highly significant. Argentina bears, without a doubt, its own manifest destiny, even exceeding the hopes of its leaders and people. "Indeed—he goes on to say—the material greatness of the Argentine people carries in itself the factors that will lead its behavior towards a more expansive policy, its intellect towards an imperialist doctrine, its emotions towards the collective feeling of imperialism . . . .If Argentina and Australia continue their extremely fast material development, which require both population growth and intensity of work, they will have some weight in international politics. In

---

[52] Augusto Bunge, "Los fundamentos biológicos de la moral," *Revista de Filosofía, Cultura, Ciencia, Educación* (Buenos Aires), 1, no. 4 (July 1915), 69.

[53] Juan B. Justo, *Teoría y práctica de la historia* (Buenos Aires, Libera, 1969), pp. 5, 13-14; original edition, 1909.

[54] José Ingenieros, *Principios de Psicología*, 6th ed. (Buenos Aires, L. J. Rosso, 1919), p. 30. The work was originally published in installments in *Argentina Médica* in 1910; the following year it was published in a special volume of *Archivos de Psiquiatría y Criminología* under the title of *Psicología Genética*. In the Spanish, French and German versions it is called Principles of *Biological* Psychology.

that case, the oversight of the other South American and oceanic nations will fall to them, an evolutionary step that will turn them into new foci of imperialist action." These perspectives seem exaggerated: "After a young United States and an adolescent Japan, it is probable that Argentina and Australia will awaken to imperialism and acquire decisive influence in the international political arena." Ingenieros sees "intensification of nationalist pride" in the social psychology of the Argentine people at the nation's centenary, as expressed in Ricardo Rojas' book *La Restauración Nacionalista*, published in 1909. At that time, only Chile and Brazil could challenge Argentina—according to the author—for South American hegemony. "Chile is a highly militarized country, with ideals of domination and conquest, prodded by compelling territorial needs. . . Brazil, on the other hand, has two major and highly respectable advantages over Argentina: the extent of its territory and superiority in numbers." But "Chile has neither territory nor fertility. Brazil lacks the proper climate and the proper race. Argentina possesses all four variables: a vast territory, a fertile soil, mild weather, and a white race." The ghost of a continental war evoked by such a prediction does not move Ingenieros. "It is very clear that the ideal for the Argentine people is in peace. . . . it only needs to let some years elapse for its distinction to become insurmountable. Territorial extension, fecundity, a white population, and mild weather, all predestine Argentina to a position of guardianship over the other nations of the continent."[55]

Positivism brought Pedro Scalabrini Ortiz, born in Como during the revolutionary passions of 1848, to our country. He taught at the Teachers' School in Paraná, founded by Sarmiento in 1870, an institution that became a proper environment for "normal positivism," where the voices of Guglielmo Ferrero and Enrico Ferri were heard. Scalabrini first won fame as an expert in public law with his *Concordancia del Derecho Público Argentino con el Derecho Público Norteamericano* (1875). Much less known is his paleontological research in Paraná, some of which comes out in his *Cartas científicas* of 1887, dedicated to governor Eduardo Racedo, and his tenure as director of natural science museums in Paraná and Corrientes.

But our interest in Scalabrini refers to a another aspect of his work, as a synthesizer of current philosophical and scientific movements under the title *Materialismo,*

---

[55] José Ingenieros, *La evolución sociológica argentina (De la barbarie al imperialismo)* (Buenos Aires, J. Menéndez, 1910, part 3: "El devenir del imperialismo argentino," pp. 10, 94, 98-101, 103, 105. In a later work, *Sociología Argentina*, 7th ed. (Buenos Aires, L. J. Rosso, 1918), Ingenieros first includes a revised text essentially the same as the one presented in *La evolución...* but extends the argument into an interesting part 4 on "La formación de la raza argentina." As was also true of Bunge, Ingenieros' overly-enthusiastic early propositions were later softened. See, on this score, the "Warning Note" to the sixth edition of *Crónicas de viaje* (1919), where he confesses: "I have discovered some traces of the only intellectual trend to which I was sensitive during my youth; and it is no wonder that, while under its influence, I was quite aware that Nietzsche's visionary prose was a result of his mental alienation." See the text in Ingenieros, *Obras completas* (Buenos Aires, Mar Océano, 1962), II: 81. Cf. Oscar Terán's introduction, "José Ingenieros o la voluntad de saber," in Ingenieros, *Antiimperialismo y Nación* (Mexico, Siglo XXI, 1979), tracing the development of his ideas, shattering the characterization of Ingenieros as the archetypical "positivist," and making evident the range of simultaneous, theoretical discourse in Ingenieros.

Darwinismo, Positivismo: Diferencias y semejanzas, first published in *La Opinión* of Entre Ríos in response to an anonymous critic, who at the end of 1887 had impugned the *Cartas Científicas* from the pages of a Buenos Aires newspaper, *El Figaro*. In 1889, a complete version was published in Paraná.

This essay by Scalabrini deserves to be judged in the light of a neo-Comtian synthesis that rejected both radical materialism and social Darwinism, bending in the direction of Comtian thought. The author begins by noticing the converging and diverging lines of "these three great intellectual currents." First he denounces the tendency of materialism to consider the best theories as simple corollaries of trivial ones. Nevertheless, if radical materialism could reduce its exaggerated claims and not invade the domains of biology, history, ethics, social science, education and, most of all, philosophy, it could complete the necessary task of clearing away spiritualism. Consequently, "positivism is opposed to materialism in the name of a greater science, more human, with a more accurate method, a real, more systematic philosophy," a paean punctuated with references to Comte's *Sistème de politique positive*.

With reference to Darwinism, Scalabrini praises its theoretical foundation but criticizes the struggle for existence: "I believe that natural selection, in the sense of the perfecting of organisms, would always exist, even if the struggle for existence—the exclusive cause of selection, according to Darwinian ideology—were suppressed. The truth is that followers of positivism do not admit a 'struggle for existence' as a permanent condition of biological progress, and much less, of social progress," Scalabrini asserts, in a direct strike against social Darwinism. More than struggle, there is *work* for existence; hence, Darwinism might be in agreement with positivism, although many have said evolutionary concepts could not be further away from Comtian orthodoxy. The altruism of Comtian positivism ends in a "proven religion," the religion of the *homo homini frater* that Scalabrini believes to be "a result of the evolution of mankind."[56]

As José Luis Romero has observed, "Perhaps it was because they shared a solid philosophy of life that the generation of the 1880s was such an efficacious force in the direction of Argentine life. Perhaps the history of philosophical thought could be written—as Alejandro Korn in fact wrote it—without mentioning any names, lest they suffer from 'the tedium of abstract disquisition.' Nevertheless, this generation has a special place in the history of ideas, because it is unusual for any system of beliefs to root itself so powerfully at the core of an elite, or to have so profound an effect on reality."[57]

This philosophy of life, social ideology or mentality—whatever one wishes to call it—which hatched diverse and even opposing lines of thought, nevertheless was galvanized by a common belief in progress as the engine and *ultima ratio* of history.

[56] See Marcelo Montserrat, "La presencia evolucionista en el positivismo argentino," in *Quipu*, 3 (1986), 91-101.

[57] José Luis Romero, *El desarrollo de las ideas en la sociedad argentina del siglo XX* (Mexico, F.C.E., 1965), p. 14.

This ideology shaped its identity in reaction to the *Syllabus errorum* of 1864 in which Pius IX firmly stated his opposition to "progress, liberalism, and modern civilization" (Proposition 80). It was a blueprint for building and educating, a territory to conquer and populate, a new material and intellectual frontier to define, a mental horizon of the liberal oligarchy of the 80s. Not tight enough to prevent it from suffering internal schism caused by its own critics and reformers, but solid and consistent enough in its core. It was an attractive horizon, an encouraging utopia about the future conceived as eternal progress, a formula that also engaged some leaders of the emerging socialist opposition.

But by the end of the century, European intellectual life was itself rent with conflicts. In 1889, while the Eiffel Tower presided over the Universal Exhibition, two books evocative of the times were published almost simultaneously: Henri Bergson's *Essai sur les données immédiates de la conscience* and Paul Bourget's novel *Le disciple*, "in which the great refuter, the deliberate, almost inhuman analyst, is humiliated before the unfathomable mystery of fate." And the Vienna of Lehár turned slowly into the Vienna of Freud, Kraus, Mach, and Musil.

Already in 1895, Ferdinand Brunetière had written a provocative article in the *Revue de Deux Mondes* titled "After a visit to the Vatican", in which he attacked Renan's *L'avenir de la Science,* written in 1848 at the request of Berthelot but not published till 1890, not long before his death. If Brunetière proclaimed "the bankruptcy of science," Berthelot replied with unconditional support for scientism, even though the edifice of European rationality had begun to show some cracks. In Argentina, it was not long before the signs of a cultural turn became evident. Ideologically, they were expressed in the nationalism of Ricardo Rojas and Manuel Gálvez, featured in the short-lived magazine, *Ideas*, under the light of Rodo's "Arielism" or the inspiration of Ganivet or Barrès.[58]

Paul Groussac, a very lucid spirit indeed, had already warned about the "paradoxes of the social science" in an article written in 1896, with a epistemological thrust that gives the essay a very modern feel. His subtle observations about the abuse of organicist methodology in the social sciences, oblivious that "such approaches were metaphorical and provisional," bolstered a discourse which held that "the flagrant barrenness of political and social science—particularly in economics—is based on a fundamental methodological error: it had been theoriezed prematurely, before completing the process of establishing facts after years of observation, and only then to deduce, wisely and cautiously, circumscribed and provisional generalizations. The great flaw of human science is fatuousness or the hopeless yearning for the unattainable. Its worshipers have not noticed—Groussac continues—that the vast and rich theories of modern physics are the result of hundreds of years of work, "and to imitate them, they start building the

[58] Carlos M. Payá and Eduardo J. Cárdenas, "El primer nacionalismo argentino," in *Criterio* 48 (1975), 585-592. By the same authors, *El nacionalismo argentino en Manuel Gálvez y Ricardo Rojas* (Buenos Aires, Peña Lillo, 1978).

house from the eaves down. . . It is widely admitted nowadays, that a branch of human wisdom does not quite deserve to be called a science, unless the phenomena it studies can be tested by means of calculus. . . But in the political and moral fields, we are barred from such mathematical accuracy. With respect to biology, 'social science' represents an organism whose tissues are composed of different cells. Viewed mathematically, this kind of problem would lead to 'indeterminate' equations and to a higher degree of resolution. Thus, in certain deep valleys, real sounds confused with echoes coming from different directions, making it impossible to discover their spatial origins. *Ignorabimus.* In social studies, we can only hope for increasingly better success at inference."[59]

In the Argentine academic world, nothing compares with Alejandro Korn's singular career. He completed his medical doctorate study in 1883 with a thesis on *Locura y Crimen* and was appointed in 1897 to a directorship at the Melchor Romero Hospital, where he remained for two decades. He was also professor in the Faculty of Philosophy and Letters of the University of Buenos Aires, later on head of the History of Philosophy Department. Beliefs then current began to fade away thanks to his pedagogical efforts, along with those of Rodolfo Rivarola in Buenos Aires and Coriolano Alberini in La Plata. So it was a physician who finally set limits on positivist biology, in spite of its anti-mechanist, anti-intellectual and anti-organicist traditions, as Ricaurte Soler so exaggeratedly liked to put it.[60]

In fact, rather than the "crisis of progress" that Babini sensed in predominance of applied over pure science—stagnant, if not yet decadent, in the 1890s,[61] the crisis is better explained by its insertion in the final stage of biologist progressivism. Whiffs of fresh air traveled around the world and an Argentine gentleman, Ernesto Quesada, who happened to be both interested and interesting, scanned the horizons of the Old Continent in search of a cultural *aggiornamento* for our country.[62]

The secure model that biologist progressivism had offered for historical, political, social and ethical analysis continued to exist for some time, especially in its Socialist version. But it was slowly to vanish, contemporaneously with the crisis of the

---

[59] Paul Groussac, "La paradoja de las ciencias sociales," *La Biblioteca*, 2 (1896), 309-320, on pp. 309-310, 319-320.

[60] Osvaldo Loudet and Osvaldo Elías Loudet, *Historia de la Psiquiatría argentina* (Buenos Aires, Troquel, 1971), pp. 133-141. See also Ricaurte Soler, *El positivismo argentino*, pp 196-197, 246-249; and Hugo Vezzetti, *La locura en la Argentina* (Buenos Aires, Folios Ediciones, 1983).

[61] José Babini, *Historia de la ciencia en la Argentina*, pp. 75-76, and "La crisis científica del 90," *Revista de Historia*, 1 (1957), pp. 86-88.

[62] Ernesto Quesada, *La enseñanza de la historia en las universidades alemanas* (La Plata, Facultad de Ciencias Jurídicas y Legales, 1910). This thick volume was based on thorough research requested by Rodolfo Rivarola, Dean of the Facultad de Ciencias Jurídicas y Legales at La Plata University, and carried out at twenty-two German universities during the winter semester of the academic year 1908-1909. Part 4, which is devoted to conclusions, is particularly interesting because Karl Lamprecht's institute in Leipzig is proposed as a model. See the reference on pp. 983-985 to Ludwig Gumplowicz and his social and racial Darwinist school.

conservative political order. The rational and harmonic universe once dreamed by Spencer, in which Argentina's imperial destiny held a reserved seat—as Ingenieros wanted—was itself transformed into a ravenous irrationalism. Once again, the dream of reason had created monsters.

Universidad de San Andrés

THOMAS F. GLICK

THE RECEPTION OF DARWINISM IN URUGUAY

RANCHERS DEBATE DARWIN

The reception of Darwinism in Uruguay followed the familiar Latin pattern of a debate between positivist and religionist intellectuals in the late 1870s and 1880s, with a significant and interesting exception. Before the intellectual debate began, another debate over Darwin's merits had already taken place among the cattle breeders who were members of the Asociación Rural—the Rural Association. That organization had been founded in 1871 with the intent of stimulating the modernization of the agrarian sector. Among its 165 founding members were members of both political parties, cattlemen, industrialists, and lawyers, all of whom subscribed to the objectives of the Association. Fifty-three of the founders were foreigners, including twenty-one Englishmen and ten Frenchmen. The Association favored the "concourse of all ideas," and its statutes prohibited any religious or politically motivated manifestation.[1] This explains, I believe, the surprisingly open debate over Darwinism in the pages of its *Revista*, in its first twenty volumes. The Association's library contains an almost complete collection of Darwin's works, in the same French edition by Reinwald that influenced all sectors of the Uruguayan intelligentsia.[2]

In the very first volume of the *Revista*, that of 1872, two members—Víctor Las Cazes and Lucas Herrera y Obes—debated whether selection was a sufficient way to upgrade the national herd. By selection, they understood, naturally, the artificial or methodical selection that breeders practiced. According to Las Cazes, the Uruguayan cattle industry was too primitive to practice crossing and selection, although the indicated method, was also unlikely to yield a productive result, "unless we

[1] José Pedro Barrán and Benjamín Nahum, *Historia rural del Uruguay moderno (1851-1885)* (Montevideo, Banda Oriental, 1967), pp. 330-332, 388.

[2] See the inventory of Darwinian books in Uruguayan libraries in Thomas F. Glick, *Darwin y el darwinismo en el Uruguay y en América Latina* (Montevideo, Universidad de la República, 1989), Appendix, pp. 107-116.

Thomas F. Glick, Rosaura Ruiz, and Miguel Ángel Puig-Samper (eds.), The Reception of Darwinism in the Iberian World, 29-52.

simultaneously improve the conformation and [feeding] regime of the animals," which had to be much more intensively pursued.[3]

Herrera's conception was similar, as he believed that "the only way to obtain cattle in continuous relationship with the nutrients that the country provides, is to impose a consistent and intelligent selection upon the animals." That selection could indeed be put into practice immediately:

> I note that Ch. Darwin in his *Origin of Species* is in complete agreement with what I have just suggested. Indeed, he says: "The improvement [of races] is by no means generally due to crossing different breeds; all the best breeders are strongly opposed to this practice, except sometimes amongst closely allied sub-breeds."[4]

This is the first mention of Darwin in the *Revista* and thus requires comment. First, Herrera had read the *Origin* in French translation (for "breeder," he uses the term *educador*, from French *éléveur*, for example). Second, he gives the impression that Darwin was already known in Uruguay. And third, he leaves no doubt as to what was the crux of the relationship between Darwin and the cattlemen: the great naturalist did not believe in the efficacy of crossing. Herrera is precise: "we can see in his chapter on variation in domestic species that in England the races of horses and sheep have been the result of a well-understood selection."[5]

The debate between selection and crossing was driven by cost. According to Herrera, selection was the better method and crossing, excessively costly when one considered the requirements of adaptation to the local range:

> Crossing can be practiced on a defective, native race with two different aims: either the transformation of the native race into the improved foreign one; or simply the correction of some of its defects.
>
> For the first objective, you need to have the same elements that are necessary for the sustenance of the race introduced; the transformation will take place more quickly, the more alike are the diet, the climate, and the [breeding] system applied. It is well known that we cannot, in economic terms, think of this kind of operation for the time being, with improved European races and our cattle, because it is contrary to the state of our agriculture.[6]

By this time the dynamic of improvement had already tilted in favor of crossing. Carlos Genaro Reyles had begun importing Durhams in 1870.[7] The success of the operation remained to be demonstrated, but that was just a question of time. In the meantime, doubts remained. In the Rural Association, Reyles was the crossing spokesman, while Faustino J. Méndez championed selection as "the most convenient

---

[3] Víctor Las Cazes, "Especie bovina," *Revista de la Asociación Rural* [hereinafter cited, RAR], 1, no. 4 (1872), 12-14.

[4] Lucas Herrera y Obes, "Especie bovina," RAR, 1, no. 6 (1872), 21-24, citing Darwin, *Origin of Species*, 1st ed. (London, John Murray, 1859), pp. 31-32.

[5] Herrera y Obes, "Especie bovina," p. 23.

[6] Herrera y Obes, "Especie bovina,", RAR, 1, no. 5 (1872), 16-19, on p. 17.

[7] On the history of crossing, see Alba A. Mariani, "Los comienzos del proceso de mestización ganadera," in *Cinco perspectivas históricas del Uruguay moderno* (Montevideo, Fundación de Cultural Universitaria, 1969), pp. 85-121. Reyles had a "scientific cabinet" is his study, which included fossil mammals; it is described by his son Carlos in his novel, *Beba*.

and economical system" and "because it is within the reach of all, it works on any kind of ranch."[8]

In May 1878, Enrique Artagaveytia, horse breeder and outspoken partisan of selection, organized three lectures at the annual meeting of the Association on the subject "Improvement of Bovine Stock." The subject of the first lecture was selection, crossing, and the formation of new races; the second, on the causes of degeneration in cattle breeding stock; and the third—the one that aroused the most interest—was a debate between Méndez and Reyles.[9] I have been unable to find concrete references as to what was said, but five years later the debate was remembered as a memorable event:

> Just a few years ago. . . right at the dawn our rural science, there was a great debate among cattlemen. Its motive was really interesting. Would there be selection or crossing? In this Darwin, the scientific authority on these matters, was brought to bear, as well as other European and American experts in this science: breeders, *éléveurs*, zootechnicians[10] in a word. The debate began in the Rural Association, first in its journal, then in lectures, reaching the point where it was brought up in general meetings where all of our principle ranchers gave their opinions. Two outspoken partisans stood out, one in favor of selection, the other of crossing.
>
> With the exposition of each doctrine, the debate concluded, and the *leaders*[11], with no clear victory on that occasion, left the meeting with the assembly divided into selectionists and those who favored crossing. But, the *leaders*? We must applaud them, for once they defended their causes with pen and word, they set out to prove their points with deeds. One practiced selection on the Uruguayan coast, the other crossing on the Río Negro. Let us display similar patience while we await their Stud-Books.[12]

Meanwhile, the debate continued in the pages of the *Revista*. As late as 1877, Domingo Ordoñana, the leading ideologue of the ruralist movement, took the opportunity to comment on Darwin in the course of some "zoological observations." According to Ordoñana, the problem lay in the nature of species, which were "fixed and invariable."[13]

> The partisans of the mutability of species have now reached the point where they believe that new species can be formed according to some stupid logic, and the learned Darwin who has *evolved* natural history to the point where he has confused races with species, also has participated in the same fantasy. Lately he has told us that, even though he immersed himself in zootechnics, he has found no observation capable of enlightening him.[14]

Crossing does not produce new and intermediate varieties; the inferior race is absorbed. Neither does selection lead to anything new; it simply multiplies individuals, either modified or perfected, "but whose modifications have the objective of augmenting their utility." And, in spite of selection, the *type* of the race is maintained intact and invariable. The zootechnicians who follow the Frenchman André Sanson and the Spaniard Navarro support the natural plan "which maintains and subjects species

---

[8] See the "expositions" of Méndez and Reyles in RAR, 6 (1877), 312-316 (Méndez) and 315-316 (Reyles). Neither cites Darwin.

[9] RAR, 7 (1878), 132. Artagaveytia is identified as a selectionist in RAR, 12 (1883), 591.

[10] I prefer the term "zootechnician" to veterinarian. These men studied animal science attuned to the needs of breeders.

[11] English in the original.

[12] Un aficionado XY, "Selección y cruzamiento," RAR, 12 (1883), 134-138, on p. 134.

[13] D. Ordoñana, "Consideraciones zootécnicas," RAR, 6 (1877), 431-434, on p. 431.

[14] *Ibid.*, p. 432.

and varieties within their limits or zoological type, only permitting the development of natural aptitudes, without surpassing them."

Here, without knowing it, Ordoñana had stumbled upon an ideological problem, inasmuch as Sanson, whose influence was great in all Latin countries, was unable to distinguish between natural and artificial selection, completely misunderstanding Darwin. For Sanson, the species was "a distinctive form," a fact he had confirmed in experiments with Merino sheep (an experiment mentioned by Ordoñana to illustrate the absorption that is the end-product of crossing).[15] Therefore, Ordoñana was simply being logical when he expressed his

> surprise at contemplating the road that Mr. Darwin's doctrines had taken him in his *Origin of Species* taking as a pretext variations that have no meaning in zootechnics, as a prologue to change.
>
> If Mr. Darwin had been a cattleman he would not have launched half of the hypotheses that plague his book, because all of them are contrary to the facts encountered in practice.

Of course, Ordoñana had it reversed. The observations on which Darwin had based his concept of artificial selection and, therefore, natural as well, had been gleaned from the same English breeders who were so frequently cited in the pages of the *Revista*. If Darwin had been a *French* cattleman. . .

A similar point was advanced less stridently by René Sacc, a Swiss zootechnician of the same tradition as Sanson who had been contracted by the Uruguayan government as an agronomist. Sacc at least had an appreciation of what aspects of Darwin's writings were useful in their application to domestic animals:

> The immortal Darwin, studying the modifications which domestic animals undergo under the influence of man, has observed that in doves and dogs, these modifications are not limited to the exterior appearance of the individual, but they also affect the interior, even modifying the number and thickness of bones. In that he is correct, and I can confirm his observations with my own of feather-footed chickens that present the incredible trait that the number of their phalanxes diminishes correlative to the feathers that adorn their feet.[16]

Sacc also agreed with Darwin that crossing only gives a temporary result: characters of type always reappear. But when he considered the theory of descent, Sacc was incapable of interpreting his own observations any further:

> Darwin, who is the most profound and lucid observer of our times, embarked from a false starting point, basing his notion of the variability of species on the modifications which are obtained under the influence of man. For these modifications disappear once the intervention of man ceases.[17]

From here Sacc is led back to Linnaeus and the fixity of species. He ends his essay complaining about the interminable debates over Darwin that had converted "the majestic field of science into a small partisan question."

Up to this point we have observed a variety of positions: crossing will transform the creole stock (Reyles); artificial selection would do it (Méndez, Herrera y Obes); neither will work because types are fixed by definition (Ordoñana). At this point a man of

---

[15] On Sanson's anti-Darwinism, see Yvette Conry, "L'Introduction du darwinisme," *Rivista di Filosofia*, 73 (1982), 72-73.

[16] Dr. Sacc, "El carnero," RAR, 7 (1878), 276-286, on p. 276.

[17] *Ibid.*, p. 283.

science joined the debate, on the side of selection: José Muñoz Romarate, the first graduate of the Faculty of Medicine in Montevideo and a Darwinian like his mentor José Arechavaleta (see below). Muñoz was also the first participant in this debate who understood the significance of artificial selection in Darwin's argument. Everything depended on variation, even though its laws were very imperfectly known. Breeders are very conscious of the complexity of the laws, as well as of the power of inherited tendencies such as albinism. The key to the problem of domestication is "the selective power of accumulation that man enjoys: *it* determines the nature of variations; man favors them to an extreme and in a determinate direction."[18] He goes on to introduce the line of English breeding theory that regards the organization of the animal as "something plastic which can be modeled as one pleases."

In his discussion of natural selection, Muñoz replies to the objections of French zootechnicians for whom natural selection was a conscious activity of animals and cannot be applied to plants. As Yvette Conry has noted, for Sanson and his school, natural and artificial selection were isomorphic categories, both requiring a selector. Taken literally, Muñoz observes, "natural selection" is a contradiction in terms and must be understood as a metaphor.

Turning to the nature of artificial selection, Muñoz observes that while breeders select only the most important characters, given the limited time any owner has in which to change his herd through selection, natural selection is always operating. Moreover, "nature is not concerned with appearances" but rather its power extends "to internal organs, both in cases of slight organic differentiation and in those involving the entire living mechanism." Natural selection is therefore a more finely tuned instrument than is artificial.[19] Now he introduces Darwin's argument on pigeon beaks to demonstrate that artificial selection, carried to the extreme, can produce two distinct sub-races, with no intermediate forms: this is the same law of divergence of character that operates in natural selection. For Darwin, "the grouping of living forms around effective centers from which they radiate and diversify is explained by heredity and by the complex action of natural selection, which implies divergence of characters."[20] Muñoz compares natural to both unconscious selection (when a breeder preserves the best individuals and destroys the inferior ones, without any explicit intention to improve the stock) and methodical selection (where the breeder intends to improve the stock).[21] Nevertheless, the breeder cannot ignore the effects of natural selection: "Even in the case of domestic animals natural selection intervenes in specified limits, beyond the action of man and even against his will."[22] Darwin had explained how unconscious selection functions, and to carry it out "it is necessary to prevent the crossing of different races."[23] Muñoz's contribution, from a historical perspective, must have clarified selection's incompatibility with crossing. The two methods are in no way complementary: either one or the other.

---

[18] José Muñoz Romorate, "De la selección como sistema de mejora en los animales domésticos," RAR (1879), 284-287, 298-301, 321-326, on p. 286.
[19] *Ibid.*, p. 299.
[20] *Ibid.*, p. 300.
[21] *Ibid.*, p. 301.
[22] *Ibid.*, p. 322.
[23] *Ibid.*, p. 323.

By this time (the first five years of the 1880s) the general Darwinian debate was in full flower, involving all strata of society. All that was lacking was for someone to make the case that natural selection had already operated on the creole herd, adapting it to the peculiarities of Uruguayan pasture. That argument was in fact made, by an anonymous author, in 1883. His argument is interesting since it no doubt represented a common opinion among cattlemen, particularly those who did not wish to invest in imported English stock:

> In their enthusiasm, Darwin's new disciples only expound one aspect of the question—that of fixity, constancy of blood lines, and thus the question of the race one is trying to perfect so that it might be stable and permanent in a fixed locality with its conditions of climate, forage, and so forth.

Therefore, there was no reason to import new breeding stock, becaue if the first breeders appreciated the characters they desired, they achieved them "by means of the life in which they were developed:"

> Nor need we leave the country to confirm it. Are the cattle of Minas and Tacuarembó anything like those of the other Departments? They are nonetheless of the same origin. But after so many generations the cattle of those Departments acquired the special conditions of cattle of hilly regions, and are therefore better for milk than for beef. That is a fact.[24]

Still, the anonymous author, although favoring selection, opted for an agnostic stance with regard to the possible advantages of crossing.

The same year there took place a debate among *estancieros* that began with "Philippic" against crossing written by Félix Buxareo Oribe, rancher and zootechnician.[25] According to Buxareo, the bottom line of crossing was that stocks had to develop in conformity with the environment of a specific locality:

> Each race is the result of local influences on a population of a species, a population which, under such influences, is modified, adapting itself to the climate, the temperature, to the feeding regime, to the kind of work of the environment in which it must exist and reproduce. It concludes by assuming a particular character which is maintained and perpetuated, not only through the permanent existent and continuous action of these local influences, but also by hereditary transmission.[26]

He thought that Durham bulls, once introduced, would lose their pristine characters over a period of a few generations. The proof of this was the perturbation introduced into the creole horse herd "with that bastard, fictitious, useless and costly animal called the English horse."

Buxareo was answered by two supporters of crossing. The first, Alfredo Herrera (a convert from selectionism) had concluded that to reject crossing was to oppose

---

[24] Un aficionado, "Selección y cruzamiento" (note 12, *supra*), p. 135.

[25] Félix Buxareo Oribe, "Sobre el cruzamiento de razas animales," RAR, 12 (1883), 39-41. Buxareo was an *estanciero* who became a zootechnician. There are a number of allusions to Darwin is his manual, *Bovinotecnia* (Barcelona, Tip. Católica, 1898). His definition of species begins with Cuvier and ends with Quatrefages, although he seems not to have grasped the latter's association of the Biblical species with the Linnean family. All of his definitions of species are based on similarity of character, which does not permit transformation. In his discussion of selection, Buxareo defines natural and artificial selection, pointing out the advantages of the latter because it required no costly imports (pp. 19, 80).

[26] Buxareo, "Cruzamiento," p. 40.

progress.[27] The second was Carlos Reyles, who declared himself "diametrically opposed to the partisans of natural selection." He meant "artificial selection," of course, but by this time the terms had become confused and interchangeable. He says he had been a backer of natural section, "like everyone else," but experience had convinced him of the benefits of crossing, inasmuch as after twenty-two years of experience at crossing Creoles with Durham bulls, their blood had not deteriorated "in the least."[28] Reyles suggests that Buxareo perform an experiment, purchasing some "select, red-colored Creole cows—I mention the hair because it is essential for the beauty of cattle that the cows be of this color," and cross half with Durhams.[29] It is interesting that Reyles insists on the importance of color. Before the diffusion of Mendel's laws, and in view of the importance that both Darwin and Wallace assigned to the importance of the coloring of animals, Reyles had assigned this trait a significance which, in fact, it didn't really have.[30]

In another article on zootechnical matters published the same year, Ordoñana explained that the degeneration of races does not occur in cattle in their country of origin when they breed only with others of the same race—"this is contrary to the doctrine of the illustrious Darwin who was a naturalist and not a zootechnician."[31] In 1885, Ordoñana embarked on a European tour, sending back some interesting travel letters for the pages of the *Revista*. It is instructive to note that in a letter dated in London in September 1885, the veteran anti-Darwinian comments on British Imperial policy in Darwinian terms: "spirits are concentrated on the zoological struggle for existence," the same expression that, later on, he appropriated to define the working class movement: "Worker strikes, which are no more than zoological struggles, continue to appear in various European countries."[32]

The great debate between selection and crossing was closed in 1886 by the same person who had begun it, Lucas Herrera y Obes.[33] Recall that in his article of 1872, Herrera had inclined towards selection, on Darwin's authority. Meanwhile however he too had begun to import Durhams, at the same time as his attitude towards Darwin had hardened. In the course of reevaluating selection, he evidently changed his views on Darwin as well:

> It is impossible to speak of improving races without establishing what is meant by "races" and whether they are capable of improvement or not. Nor can one speak of crossing without deciding if there are, or are not, impassable barriers between the different kinds of animals on which we operate. Nor can we approve or condemn selection as an improving

[27] Alfredo de Herrera, "Sobre ganado Durham," RAR, 12 (1883), 68-70. See another article by the same author, "El cruzamiento de razas," RAR, 8 (1879), 22-26; on p. 24, he mentions Darwin's comments on the Collins, a family of English breeders.

[28] Carlos Reyles, "Sobre el cruzamiento de nuestro ganado vacuno con el de raza inglesa Durham," RAR, 12 (1883), 164-165, on p. 164.

[29] *Ibid.*, p. 165. See the discussion of this exchange by Barrán and Nahum, *Historia rural (1851-1885)*, pp. 604-611.

[30] See the anonymous article, "Colores de los animales," RAR, 11 (1882), 200-201, where the interest of Darwin and Wallace in the topic is mentioned.

[31] D. Ordoñana, "Historia natural zootécnica," RAR, 12 (1883), 228-230, on p. 230.

[32] D. Ordoñana, "Correspondencia de Ordoñana," RAR, 14 (1885), 609-617, on p. 609, and 15 (1886), 193.

[33] Lucas Herrera y Obes, "La mejora de ganados," RAR, 15 (1886), 40-45.

agent without knowing, appreciating, and judging the Darwinian school, which is that which introduced this scientific term with the meaning it has today.[34]

He goes on to describe the mutability of organic types according to the Darwinians, concluding that, for them, selection is the *agent* while for anti-evolutionists it is the *instrument*. The question that remains is the definition of species, whether it is absolute or merely conventional: "It is not possible to be only half a Darwinian: it is necessary to be one, or not to be one:"

Whenever selection has been identified as an agent for the improvement of our races of cattle, I have impugned it on the understanding that the question could not be discussed except scientifically and the virtue of selection, so considered, rests on Darwinian theories which we do not accept. We reject selection as an agent, even though we do not admit it as an instrument either; that is to say, as a consequence of heredity.[35]

He had, in the end, decided in favor of the Durhams which, in the context of the present debate, implied a rejection of Darwinism. It is obvious that Herrera, once convinced of the benefits of crossing, could abandon Darwin and give full rein to his ideological inclination to oppose him, inasmuch as he no longer had any practical motive to continue in an agnostic, utilitarian stance. His brother Julio was an outspoken anti-Darwinian.[36] Lucas Herrera and Domingo Ordoñana are convenient symbols of the limits of "civil discourse" within the Rural Association: to consider the ideas in a public forum and apply them whenever they were useful was possible without the abandonment of previous ideological positions.

It is important to note that during the period 1872-1890, the *Revista* of the Rural Association was the most important medium for the diffusion of Darwinism in Uruguay. The selection/crossing debate aside, the journal published numerous articles on Darwinism, almost all of which were favorable: for example, an article on mange as a hereditary predisposition, with an allusion to Darwin's theory;[37] a note on *Toxodon platensis*, referring to a specimen collected by Darwin and described by Richard Owen;[38] an article by the Spanish Darwinian, Odón de Buen, on similarities between plants and animals, with an openly monistic conclusion;[39] an essay by Fernando Manduit on atavism in a fully Darwinian and selectionist context;[40] and, to close the cycle, a panegyric of T. H. Huxley in which he roundly asserts that "the most powerful instrument for the extension of the domain of knowledge in natural history which has been placed in human hands since the publication of Newton's *Principia* is the *Origin of Species*, by Darwin."[41] The importance of these articles transcends their practical or

---

[34] *Ibid.*, p. 41. Herrera y Obes is identified as an importer of Durham bulls by Alfredo de Herrera, "Sobre ganado Durham" (note 27, *supra*), p. 68.
[35] Herrera y Obes, "La mejora de ganados," p. 42.
[36] Julio Herrera y Obes, "Evolución," in *Escritos* (Montevideo, 1947), pp. 7-11.
[37] Eugenio Clairian, "El acarus o arador y su probable procedencia," RAR, 2 (1873), 100-109 (reference to Darwin on p. 106).
[38] J. T. B., "Paleontología. El Toxodon Platensis," RAR, 4 (1875), 1125.
[39] Odón de Buen y del Cos, "Analogías y diferencias entre animales y vegetales," RAR, 14 (1885), 203-204, on p. 204: "Insofar as concerns the great laws of the struggle for life, heredity, and evolution, there is only the slightest difference between plants and animals; atmospheric agents might vary and induce variations in a plant, which either accommodates its organism to exterior influences or dies."
[40] Fernando Manduit, "Atavismo," RAR, 14 (1885), 570-573 (reference to natural selection on p. 572).
[41] T. H. Huxley, "El origen de las especies," RAR, 18 (1889), 127-129, on p. 129.

topical interest inasmuch as, in the words of one rancher, the *Revista* of the Association constituted "the reference book of young ranchers."[42]

THE NEW BIOLOGY IN URUGUAY

We can date the first discussion of Darwin among Uruguayan positivists to 1874 at the earliest. It has been claimed that the first supporters of Darwin, in 1875 and 1876 respectively, were José Pedro Varela (1845-1879) and Angel Floro Costa (1838-1905).[43] That is the impression that one gets from reviewing the student and positivist journals of the 1970s; I find no mention of Darwin there before 1875.[44] No Darwinian titles appear in a manuscript catalog from the National Library in 1870.[45] But from this evidence one can conclude only that in university circles Darwinism and positivism made their appearance simultaneously. A search of Darwinian titles in other Uruguayan libraries up to 1900 produced a strong predominance of Darwin—both in titles (28) and copies (56) over Haeckel (16 titles, 24 copies), and Huxley (9 titles, 16 copies). Uruguayan intellectuals preferred to read Darwin directly, rather than in popularizations by others and read Darwin in French, as they did the works of other Darwinians like Haeckel, Huxley, and Gegenbaur, as well as Spencer's *Principles of Biology.* Only Varela had a collection of Darwin's works in English, but he too seems normally to have used French translations, which he cites in his works.[46] So my search confirms the observation of Ergasto H. Cordero, with reference to the Reinwald editions, that "it was in those works that young people imbibed the new theories."[47]

Darwinism was more widely diffused in university courses and public lectures than through the written word. At the Ateneo del Uruguay, an intellectual club where the most important public lectures were presented, "Of those who lecture here, the majority

---

[42] Alfredo de Herrera, "Sobre ganado Durham," p. 68.

[43] Arturo Ardao concludes that positivism first appeared in Uruguay somewhat before 1875 and that the first mention of Darwin was by Varela, in *La legislación escolar* (1876); *Espiritualismo y positivismo en el Uruguay* (México, Fondo de Cultura Económica, 1950), pp. 79-80; and *Etapas de la inteligencia uruguaya* (Montevideo, Universidad de la República, 1971), p. 109.

[44] The first mention occurs in L. Figuier, "El hombre primitivo," *La Voz de la Juventud*, 2 (1875), 10-11, 19, 22-23, 28. Darwin and Quatrefages are mentioned on p. 23, with both names misspelled (Darwn, Buatrefages).

[45] Biblioteca Nacional, list dated March 19, 1870. It is odd that among the 111 natural science titles there is not even one evolutionary volume, stranger yet because José Arechavaleta was a member of the governing board of the Library at this time. The list is reproduced in *El Club Universitario*, 2 (1872), 301-303 (283 volumes, mainly in French).

[46] In spite of his collection of Darwin in English, Varela used French translations: of the *Origin*, that of Royer, with its Lamarckian annotations, and of *The Descent of Man* (Reinwald edition, with Carl Vogt's introduction). See Varela, *El destino nacional de la Universidad*, 2 vols. (Montevideo, Colección de Clásicos Uruguayos, 1965), I, 185, 199. Cf., ibid., p. 180, n. 1, where he provides Spanish titles, and p. 218, n. 1, where he gives information on the circulation of scientific books in Montevideo: both he and his opponent Carlos María de Ramírez depended on the same copy, in French, of Spencer's *Social Science*, owned by a common friend. "That friend," Varela explains, "obliged me to return the Spencer so he could lend it to Dr. Ramírez, which would have deprived me of a powerful weapon, since there are only a few copies of this book in Montevideo, had not another gentleman done me the kindness of lending me the same work, only in English." Clearly, Varela was more at ease reading works of English science in French.

[47] Ergasto H. Cordero, "Dos aspectos de la vida científica de Arechavaleta," *Revista Nacional*, 15, no.44 (1941), 250-255, on p. 252.

are evolutionists," according to the professor of natural history, José Arechavaleta. "Dr. Manuel B. Otero and Srs. Susviela Guarch, Felippone, and Regúnaga are outspoken partisans of [evolution]."[48]

In the Faculty of Law, Martín C. Martínez's chair of Natural Law was a center of Darwinian discourse. In his inaugural lecture of 1882, Martínez commented on the recent biological revolution: "It has been scarcely twenty years since that kind of natural history, which had been reduced to the humble classification of species, disappeared under the impulse of the greatest of scientific revolutions." That revolution had been so potent that it also caused the reformation of other sciences, such as psychology: "It is clear that the laws of heredity and natural selection...eliminated forever the empty system of [mental] faculties with which it had attempted to build an entire explanatory system of psychical life."[49]

The progress of the history of science, in Martínez's view, was linked to the progress of society generally. In his courses, Martínez taught law as a variant of sociology. In the curriculum for his 1885 course, for example, he develops the influence of ambient societies, including "the effects of the struggle for existence," on the development of social institutions. He also explained that the social and economic changes historically introduced in the organization of property respond to the "influence of natural selection." The course ended with his conception of philosophy of law, pointing out "modifications introduced by evolutionary theory in the utilitarian concept of law in the direction of explaining its historical development and the obligatory character of its prescriptions."[50] Martínez, who was both a Spencerian positivist and a Social Darwinist, characterized himself as an "explanatory naturalist," insofar as he believed that Darwinian mechanisms explained social phenomena in a literal, and not merely figurative, way. We can presume that his presentation of Darwinian theory was faithful to its biological meaning, as Martínez understood it.

In the chair of Constitutional Law, held since 1873 by the anti-Darwinian Justino Jiménez de Aréchaga, the students took part in well publicized debates on Darwinism, according to the direct testimony of one of the participants, Juan Antonio Ramírez, who recalled debating the *Origin of Species* with fellow student Carlos Vaz Ferreira.[51] Both Martínez and Aréchaga directed theses sustaining positions contrary to their own.

In the curriculum of the preparatory course in General Geography in 1884, Antonio M. Rodríguez taught anthropology from an evolutionary perspective:

---

[48] José Arechavaleta, "La teoría de la evolución es una hipótesis?" *Anales del Ateneo del Uruguay*, 1 (1881), 121-131, on p. 121. Florentino Felippone, student of Arechavaleta, was Professor of Chemistry. In 1988, I purchased the Ateneo's discarded copy of Carl Gegenbauer, *Manuel d'Anatomie Comparée* (Paris, Reinwald, 1874) from a second-hand book dealer. During the military dictatorship of the 1970s, the Ateneo purged its library of most of its Darwinian titles.

[49] Martín C. Martínez, "Discurso inaugural," cited by Juan Antonio Oddone and M. Blanca París de Oddone, *Historia de la Universidad de Montevideo. La Universidad Viejo, 1849-1885* (Montevideo, Universidad de la República, 1963), pp. 405-410, on pp. 406-407.

[50] Martín C. Martínez, "Programa de Derecho Natural," *Revista de la Sociedad Universitaria*, 3 (1885), 5-11, on p. 11.

[51] See Juan Carlos Gómez Haedo, "Justino Jiménez de Aréchaga," *Revista Nacional*, 1 (1938), 69-77, on p. 76; and Oddone, *La Universidad Vieja*, p. 264.

> General anthropology—State of humankind considered globally—Monogenism—Polygenism—Transformism and Selection[52]—Applications to man.

The text in that course was the *Antropología* of Paul Topinard, a French Lamarckian.[53]

Two of the earliest professors of the Faculty of Medicine (founded 1876) were eminent Darwinians: the professor of anatomy, Julio Jurkowski, and the professor of physiology, Francisco Suñer Capdevila, both of whom debated Carlos María Ramírez in 1877.[54] Suñer (1848-1916) was a Catalan deputy to the Cortes of the Spanish First Republic who fled to Uruguay after its fall.[55] The monistic and Darwinian orientation of his course is detailed in the first curriculum of the Faculty of Medicine. In the very first lesson students had to consider, among other topics, "the parallel between inorganic bodies and organized beings and between animals and plants" and theories about life. The second lesson introduced the Haeckelian theme of simple bodies in nature before proceeding to cellular evolution.[56] More characteristically monist themes appear in a succeeding lecture: "intellectual faculties," including "the influence of organization on intellectual faculties. Intelligence of animals" and the origin of instincts.[57] The debate over "moral faculties," whose source was the chapter so titled in *The Descent of Man*, was the subject most debated by positivists and idealists (*espiritualistas*, conventionally), inasmuch as the latter sought to preserve a non-material basis for human morality. The last lesson in Suñer's course was an openly Darwinian discussion, ending on Huxleyian note:

> Lesson 52. Development of the individual after birth. Ages. Temperaments. Death. Human races and their characteristics. Are they distinct species or branches of the same species? Animal species. Origin of species. Ascending progress of organisms. Spontaneous generation. Man's place in nature.[58]

Jurkowski (1843-1913), although an evolutionist, did not teach anatomy from an evolutionary perspective. Evolution is absent from his anatomy program, as is any comparative perspective, an oddity among Darwinian anatomy professors in this period. As a text he used Sappey's *Tratado* in preference to the evolution-oriented manuals of Testut or Gegenbaur.[59] Either Jurkowski's Darwinism was rhetorical only and did not affect his instruction, or he used Sappey because there were no Spanish translations of the evolutionary textbooks, or he judged that descriptive, rather than comparative, anatomy was more appropriate for his students.

---

[52] *Elección*, suggesting that he had used Royer's translation of *Origin*.

[53] Archivo General de la Nación (AGN), Archivo de la Universidad de Montevideo, 1883, expediente no. 20. Aula de Geografía General, Program del Primer Año, p. 12.

[54] Blanca París de Oddone, *La Universidad de Montevideo en la formación de nuestra conciencia liberal* (Montevideo, Universidad de la República, 1958), pp. 140-141, n. 64, provides extracts from lectures by Jurkowski and Suñer from the *Revista Científica Literaria*, 1 (1877).

[55] Suñer Capdevila is often confused with his brother of the same name. See Thomas F. Glick, "La 'Idea Nueva': Ciencia, política y republicanismo," in B. Ciplijauskaité and C. Maurer, eds. *La voluntad de humanismo: Homenaje a Juan Marichal* (Barcelona, Anthropos, 1990), pp. 57-70, on pp. 59-61.

[56] *Programa de la Facultad de Medicina* (Montevideo, 1876), p. 6 (Programa del Aula de Fisiología).

[57] *Ibid.*, p. 11 (lesson 47).

[58] *Ibid.*, p. 12.

[59] *Ibid.*, pp. 3-5. Ph. C. Sappey, *Tratado de anatomía descriptiva*, Rafael Martínez y Molina, trans., 2nd ed. (Madrid, Carlos Bailly-Baillière, 1878).

JOSÉ ARECHAVALETA

Arechavaleta (1838-1912), the founder of modern biology in Uruguay, was born in 1838 in the village of Urioste, near Bilbao. He emigrated to Montevideo at the age of 18. There he studied botany and entomology with Ernesto J. Gibert, French naturalist and a refugee from the Revolution of 1848. On the basis of these studies, Arechavaleta obtained a pharmaceutical license, his only professional credential. As a young man, Arechavaleta took part in a scientific salon or *tertulia* which met in the back room of the Las Cazes family pharmacy—meetings frequented by Gibert, Teodoro Vilardebó and other physicians and naturalists. Years later, Arechavaleta's own pharmacy was the site of a famous tertulia of positivists whose participants included Varela, the Ramírez brothers, and Carlos María de Pena.[60] Inasmuch as Arechavaleta's earliest publications date from the 1880s we do not know when he first became a Darwinian, although among the evolutionary books in his library, a few date from before 1870.[61]

In an important lecture read at the Ateneo del Uruguay in 1881, Arechavaleta in replying to Prudencio Vázquez y Vega's assertion that evolution was merely a hypothesis, identified himself as an evolutionist of the school of Ernst Haeckel.[62] For Arechavaleta, evolution was a fact amply demonstrated by evidence from comparative morphology and physiology, anatomy, embryology and, more specifically, studies of rudimentary organs, geological succession, and the geographical distribution of species.[63] He expounds Haeckel's three doctrines of the general theory of evolution, and those of descent and selection, emphasizing the fact that evolution lacks any plan or design and that evolutionary processes are as valid for one-celled organisms, as they are for complex ones.[64]

It is clear that in this period Arechavaleta conceptualized his research program in terms derived from his reading of Haeckel. Of this there is an indirect testimony of one of his students during the 1870s, Manuel Tardáguila. In a review of the history of entomology in Uruguay, Tardáguila observes that Lamarck and Darwin "have completely transformed scientific studies and their classification, an inevitable

---

[60] On Arechavaleta, see Antonio Peluffo, "Arechavaleta: el investigador, el maestro, el hombre," *Anales de la Facultad de Química*, 6 (1960), 7-22; Peluffo, "José Arechavaleta," *Revista Nacional*, 1 (1938), 121-129; Telésforo de Aranzadi, "Don José Arechavaleta y Balpardo," *Boletín de la Real Academia Española de Historia Natural*, 13 (1913), 528-545; Joaquín de Saltearin, "José de Arechavaleta," *Revista Histórica*, 9 (1918), 77-95; Cordero, "Dos aspectos" (n. 47, *supra*); Ardao, *Espiritualismo y positivismo*, pp. 129-135.

[61] Arechavaleta's personal library was incorporated after his death into that of the Museo de Historia Natural in Montevideo. Among the earliest titles are Richard Owen, *The Zoology of the Voyage of the H. M. S. Beagle*. Part I, Fossil Mammals (London, Smith Elder, 1840), and Louis Buchner, *Confèrences sur la théorie darwinienne de la trasmutation des espèces* (Paris, Reinwald, 1869). He owned more than one Darwin title, in the Reinwald French edition, from the 1870s, as well as copies of Vogt's translation of Gegenbaur, *Manuel d'Antomie Comparée* (see n.48, *supra*) and Huxley's *Éléments d'anatomie comparé des animaux invertébrés* (Paris, Adrien Delahaye, 1877). His books are identified by the stamp "J. Arechavaleta". His copy of Haeckel, *Psychologie cellulaire* is inscribed "Al profesor don José Arechavaleta, mi distinguido maestro y gran amigo. J. Regúnaga. Buenos Aires, enero de 1880." His copy of Huxley, *L'Ecrevisse* lacks the stamp but is inscribed, "Al Sr. Don José Arechavaleta, su aftmo. amigo y comp[añer]o. A. Vásquez Acevedo. Ag. 24 1880."

[62] Arechavaleta, "La teoría de la evolución es una hipótesis?" (n. 48, *supra*); Prudencio Vázquez y Vega, "Crítica de la moral evolucionista," *Anales del Ateneo del Uruguay*, 1 (1881), 201-222, on p. 215.

[63] Arechavaleta, "La teoría de la evolución es una hipótesis?", p. 125.

[64] *Ibid.*, pp. 123-124.

consequence of the two great laws of evolution and descent" that they formulated. From this, he concludes that "the natural classification of Haeckel, based on those laws...transforms [zoology and botany] completely." Tardáguila had studied botany and zoology with Arechavaleta in 1874 and 1875, when the latter had imparted to his students the new rules of natural taxonomy as outlined by Haeckel.[65]

According to natural classification, based on the findings of embryology, comparative anatomy and paleontology, Haeckel, in his *Naturliche Schopsfungs-geschichte* of 1868, a genealogical tree of twenty-two human ancestors. In the first stage were found moneras, primordial ancestors of all animals, lacking—according to Haeckel—a nucleus. Years later it was ascertained that Haeckel had invented these purely hypothetical ancestors by analogy with ontogenetic states of embryogeny (an analogy based on a misunderstanding of the fate of the nucleus of a fertilized egg).[66] Haeckel's hypothesis stimulated the "discovery" of various moneras, including the famous—or notorious—*Bathybius* of T. H. Huxley. As a pure Haeckelian, Arechavaleta repeated the feat: in the mud of a swamp in Carrasco, Arechavaleta believed he had found a Bathybius, "a supremely simple organism, a protoplasmic mass, entirely naked and without a nucleus," that he baptized *Helobius Oterii* in honor of his friend and fellow Darwinian, Manuel B. Otero.[67] The discovery itself is of less interest than its significance within the Haeckelian universe of Arechavaleta: the importance of the moneras—as Arechavaleta himself observed—is that their existence would reduce the distance between organic and inorganic beings—a crucial point for monists who wanted to do away with that distinction.[68]

Moreover, even though Arechavaleta admitted that even Huxley had, by that time, abandoned his *Bathybius*, it was nevertheless clear that for him to accept or reject moneras was the equivalent of accepting or rejection evolution itself.[69] He viewed his discovery in Carrasco as providing evidence for Haeckel in his moment of crisis: "In my opinion, then, *Helobius Oterii* should occupy the first step of the ascending scale, alongside *Protobathybius* and *Bathybius*." [70]

It was as a pedagogue that Arechavaleta's influence in the reception of Darwin in Uruguay was most felt. His role as a teacher has been concisely described by Telésforo de Aranzadi:

> with his teaching he awoke in an entire generation of studies love for the study of the observational sciences, for in an epoch when the laboratory and the microscope were

[65] Manuel Tardáguila, "Nuestra historia de entomología," *Boletín de la Sociedad Ciencia y Artes*, 8 (1884), 257-259, on p. 257. Tardáguila, who received a B.A. in 1871 and J.D. in 1882, appears in 1883 as a student of medicine when he applied for the position of interim professor of botany and zoology in the preparatory program of the University. He had been examined in General Botany in 1874 and 1875, in Zoology in 1874, and in Medical Botany, also by Arechavaleta, in 1881 (AGN, Archivo de la Universidad, expediente no. 52 of 1873) . In 1874 Tardáguila appears with other students of Arechavaleta in a request for railroad tickets from Montevideo to Juan Chazgo "for the purpose of herborizing with their professor" (AGN, Archivo de la Universidad, expediente no. 59 of 8174 (22 October).

[66] See the discussion in Stephen Jay Gould, *Ontogeny and Phylogeny* (Cambridge, Harvard University Press, 1977), pp. 170-173.

[67] José Arechavaleta, "Apuntes sobre algunos organismos inferiores," *Anales del Ateneo del Uruguay*, 3 (1882), 41-46, 250-255, on p. 44.

[68] *Ibid.*, p. 43.

[69] *Ibid.*, p. 251.

[70] *Ibid.*, p. 253.

> unknown as instruments of teaching, in which all was theory and speculation, it was he who, making his students see and observe the fundamental phenomena of biology, speaking to them of evolution and making them translate Darwin and Haeckel, definitively determined the orientation of the future studies of many of them towards the biological sciences.[71]

For the use of his zoology students at the Ateneo, Arechavaleta translated from the French a chapter that Alfred Giard, had written to accompany his translation of a book of Huxley's that he used as a textbook in his chair at Lille. In a translator's introduction, Arechavaleta observed that "The considerable advances of the natural sciences in these last few years is owing to the evolutionary theory contained and developed in the immortal book, *Origin of Species*." The Ateneo, wishing to keep abreast of the scientific movement, had founded chairs for its instruction. In his, that of zoology, he lectured "according to the new principles, which I have professed for many years."[72] Giard's text was a composite of Darwinian and Lamarckian notions. He presents three corollaries to Malthus's' "theorem" of population: first, the struggle for existence; second, the Lamarckian "law of adaptation," that is, the inheritance of acquired characteristics; third, natural selection."[73] Although neo-Lamarckian mechanisms were everywhere in vogue at this time, still Arechavaleta's choice of a text is odd, because nowhere else in his writings is there even a hint of Lamarckian explanation. After 1888 Giard began to emphasize Lamarckian mechanisms even more and to criticize what he regarded as the hypervaluation of the Darwinian struggle for life.[74]

But Lamarckism was only one aspect of Giard's evolutionism. He is today remembered more for his contribution to Haeckel's theory of recapitulation. Indeed, the pamphlet translated by Arechavaleta continues with Haeckel's biogenetic law and Fritz Müller's principle that the series of phases that an embryo presents in the course of its development will be abbreviated or condensed. The result of such condensation will be a more rapid evolution, so that variations advantageous in the struggle for life will come into play.[75] The discussion of Müller's ideas takes up five pages of the

---

[71] Aranzadi, "Don José de Arechavaleta" (n. 60, *supra*), p. 536. Cf. Peluffo, "Arechavaleta," p. 19: Arcechavaleta "translated for his students whole chapters of works on transformism: Haeckel, Spencer, Russell (*sic*; he meant Wallace), Weismann, and others" Another student of Arechavaleta recalled: "He was the first among us who dared to speak of Darwin and Pasteur with respect and admiration. I say this because, compared with the archaic dogmatism of the arguments of those times, anything that was not metaphysical speculation or scholastic dialectic was considered an act of irreverence and the person who so held, a vulgar empiricist" (Joaquín de Salterain, "José de Arechavaleta," p. 80).

[72] Alfredo Giard, *Un capítulo sobre los principios generales de la biología* (Montevideo, 1879), p. vii. Among the copies of this pamphlet preserved in the National Library is one that Arechavaleta signed for Angel Floro Costa. Giard's book was broadly based on Huxley's "Notes on the Invertebrata for the Use of Students of Zoology" (1874).

[73] Giard, *Capítulo*, pp. 29-30.

[74] On Giard's Lamarckism, see Peter J. Bowler, *The Eclipse of Darwinism* (Baltimore, Johns Hopkins University Press, 1983), pp. 111-112. It is interesting to compare Giard's corollaries with those adduced by Carlos Berg in 1891: "Growth with reproduction; transmission by heredity almost included in the reproductive process; variation through the direct or indirect influence of external conditions and the use and disuse of organs; rapid multiplication which must produce a struggle for existence and which implies natural selection which, in turn, determines divergence of character and the disappearance of less perfected forms (individuals, etc.) ; "Elementos de zoología," *Anales de la Universidad* (Montevideo), 1 (1891-92), 220-252, on p. 249.

[75] On Müller and condensation, see Gould, *Ontogeny and Phylogeny*, p. 101. In 1887 Giard made a important contribution to the theory, clarifying the embryological mechanisms of neoteny and progenesis (*ibid.*, p. 227).

pamphlet's thirty-seven. Why would Arechavaleta devote so much space to complicated mechanisms of recapitulation that few of his secondary level students could follow? He must have thought that in view of the reigning confusion over the mechanisms of evolution, Haeckel's biogenetic law offered the greatest hope of clarity.

POSITIVISTS VERSUS RELIGIONISTS

In Uruguay, just as in the majority of Hispanic countries, Darwinian mechanisms of evolution were not debated. What was put in play were systems of values and social diagrams. Darwin was a symbol conveniently appropriated as a weapon in the ideological battles that divided nineteenth-century elites. Darwinism was introduced as a theme of debate in academic circles at the same time as was Spencerian positivism, in 1875. A decade late, José T. Piaggio evoked the mood in the University Society:

> Ideas succeeded one another with each change of speaker. Scarcely had the echoes of a proponent of positivism died down when the semi-eloquent voice of a young Cartesian resounded in the hall. That was a vortex of ideas in play. . . some laughed at Spencerian doctrines and so many things were spoken about Darwin![76]

The first major episode was the debate between Carlos María Ramírez and José Pedro Varela in the fall of 1876.[77] Ramírez was correct when he asserted that Varela's introduction of Darwin in a critique of the University in his book *La legislación escolar* had a broader motive: "And now that I've named Darwin, I ask myself whether or not resistance or acceptance of Darwin has not become an unequivocal criterion of the retardation or precocity of whatever system of legislation among cultivated nations."[78]

Ramírez himself tried to block out an agnostic position while the scientists clarified the matter, inasmuch as—as he admitted, "I am inclined to believe that we do not descend from Adam and Eve; and I do not see that, in order to convince myself of our genealogical relationship with monkeys—in a very remote past—I need perceptibly change what I think about the present nature and future destiny of the human race."[79] Varela had to back off somewhat, explaining that he had alluded to the method of teaching philosophy, not the content, inasmuch as he was an advocate of the freedom of instruction.[80] Whether accurate or not, Ramírez's perception is an index of the strong polarization over Darwinism, that left no space for calmer spirits (what few there were) like Ramírez.

The next marker in positivist/idealist debate was the "conversion" of Gonzalo Ramírez (1846-1911), professor of Penal Law and brother of Carlos María, in 1878. In view of the piling up of

> unchallengeable proofs which depose in favor of the grand view of the English naturalist Charles Darwin, the modern philosopher and moralist resolutely proclaims the schism of the

[76] José T. Piaggio, "Discurso," *Revista de la Sociedad Universitaria,* 3 (1885), 271-275, on p. 273.
[77] José Pedro Varela and Carlos María Ramírez, *El destino nacional y la Universidad.* Polémica, 2 vols. (Montevideo, Colección de Clásicos Uruguayos, 1965).
[78] *Ibid.*, II, 119.
[79] *Ibid.*, II, 120.
[80] *Ibid.*, II, 188-189 n. 1.

> sciences; and following the example of the Catholic philosopher, they sentence us either to renounce science or to be atheists.
>
> My profession of faith I do here consign and, sealing it with an intimate memory of the other world, I tell you with all sincerity that, without ceasing to be a humble sectarian of the doctrines of Charles Darwin, I have been able to extend for the last time the frozen hand of a beloved being, feeling the idea of a Supreme Being palpitate in my brain, enlivening my heart with the beautiful dream of immortality.[81]

Gonzalo Ramírez's honest confession caused a sensation in intellectual circles and won him, to boot, a sardonic open letter from Angel Floro Costa. Costa noted that Ramírez's profession had stimulated "a formidable imprecation" from Julio Herrera y Obes who asserted that one could not be a Darwinian and a idealist at the same time. With this, Costa was in agreement, since "science neither permits nor tolerates diplomatic maneuvers."[82] The rest of Costa's letter was a kind of plea for an end to all anthropomorphic illusions.

The conversion of Ramírez marked the end of the "spiritualist" movement that had begun in 1872 with the Club Universitario's "Profession of Rationalist Faith," a Deist tract which declared both belief in God and a rejection of all Catholic dogma.[83] The ideological polarization that seemed inevitable when Darwin was debated in Catholic countries did not permit agnosticism.

In 1878 too, there took place a polemic over the conflict between science and religion in which the bishop of Montevideo Mariano Soler defended the Catholic side and Manuel B. Otero, the positivist Darwinian side.[84] Soler's strategy in defending the Biblical account of creation consisted in demonstrating that the *sequence* of events in the *Genesis* narrative was in accord with the findings of science. The key point was the supposed proof that animal life had preceded the appearance of plants. Soler looked foolish citing whatever scientist could be adduced in support of his position, however unknown or ancient he might be ("fossil authors," according to Otero), while Otero made the mistake of invoking Dawson's *Eozoon* which, as Soler knew, had already been discredited by Darwinian naturalists.[85]

In 1880, Soler published a pamphlet attacking Darwinism from the same Biblical perspective.[86] Although Soler prided himself in presenting only scientific arguments and even though he cited a plethora of sources (he had a copy of the Spanish translation of the *Origin* in his personal library), his science was nothing more than a thin patina that covered a religious apology of a genre that had by this time become traditional. He

[81] Gonzalo Ramírez, "Clase inaugural del curso de derecho natural y penal, 1878," *Revista Nacional*, 14 (1941), 295-298, on pp. 297-298.

[82] Angel Floro Costa, "La metafísica de la ciencia. Fantasía literaria dedicada a mi compatriota y amigo el doctor don Gonzalo Ramírez," *El Panorama*, 1 (1878), 25-31, 38-43, 97-102, 129-135, on p. 26. On the polemic between Floro Costa and Gonzalo Ramírez, see Fernando Mañé Garzón, *Un siglo de darwinismo: Un ensayo sobre la historia del pensamiento biológico en el Uruguay* (Montevideo, Facultad de Medicina, 1990), pp. 43-56.

[83] On the "Profession," see *El Club Universitario*, 2 (1872), 361; and París, *Universidad de Montevideo*, p. 115. Ardao (*Etapas*, p. 110) locates the end of the first phase of positivism in 1879, with the polemic of Jurkowski and Arechavaleta against Vázquez y Vega. In my view the conversion of Ramírez was more significant, inasmuch as it demonstrated the impossibility of a middle way.

[84] The lectures of Soler and Otero are reproduced in Soler, *El Génesis y la geología* (Montevideo, A. Barreiro y Ramos, 1878).

[85] *Ibid.*, pp. 75 (fossil authors), 131 (eozoon, the imaginary fossil).

[86] Mariano Soler, *El darwinismo ante la filosofía de la naturaleza* (Montevideo, Marella Hermanos, 1880).

proposes a long list of anti-Darwinian scientists (including Hooker!) and identifies Haeckel as an exaggerated Darwinian (separating Haeckel from Darwin was a *leitmotiv* in Catholic apologetics). He cites the Catholic evolutionist St. George Mivart without first-hand acquaintance with his ideas, as well as the Mivart's Spanish follower Zeferino Cardinal González, without appreciating the conciliatory stance of this interesting and important figure.[87] In another article, Soler introduces an interesting detail: Huxley's findings on the close relationship of birds and reptiles confirms *Genesis* I.20, where Moses indicates the commonalty of the two groups, inasmuch as both were created on the same day. Huxley had added nothing new![88]

Soler was the captain of the anti-Darwinian team (together with his lay epigone, Juan Zorilla de San Martín). His *modus operandi* is instructive. He was a member of the Society of Sciences and Arts, an association devoted to scientific popularization whose most vocal spokesmen were the engineers Carlos Honoré and Melitón González, two outspoken Darwinians who in May 1882 had organized a memorial program in honor of Darwin, one month after his death. The Society had even sent a floral arrangement to Darwin's tomb. González opened the session with a brief talk, describing natural selection and presenting Florentino Ameghino's hypothesis of America as the cradle of humanity.[89] Honoré followed with a long lecture that began with an evocation of his student days in Belgium, when he had had

> the opportunity to defend in lectures the theory of the *Origin of Species* against many young people inspired by outmoded ideas. . . There were but a few of us who resolutely popularized the ideas of that fecund thinker outside of the University classrooms in which his ideas were discussed with either a marked timidity or a completely hostile spirit. It was we who sought an expansion of the official curricula so that they might encompass the subjects indispensable for the appreciation of the new theory.[90]

Honoré's recollection most likely refers to the Free University of Belgium during the decade of the 1860s. This University was a stronghold of Krausism at this time; Wilhelm Tiberghien was rector in 1867-68 and again in 1875-76.[91] He also recalled that ten years before (1872) in the Club Universitario "there predominated that scant information which persons devoted exclusively to legal concepts acquire about nature." Honoré had been one of those who lectured on Darwin there. Compare that to the present situation in 1882:

> Today we see quite a different spectacle: our libraries now offer all the works of the *poorly known* individual [Darwin] of those days. We have attended public lectures, heard spirited debates, read interesting polemics, all of which demonstrate that his ideas have already

[87] On the ideas of Mivart and González, see Glick, "Spain," in *The Comparative Reception of Darwinism*, 2nd ed. (Chicago, University of Chicago, Press, 1988), pp. 307-345, on pp. 340-343.

[88] Mariano Soler, "La cosmogonia de la ciencia," *Boletín de la Sociedad Ciencias y Artes*," 8 (1884), 87-101, 122-124, 133-139, on p. 138. For Huxley's lumping of birds and reptiles in the same group, see his article "On the Classification of Birds," *Proceedings of the Zoological Society of London*, 1867, 415-472.

[89] "Sociedad Ciencias y Artes. Sesión celebrada en honor del naturalista Carlos Roberto Darwin," *Boletín de la Sociedad Ciencias y Artes*, 6 (1882), 229-236.

[90] Carlos Honoré, "Discurso sobre Darwin, sus obras y la influencia que ejercieron en la ciencia," *Boletín de la Sociedad Ciencias y Artes*, 6 (1882), 241-247, 253-260, on p. 242. In Mañé's view (*Un siglo un Darwinismo*, p. 115), Honoré was the most intelligent of Uruguayan commentators on Darwin.

[91] See Susana Monreal, "El krausismo en la Universidad Libre de Bruselas (1834-1897)," in *Krause-Ahrens-Tiberghien: Estudios y selección de textos* (Montevideo, Fundación Prudencio Vázquez y Vega, 1988), pp. 33-38.

> penetrated our spirits and that Uruguayan society has not been indifferent to the change of ideas which has operated in knowledge of General Laws which guide the organic world.[92]

Honoré, who had done some meteorological research, observed that Darwin's studies of the glaciers of Chile and Patagonia had lighted the path of his own studies of glaciers.[93] Asserting that there was no doubt that "the open book in which Darwin had deciphered the most essential elements of the Law of evolution had been the soil of Uruguay and Argentina," he went on to discuss the evidence that Darwin had gathered in South America. The lecture ends with an evaluation of the stimulus that Darwin had given to various scientific disciplines: geology, morphology, embryology, and the social sciences.

Beginning with the seventh number of the next volume of the *Boletín* of this Society (February 17, 1883), Mariano Soler's name appears as a member of the editorial board; and, in the same number, there appears another version of his critique of Darwinism.[94] Accompanying Soler's article is an editorial note by Melitón González explaining that "Without endorsing the ideas presented, we want to stress again the responsibility of the Sociedad Ciencias y Artes and rectify one of the concepts mentioned by the author at beginning of his article." Soler had written that inasmuch as Darwinism had been defended in the *Boletín*, there was also an obligation to present a refutation. González replied that the Society neither defends nor attacks the Darwinian theory. Nevertheless, "we must point out that we do not agree with the synthesis that [Soler] presents of the system; a synthesis which serves him as a premise to attack it." He provides some examples of errors the bishop had made and reserves the right to refute his refutation.[95]

Soler replied that González had scruples when it came to scriptural doctrines but not when "the same *Boletín* has repeatedly printed completely materialist doctrines."[96] In another short article, Soler asks González to refute him, but that he could refute that refutation inasmuch as González had formed an erroneous concept of the position of Soler, who had not attacked the doctrine of Darwin concretely, but rather that of "Darwinism in its evolutionary and transformist form, expanded by Haeckel, Huxley, and Vogt. . . . The statutes [of the Society] do not extend to the President [González] the right to print in its *Boletín* a more or less scientific and extensive critique of any theory whatsoever, and then describe it as false and inexact, in a dictatorial way, and in the guise of a chronicle."[97]

The Catholic Club of Montevideo was established by Soler in part to establish a platform from which to take the offensive against materialism and Darwinism. I found no books by Darwin in the library of the Club, but it did have copies of books by Spanish clerics who struck a conciliatory note, like Zeferino González and Miguel Mir. Unlike the Spaniards, however, Soler was not a conciliator. The obituary of Darwin that Juan Zorrilla published in the daily, *El Bien Público,* was one of the most violent

---

[92] Honoré, "Discurso," p. 243.

[93] *Ibid.*, p. 244.

[94] Mariano Soler, "Crítica del darwinismo bajo el aspecto de las ciencias experimentales y de la filosofía de la naturaleza," *Boletín de la Sociedad Ciencias y Artes*, 7 (1883), 428-431 and following.

[95] "Crónica. El darwinismo," *Boletín de la Sociedad Ciencias y Artes*, 7 (1883), 431.

[96] Mariano Soler, "Crónica. Con ocasión del darwinismo," *Boletín de la Sociedad Ciencias y Artes*, 7 (1883), 443.

[97] Mariano Soler, "Tomamos la palabra al señor González," *Boletín de la Sociedad Ciencias y Artes*, *loc. cit.*

and no doubt reflected Soler's views as well.[98] Zorrilla began with a spectacular error: "The celebrated Darwin has just died in the United States."[99] It is difficult to take Zorrilla at face value in view of his inability to identify Darwin's nationality. "Darwin was famous," he concluded, "as one of those who wounded humanity in its most vital and noble fibers: in its dignity, in its divine origin and its sublime and immortal destiny." He accuses Darwin of having deliberately sought in his research weapons against the faith, an accusation that even his Catholic detractors in Spain never dared to say.[100] Spanish Catholics, to the degree that they controlled public education, could well display equanimity upon contemplating the death of their great enemy Darwin, or so they perceived him. Zorrilla and Soler, on the other hand, were much more combative, inasmuch as they had to recover a hegemony which they had already lost.

VÁZQUEZ ACEVEDO'S UNIVERSITY: A POSITIVIST DICTATORSHIP?

In 1881 Martín C. Martínez, Eduardo Acevedo, and Prudencio Vázquez y Vega all received the doctorate in law. The first two, both positivists, drew up a new philosophy curriculum which the last named—an idealist—attacked harshly for its Darwinian and materialist contents.[101] The program (the version approved was emended to reflect Vázquez y Vega's critique) is the clearest document of the intellectual program of the university Darwinians. The first part of the program is a course on psychology, whose principal objective was to ascertain the causal role of natural selection in the evolution of the senses.[102]

That was not the worst of it, however, from the point of view of Vázquez y Vega and other anti-materialists. Most objectionable was a section on evolutionary ethics, based on Spencer and on Darwin's chapter on "Moral Faculties" in the *Descent of Man*, where Darwin asserts that "sympathy" (a term straight out of Rousseau and/or Adam Smith) is an instinct that endows groups of humans with social solidarity, and the groups better so endowed are advantaged over other groups in the struggle for existence.[103] Egoism and altruism are also traits that likewise advantage their carriers.[104]

---

[98] "Darwin," *El Bien Público*, April 30, 1882.

[99] The error was so gross that an editor of an anthology of the writings of Zorrilla had to add a footnote to explain Zorrilla's lapse: *Juan Zorrilla de San Martín en la prensa. Escritos y discursos*, Antonio Seluja Cecín, ed. (Montevideo, Comisión Nacional del Homenaje del Sesquicentenario de los Hechos Históricos de 1825, 1975), p. 47.

[100] Cf. The obituaries of Darwin published in traditional Catholic media, reproduced by Diego Núñez Ruiz, "La muerte de Darwin en la prensa española," *Mundo Científico*, 2 (1982), 396-404, on pp. 398-399.

[101] The texts of the proposal, the polemics over it, and the program finally approved are reproduced in full in María Teresa Carballal de Torres, "La reforma positivista del programa de filosofía, en 1881," *Cuadernos Uruguayos de Filosofía*, 3 (1964) , 203-290.

[102] *Ibid.*, p. 211: "Sense of sight. . . Is it possible to explain with the help of natural selection the development of this sense based on the optic pigmentation of the lower animals? Sense of hearing. . . How does evolutionary theory explain the development of this sense in the animal kingdom? (p. 212): Sense of smell. . . . Is the sense of smell in civilized man a rudimentary sense that is tending to disappear? Darwin's view. . . genetic sense . . . Development of this sense in the animal kingdom. How does evolutionary theory explain that development?"

[103] *Ibid.*, p. 230: "The great influence which, according to Darwin, sympathy has had in the genesis of morality." Cf. Prudencio Vázquez y Vega, "Crítica de la moral evolucionista," *Anales del Ateneo del Uruguay*, 1 (1881), 210-222.

[104] "Reforma positivista," p. 231.

When Vázquez y Vega attacked their syllabus, Azevedo and Martínez replied that

> The study of evolutionary theory offers the advantage of unquestionable practical utility. Darwin's grand system represents the most powerful inductive effort ever made throughout history and, studying it, students are made cognizant of mental operations and their utility with greater facility and precision than could be obtained learning abstract theories and rules by rote memory from a text book.

Still, the reformers were prepared to defer to their opponent's wounded feelings and "in order to remove "even the suspicion of partiality," they resolved "to excise from the curriculum the enumeration of those arguments of Spencer that have so exercised the spirit of Dr. Vázquez y Vega."[105]

At graduation ceremonies, students were invited to make statements of academic conscience, the best of which were presented publicly. Darwin turns up in two such pronouncements for 1882:

> In all the annals of science there is no more colossal reform than then one at work in the present century, ever since the immortal Charles Darwin launched his great theory on the origin of species. From biology it has extended itself to all other branches of human knowledge.
>
> Lorenzo Barbagelata, May 23

And perhaps in the contrary sense:

> England is not an innovative nation; thus it has not been able to bring about a scientific revolution. This is why Darwin is the Amerigo Vespucci of anthropology, while in the social sciences Spencer adumbrates his reflection. The true revolutionaries were Lamarck and Auguste Comte... the Latin race is a race of Gods.
>
> Isidro Revest, May 22[106]

Revest's formulation is interesting as a call for both evolutionism, and positivism in the French style that Varela had attacked in *La legislación escolar*.

In a similar contest in 1886 one of the student statements provoked the ire of the editors of the Protestant magazine, *El Evangelista*:

> To deny the origin of man from a lower form, as Darwin has demonstrated in his transformist theory, is to deny the law of evolution [which is] admitted as true by the entire scientific world.[107]

For evangelicals, Christianity admitted no changes of species.

The Darwinian tone of the University was set by its leader Alfredo Vázquez Acevedo, elected rector on a positivist list in 1880 and who served in that position until 1889 with the exception of two terms of two years each.[108] Anti-Darwinians viewed him as the orchestrator of an evolutionist take-over of the University, a view which the rector did nothing to oppose. Indeed in his 1885 graduation speech, Vázquez Acevedo reflected on the role of evolutionism in the University and in Uruguay:

> In few countries has the modern theory of evolution made as rapid progress as in our small republic. While the old nations of Europe hobble the truths that the eminent Darwin has

---

105 *Ibid.*, pp. 252-253. On this polemic, see also Mañé Garzón, *Un siglo de darwinismo*, pp. 99-106

106 AGN, Archivo de la Universidad de Montevideo, expediente 40 bis of 1882.

107 "Descendemos del mono?" *El Evangelista*, 9 (1886), 333.

108 On Vázquez Acevedo's election, see Ardao, *Espiritualismo y positivismo*, pp. 175-178; and París de Oddone, *La Universidad de Montevideo*, pp. 82-84.

> taught, we dare promote them, carrying their explanations and philosophical consequences farther than the English scholar himself.[109]

These lines were cited in a congressional debate in 1886 in order to show that the rector had converted the university into a positivist dictatorship. Such was the accusation of the idealist deputy Carlos Gómez Palacios in the session of June 11 when he presented three "propositions:"

> (1) That the University is a philosophical sect, a materialist sect, where the only system taught exclusively is materialism .
>
> (2) That the Rector's personality constitutes a dictatorship, that he exercises, with the professors, deans, and members of the council as Satraps.
>
> (3) That the professoriate is notoriously incompetent.[110]

Gómez Palacios urged the recovery of the "true principles of science," in a sense contrary to the Darwinian direction indicated in the Rector's speech, concluding that was no freedom of instruction in the University inasmuch as the only philosophy imparted there was positivism, or "disguised materialism."[111]

Next to speak in this debate was the deputy Lamas who also weighed in with a conspectus of the principles of the University as he saw them, including egoism and Spencer's struggle for life:

> In this narrow sphere they confine all the forces of human evolution. I hope it is nothing more than that which Buffon assigned to the lower species, when, in denying them free will, he said that they only functioned through desire and repugnance, which is the same as pleasure and pain.[112]

Inasmuch as the idealists were marginalized in the University, they had no other recourse except to use the Congress as a tribune for reclaiming the position they had lost. I do not think, however, that their testimony in itself constitutes proof of a positivist dictatorship. More to the point were the words of Carlos Honoré, positivist deputy, when he alleged that there reigned in the University a philosophical intolerance that ought to be replaced by "true positivism" of the tolerant kind.[113]

A better criterion for assessing the state of the University had been introduced in the debate of the previous year by the deputy Mendoza, according to whom "it is necessary to ascertain whether there is freedom of instruction in the University... if a spiritualist student is obliged, by force, to give a Spencerian examination," that is, to respond to questions in conformity with Spencer's ideas.[114] On this criterion it is doubtful that positivists ruled in a dictatorial way. We noted, for example, the atmosphere of free

---

[109] Ardao, *Espiritualismo y positivismo*, p. 221 (Discourso de Vásquez Azevedo en la Colación de grados de 1885). Cf. the paraphrase by Antonio María Rodríguez delivered in the Uruguayan House of Representatives in 1886: "that our country, so small in area, was nevertheless one of those which received with greatest enthusiasm and affection the most important scientific advances of our epoch: the evolutionist doctrine, the most important one of the contemporary period, found here a greater number of followers than are found in many European countries, and it is better known and more studied...than it is in many retrograde countries of Europe;" *Diario de Sesiones de la Cámara de Representantes*, 79 (1886), 275.

[110] *Diario de sesiones*, p. 280.

[111] *Ibid.*, pp. 282, 295.

[112] *Ibid.*, p. 313.

[113] *Ibid.*, p. 345.

[114] *Diario de sesiones*, 73 (1885), 364.

debate in Aréchaga's classroom. Similarly we might note that Martínez, the staunch Darwinian, approved the anti-evolutionist thesis of Jorge Arias in 1884, the same year in which Manuel Herrera's anti-Darwinian thesis was also approved.[115]

SOCIAL DARWINISM

Here we examine the thought of Martín C. Martínez and Eduardo Acevedo, both of whom emphasized natural selection as the principal mechanism (for Martínez, the only mechanism) of social evolution. Both defined their concepts in doctoral theses defended in 1881. Acevedo begins his treatise, *El gobierno municipal*, by considering the lack of specialization in simple animals. Thus, in the monera, "Every part of this rudimentary organism feels, every part breathes, every part digests." There is no specialization. But as one reviews the zoological scale from the monera to man one observes the increasing division of labor. Sociology offers analogies: there is no division of labor among primitive tribes: Evolution explains the transformation from rudimentary to civilized society, "demonstrating that in the struggle for existence those beings win out whose physiological functions are better distributed and those societies in which the principle of the division of labor has been most wisely applied."[116] In prehistoric societies, the struggle for existence was brutal. Then the first steps towards division of labor were consolidated through natural selection which favored groups organized in tribes over isolated hunters. Eventually democracy prevailed over tribal despotism, because it represented a greater division of labor:

> In the struggle for existence, natural selection consolidated political and civil liberty because only such laws could communicate to peoples the power required to destroy the effect of the various agents of destruction that are opposed to the development of all organisms… natural selection leads to such a grand and fecund result and demonstrates that it is possible to explain human progress—the most important phenomenon in sociology with only the aid of the general laws that govern all beings.[117]

His thesis that local government represented the segregation of functions must be understood in this biological context: that is, a greater division of labor than that represented by the centralizing Uruguayan government. Acevedo's archly Darwinian dissertation, by the way, was directed by the anti-Darwinian Aréchaga.

Martínez's dissertation on "territorial property" presented a similar type of historical argument viewed phylogenetically. For Martínez

> The natural sciences have rehabilitated history. The study of organisms from intrauterine life on, the history of the layers of the earth studied according to their order of appearance, man himself studied from the embryological stage to that of the child, primitive and

[115] Jorge Arias, *Consdieraciones acerca de la escuela de la evolución* (Montevideo, Universidad Mayor de la República, 1884), a thesis praised by Protestants as "a healthy reaction of enlightened youth against the philosophical modernism that invades everything;" "La tesis del doctor Arias," *El Evangelista*, 7 (1884), 109-110, on p. 109. Manuel Herrera, *La evolución en las ciencias jurídicas* (Montevideo, Universidad Mayor de la República, 1884), a thesis criticized by positivists as "reactionary, an intent to recapture for metaphysics the dominion that the theory of evolution has conquered in the world of ideas;" review in *Revista de la Sociedad Universitaria*, 2 (1884), 329-331, on p. 331. On the theses of Arias and Herrera, see Mañé, *Un siglo del de darwinismo*, pp. 139-141.

[116] Eduardo Acevedo, *El gobierno municipal* (Montevideo, 1882), p. 6.

[117] *Ibid.*, pp. 9-10.

> civilized, illuminated the problems of the origin of species and of man, of his purpose in life, of his social laws and psychic nature.[118]

Martínez's recapitulationist bias permitted him to identify the property system of present-day Tasmanians (preserved as "lower organisms") with those of our own Stone-Age ancestors.[119]

Nomadic tribes, he continues, replaced hunters owing to the operation of natural selection. In a more recent period, slavery became a powerful agent of progress, inasmuch as it represented an advance in the division of labor. And so he continued on through successive stages of human history: sedentarization and invention of agriculture, village communities, finally families.

> The most active cause of this historical development has been natural selection. As nomads triumphed over hunter hordes, sedentary peoples triumphed over nomads. Sometimes nomads conquered sedentary societies; but besides the fact that this happened when barbarians had acquired the habits of peoples they later conquered, their conquests were destructive whirlwinds, but nothing more than that. They either disappeared or else adopted the customs of the peoples they dominated. By contrast, the conquests of sedentary peoples were permanent, and with them they brought their customs, including the way they appropriated land.[120]

In another essay of the same period, Martínez explores the evolutionary meaning of warfare.[121] The central note is Haeckel's notion that evolution had transformed death into the source of life. Through war, stronger races have replaced weaker ones.[122] The rigid and brutal legal codes of Antiquity constituted a form of selection: through them, less brutal societies were produced. In our days, however, war has been converted into a cause of retrogression and ruin, inasmuch as, owing to technological innovations, the strong perish along with the weak. Moreover the prosperity of a world empire, like England, came to depend on peace, lest the international market be disturbed. He concludes that industrialism had replaced warfare as the main focus of the struggle for existence: "Our democratic regime is nothing more than the substitution of one form of struggle for another: competition replaces struggle through warfare; debate is the substitute for persecution.[123]

I have focused on the ideas of Martínez and Acevedo because they were the Uruguayan representatives of classical Social Darwinism. Organic models had become the normal mode of social explanation. Social recapitulationism was widespread, too, as in Carlos María de Pena's Haeckelian pedagogical notions: "The education of the child ought to harmonize, in the manner and order followed, with the education of the human race, viewed historically. The genesis of knowledge in the individual should follow the same path as that of the genesis of knowledge in the race."[124] There was also, of course, strong dissent in the liberal camp from this kind of reasoning. From the older idealist

---

[118] Martín C. Martínez, "La teoría evolucionista en la propiedad territorial," in his *Estudios Sociológicos* (Montevideo, Colección de Clásicos Uruguayos, 1965), pp. 3-51, on p. 30.
[119] *Ibid.*, p. 33.
[120] *Ibid.*, pp. 45-46.
[121] Martínez, "La concepción contemporánea de la guerra," *Estudios sociológicos*, pp. 80-112.
[122] *Ibid.*, p. 86.
[123] *Ibid.*, pp. 103-104.
[124] Carlos María de Pena, "Pro Herbert Spencer: Influence en el Uruguay de algunas ideas de Spencer sobre educación," *Anales de Instrucción Primaria*, 1 (1903), 497-507, on p. 499.

point of view, Carlos María Ramírez, after reviewing the cruel markers of Spencer's social vision, had to demur, in a footnote: "Some of these phrases are literal, as incredible though it might seem."[125]

CONCLUSIONS

It seems to me quite clear that the Latin countries present a different pattern in the reception of Darwinism than do those of England and the United States. In the English-speaking countries the crucial variable in the reception of Darwinism in any one locale was the prevailing dynamics of interaction there among the various Protestant denominations. That is, reception was clearly decentralized. In Latin, Catholic countries, however (and here, I mean France, Spain, Italy and all the countries of Latin America) the centralized mode of social control—whether by state or church—would seem to be the principal variable. Inasmuch as the vast majority of population those countries were Catholic, the dynamic of religious differentiation does not attain *per se* (although of course secularization was a key issue). Centralization was crucial, however. The fact that Col. Latorre, the positivist dictator, delivered public education into the hands of an outspoken Darwinian, José Pedro Varela, in large part explains the ease with which Varela put across his positivist educational reforms. Catholic presidents, whether of the parliamentary or the authoritarian variety, had similar successes in opposing Darwin.

Boston University

[125] Varela and Ramírez, *El destino nacional y la universidad*, I, 42 n. 1.

PEDRO M. PRUNA GOODGALL

BIOLOGICAL EVOLUTIONISM IN CUBA AT THE END OF THE NINETEENTH CENTURY

There seems to be a by now somewhat receding tendency to consider biological and social evolutionism as a single entity.[1] A more sober view of the matter would be perhaps to regard these two conceptions—which stem from different roots—as separate elements interacting within a single "evolutionistic movement," a cultural current that prevailed or at least attempted to do so in the intellectual milieu of many European and American countries during the last three decades of the nineteenth century. It is quite evident, and this is especially true of England and the United States, that the intellectual mood favoring an evolutionary approach to virtually all phenomena strongly contributed to the acceptance of Darwin's theory alongside Spencer's; but, in other countries, this reception was linked also to a resurgence of Lamarckism, the diffusion of Haeckel's *monism* and/or Hegel's evolution of the *Spirit*, as well as other similar doctrines (such as Krausism in Spain). It was certainly difficult to find "thoroughbred" Darwinians at the time: "To confuse matters for the historian, the exponents of widely differing interpretations called themselves 'Darwinians' because they acknowledged Darwin's lead even if they did not accept all the details of his theory."[2]

In several European countries, as well as in the United States, the introduction of Darwinism (*sensu lato*, as explained before) was perhaps related to the increasing importance and the growing respectability of the scientific community, only recently organized at the time, into societies for the advancement of science. In Latin America, however, the introduction of Darwinism accompanied the diffusion of positivism, a political and ideological movement which prevailed among liberal intellectuals from the 1870s on as a reaction against the remnants of the old colonial regime, symbolized by the virtual monopoly of ideology through the indirect control of institutions of higher learning and the exercise of censorship by the Catholic Church. The case studies of

[1] See, for example, John C. Greene, *Science, Ideology and World View* (Berkeley, University of California Press, 1981).

[2] Peter Bowler, *The Norton History of the Environmental Sciences* (New York and London, W. W. Norton, 1992), pp. 325-326.

Thomas F. Glick, Rosaura Ruiz, and Miguel Ángel Puig-Samper (eds.), The Reception of Darwinism in the Iberian World, 53-64.

Mexico, Brazil and Argentina[3] are good examples of the association of French or English positivism with evolutionism. The Cuban case follows the general pattern common to most Latin American countries, but is perhaps somewhat more complicated, among other reasons because Cuba remained a Spanish colony until 1898[4].

1. PROLEGOMENA

There seems to be no reference to evolutionism in Cuban scientific literature before 1858, and even then the reference was only slight and indirect. The Cuban naturalist Felipe Poey (1799-1891), a professor at Havana University, was familiar with the terms of the public debate between Georges Cuvier and Etienne Geoffroy Saint-Hilaire, which took place in Paris in 1830, a time when Poey was in France. Years later, in 1858, Poey confessed that he was convinced of the truth of Cuvier's ideas, but had found "so much philosophy in the doctrines of the contrary school" that he felt somewhat attracted to them. Yet, he preferred "to remain on Cuvier's side" unless forced otherwise by new evidence. Poey actually considered Saint-Hilaire's theory (on the transformation of morphological types under the direct influence of environment) somewhat "exaggerated."[5]

By 1861, Poey's position in this regard had not changed, in spite of strong criticism by his son, the meteorologist Andrés Poey, who was a convinced Comtean positivist and, therefore, an enemy of creationism. Moreover, Felipe Poey had somehow arrived at the conclusion that evolutionism, when applied to man, was nothing less than an attempt to justify racism and slavery; whereas truly religious beliefs were against racist doctrines. A speech he gave that year on "The Unity of the Human Species" underlined his concept of human races as temporary adaptations to climatic conditions (Saint-Hilaire's and especially Lamarck's idea, but restricted to intraspecific diversity), as varieties of the one and only human species created by God. The speech was strong enough for its place and time. It was given in the presence of the Spanish Governor General when slavery was still predominant in Cuba and was apparently interpreted by some as a sign of disrespect for the Governor and also somewhat subversive. Poey was

---

[3] Rosaura Ruiz Gutiérrez, *Positivismo y evolución: Introducción del Darwinismo en México* (Mexico City, Universidad Nacional Autónoma de México, 1987); Roberto Moreno, *La Polémica del Darwinismo en México. Siglo XIX* (Mexico City, Universidad Nacional Autónoma de México, 1989); Terezinha Alves Ferrreira Collichio, *Miranda Azevedo e o Darwinismo no Brasil* (Belo Horizonte and Sao Paulo, Editora Itatiaia and Editora da Universidade de São Paulo, 1988); Marcelo Montserrat, *Ciencia, historia y sociedad en la Argentina del siglo XIX.* (Buenos Aires, Centro Editor de América Latina, 1993). See also Thomas F. Glick, *Darwin y el Darwinismo en Uruguay y en América Latina* (Montevideo, Universidad de la República, 1989).

[4] The reception of Darwinism in Cuba has been studied by Pedro M. Pruna, "La recepcion de las ideas de Darwin en Cuba durante el siglo XIX," *Conferencias y Estudios de Historia y Organización de la Ciencia* (Havana), no. 32 (1983), pp. 1-29; reproduced in *Revista Latinoamericana de Historia de las Ciencias y la Tecnologia (Quipu)*, 1 (1984), 369-389; Thomas F. Glick, "La polemica del darwinismo en Cuba" in *Actas del II Congreso de la Sociedad Española de Historia de las Ciencias,* 1984, I, 413-418; Pedro M.Pruna and A. García Gonzälez, *Darwinismo y Sociedad en Cuba. Siglo XIX* (Madrid, CSIC, 1989); Pruna, "Darwinisme Cubain," in *Dictionnaire de Darwinisme et de L'Evolution,* P. Tort, ed. (Paris, Presses Universitaires de France, 1996), I, 894-896.

[5] Felipe Poey, *Memorias sobre la Historia Natural de la Isla de Cuba* (Havana, Barcina, 1858), II, 109.

therefore forced to recant: "if I ever used the word equality, it was referring to the immortal spirit all men possess, without discussing [the problem of] rights" and without any intention to cause the Governor General any displeasure.[6]

Seven years later, in 1868, during the first public discussion of Darwin's *Origin of Species* (which took place at the Academy of Sciences in Havana), Poey's position had changed radically. Against the opinion of the Count of Pozos Dulces, who had criticized Darwin's theory, Poey considered the "species problem" an open question and evolution, a possible answer. He promised to deliver a series of lectures on this matter and did so the next year.[7] In the following years he finally accepted evolutionism (without renouncing his antiracism) and interpreted zoological classifications in evolutionary terms in his classes at the University.

In 1867 Alvaro Reynoso, a distinguished Cuban chemist and agronomist, mentioned Darwin's *Origin of Species* (which he, as some other Cuban scientists, read in the notorious French translation by Clemence Royer) only to point out that it had "awakened the old controversy" between transformists and creationists.[8] In 1870 a paper sustaining creationist views regarding the origin of man was severely criticized by a group of positivists at the Havana Academy.[9]

In spite of these facts, one cannot really speak of the introduction of Darwinism into the common discourse of Cuban intellectuals until the late 1870s. The main reason for such a delay was that a war of independence took place in Cuba from 1868 to 1878. Since Darwinism went against the official ideology (that of the Catholic Church), it could be construed as a sign of political dissent, with all the risks that usually accompany dissent in wartime. After 1876, Spain carried out a more tolerant policy toward Cuban intellectuals, within its general effort to pacify its colony. A Spanish translation of Darwin's *Origin of Species* was published in Spain in 1876 and sold in Cuba the next year.

THE INTRODUCTION OF EVOLUTIONISM

In the same year of 1877, when the political situation in Havana was somewhat more relaxed, a small group of lawyers and writers, headed by José Antonio Cortina (1853-1884) and Julián Gassie (1850-1878) organized the Liberal (Autonomist) Party to strive for some sort of self-government under Spanish sovereignty. Many members of this party, among them probably Cortina himself, were actually in favor of political

[6] Poey's speech was never published, except for some fragments. The original manuscript is kept among his papers at Havana University. For detailed discussions of the Geoffroy-Cuvier debate see: Toby Appel, *The Cuvier-Geoffroy Debate* (Oxford, Oxford University Press, 1987), and I. E. Amlinskiy, *Geoffroy Saint-Hilaire and his Struggle against Cuvier* (Moscow, Izdatel'stvo Akademii Nauk SSSR, 1955) (in Russian).

[7] His lectures on "The Notion of Species" at the Academy of Sciences (1869) were never published and were not abstracted in the proceedings of the Academy. The original manuscript has not been found.

[8] Alvaro Reynoso, *Apuntes acerca de varios cultivos cubanos* (Madrid, Rivadeneyra, 1867), p. 252. I thank my colleague Rolando Misas, for calling this reference to my attention.

[9] *Anales de la Real Academia de Ciencias Médicas, Físicas y Naturales de la Habana*, 7 (1870), 352-361.

independence, but did not consider it immediately attainable. Cortina founded an influential journal called the *Revista de Cuba*. The second number of this publication carried an article by Gassie on Ernst Haeckel's *Anthropogeny*. The following numbers included a translation of an article by Haeckel and a review of his ideas on the origin of man.[10] Most members of the editorial board of the journal were under the influence of Hegel and found Haeckel more acceptable (perhaps more philosophical) than Darwin.[11] A close associate of this group, the poet Francisco Sellén (1836-1907), wrote at the time:

> Immense, unceasing transformation!
> Flame that cannot be extinguished!
> And we exist, palpitating, in the WHOLE.
> And all is life, and light and movement."[12]

The editors of the *Revista de Cuba* also organized several meetings to discuss the implications of Darwinism and were actively involved in the public debate on Darwinism and the origin of man which took place in the Liceo de Guanabacoa in 1879. The speakers in this debate were José Antonio Cortina, Antonio Mestre (1834-1887), then Secretary General of the Havana Academy and a renowned positivist, the philosopher Enrique José Varona (1849-1933), the first Cuban Spencerian, and others who favored evolutionism, whereas the poet Antonio Vinageras (1832-1904) and the physician and Comtean positivist José Francisco Arango (1843-1893) opposed Darwinism.[13] This meeting was strongly criticized by the Spanish colonialists' newspaper *La Voz de Cuba*, and also by a group of seventy distinguished ladies, who were outraged by the fact that the thesis on the animal origin the human species had been expounded before an audience which included women and children.

The main object of criticism by the irate ladies was Antonio Mestre's paper on "The Natural Origin of Man." Mestre presented a careful analysis of the different "proofs" of evolution presented by Darwin in *The Origin of Species* (which he read in Clemence

---

[10] Julián Gassie, "La Anthropogenia de Haeckel y el transformismo unitario en Alemania," *Revista de Cuba*, 2 (1877), 256-263; Ernst Haeckel, "La teoría de la evolución en sus relaciones con la filosofía Natural," *ibid.*, 3 (1878), 52-64; Enrique Fernández Veciana, "Haeckel y el origen del hombre," *ibid.*., 4 (1878), 295-301.

[11] They were closely linked with José del Perojo (1852-1908), a Spanish philosopher, born in Cuba, who was the editor of the important Spanish journal *Revista Contemporánea*, which also published articles on Haeckel. Perojo was the main opponent (with a strong anticlerical bend) of Menéndez y Pelayo's interpretation of the effects of the Inquisition on science in Spain, which Pelayo regarded as mild and almost negligible. See Marcelino Menéndez y Pelayo, *La Ciencia Española* (Madrid, Revista de Archivos, Bibliotecas y Museos, 1915), II, 112-142. This was one of the last episodes of what Spanish historians of science call "the polemic of Spanish science." For a rather recent periodization and appraisal of this debate, see José María López Piñero, *Ciencia y Técnica en la Sociedad Española de los siglos XVI y XVII* (Barcelona, Labor, 1979), who considers that this phase of the debate was "a mere excuse for the confrontation of the positivist, progressive and secular mentality with the conservative, clerical and chauvinistic ideology."

[12] Francisco Sellén, "Panteísmo," *Revista de Cuba*, 5 (1879), 50-52.

[13] Most of the speeches were published in *Revista de Cuba* in 1879. The same year, Vinageras published a very long poem called *El Congreso de Guinea* (first referred to by Thomas F. Glick, see n. 1). It includes the following verses (which I have translated from Spanish): "With tails! and convened/ by an old an ugly orangutan/ patriarch of the monkeys of the world/ profoundly judicious/ and born in a jungle in Borneo."

Royer's French translation). He considered that although geological, paleontological, embryological and other facts presented by Darwin spoke in favor of his interpretation, they could not be regarded, however, as "definitive proofs." Darwin's was only a "plausible theory", similar to others which had originally been so, but had finally been accepted by science (such as those advanced by Copernicus and Newton). It therefore deserved an open and wide discussion, free of the ethical and religious prejudices which usually marred debates on human origins.[14]

The Anthropological Society of Cuba (founded in 1877) also served as a vehicle for the introduction of Darwinism. A lengthy discussion on the "antiquity of man" took place in 1878 and 1879, originally promoted by Gassie, but also by Luis Montané (1849-1936), a disciple of Paul Broca, the French patriarch of positivist anthropology. Physical anthropology was developed mainly in colonialist countries and tainted with racist considerations. When transferred to countries (such as the United States, Brazil, and Cuba), where slavery had existed in recent times or was in the process of disappearing, it retained its racist bent. There was, however, a rather subtle difference between "cultural" and "biological" conceptions of racial differences in Cuba. The "cultural" approach considered the "lower" intellectual status and the habits of the black population a consequence of slavery and therefore amenable to change (toward the "higher" standards of the whites), whereas the "biological" approach deemed these traits the inborn and essentially unchangeable defects of the former slaves and black freemen. Although these two tendencies were prevalent within the Anthropological Society, some members of the Society were openly against any form of racism whatsoever. The debates on the nature of racial differences were obviously related to the task of incorporating the former slaves into the social structure created by the new dominant elite of owners of large sugar mills (requiring free labor) and extensive latifundia, after slavery was dealt a deathblow during the war of 1868-1878 and finally abolished in 1880-1886.[15]

The novel *Looking for the Link* (*En Busca del Eslabón)*, by Francisco Calcagno (1827-1903), a physician and a member of the Anthropological Society, reflects the attitude of many nineteenth century intellectuals in most "civilized" countries regarding the "Negro race." It is the tale of a round-the-world expedition, under the leadership of a Yankee captain, in search of the Haeckelian "missing link," which is never actually found. One of the characters in the novel is the black servant and former slave of a Cuban professor, but also the main jester in the plot. He—like all blacks—is undoubtedly human, the novel concludes, but the missing link will nevertheless be found somewhere in-between the Africans and the anthropoid apes. Calcagno's opinion on this matter clearly differs from that of his father in law, Felipe Poey, who may have

---

[14] Antonio Mestre, "El Origen Natural del Hombre," *Revista de Cuba*, 5 (1879), 419-433, 508-528.

[15] Discussions on these matters may be found in the proceedings of the Anthropological Society, compiled by Manuel Rivero de la Calle: *Actas de la Sociedad Antropológica de la Isla de Cuba* (Havana, Comisión Nacional Cubana de la UNESCO, 1966).

been the prototype for the Cuban professor Don Sinónimo, somewhat ridiculed in the novel.[16]

Another aspect of the controversy on Darwinism in Cuba is obviously associated with politics. The so-called "heroic period" of the Autonomist Party ended around 1885, after the death of José Antonio Cortina. During the following years, until its dissolution in 1898, the party's program was openly against the movement for independence and clearly favored Spanish domination under the limited home rule autonomists demanded. The Spanish government, however, never actually carried out reforms leading to home rule.[17] Some former party leaders (such as Enrique José Varona) and many members, who were actually in favor of independence, drew away from the party. In spite of this, the leadership of the Autonomist Party continued to believe that an all encompassing "law of evolution" would bring home rule to Cuba, as it had done to Canada, without resorting to violent struggle. The Party therefore adopted the slogan, "All through evolution, nothing through revolution." Manuel Sanguily (1848-1925), a distinguished Cuban intellectual who favored political independence, denounced the autonomists' slogan as a deformation of the true sense of the word "evolution". Darwin—he wrote—was in favor of struggle, and no one can deny that "revolution is a necessary and always fatal form of evolution." The autonomists—he argued—had no right to consider themselves Darwinists since the slogan itself was created by a person (the Count of Pozos Dulces) who was probably under the influence of Hegel, not of Darwin.[18]

Darwinism was one of the regular topics in several publications in the 1880s but it held a preferred status in the journal *El Eco de Cuba*, founded in 1885. Its editor, José María Céspedes published there several articles on evolution. He favored Haeckel's and especially Spencer's "comprehensive" interpretation of nature and society. There were, of course, critics of evolutionism, but mostly among the Spanish colonialists. Cuban intellectuals were overwhelmingly in favor of the new, Darwinian (*sensu lato*) approach and this allowed for more daring enterprises, such as the open explanation of Darwinism (*sensu stricto*) in university courses.

CLASSIFICATION

Although the final acceptance of evolutionism by many Cuban naturalists was probably influenced by the "evolutionistic mood" in national culture, which prevailed from the late 70s until 1895, a rather independent tradition within natural history may be traced which eventually favored the reception of the theory of evolution. Within this tradition, discussions dealt mainly with problems of classification and, to a certain extent, trod the path previously followed by naturalists in other countries in reexamining certain concepts of taxonomy which were considered problematic at the time.

---

[16] Francisco Calcagno, *En busca del eslabón* (Havana, Letras Cubanas, 1983). Originally published in Barcelona in 1888.

[17] Parliaments and councils of ministers were established in Cuba and Puerto Rico in 1897. They had very little authority and actual power remained in the hands of the Spanish governors. This was quite obviously a belated attempt by Spain to quell the the second Cuban war for independence (1895-1898) and to avert its extension to Puerto Rico. The war ended after the intervention of U.S. troops in 1898.

[18] Manuel Sanguily, "Las reformas politicas y el darwinismo," *El Cubano*, no. 88 (October 29, 1887).

The first such taxonomic problem was that of "distribution" according to "affinity." In other words, how should resemblances between taxa be represented? Linnaeus thought such "affinities" could be reflected on a surface: a map in which the degree of resemblance was inversely proportional to the distance between circles representing the different taxa. Most naturalists, however, preferred the more modern derivative of the ancient idea of *scala naturae*—the "series"—a chain, with or without branches, which in some cases was similar to a tree. Such ordering of taxa on the basis of morphological resemblance has been an important step in the construction of higher taxa and of the hierarchical classification itself.[19] It is usually referred to, in contemporary texts, as the formation of morphoclines.[20]

Felipe Poey considered it desirable to "combine the series or chain of Bonnet, Lamarck and Blainville, with Linnaeus' net or geographical chart, Cuvier's parallel line and MacLeay's quinary circles."[21] He, nevertheless, favored a system derived from a certain Mr. Adams' distribution of species and varieties in circles, but substituting what he called "conglobulate spheres" for circles.[22] Poey was apparently not aware that Lorenz Oken had also proposed a distribution of spheres within spheres to represent taxonomic affinity. It is generally recognized that spherical and circular distributions do not reflect evolutionistic leanings (although Oken was inclined toward some kind of transformism), and it is certainly true that Poey did not favor evolutionism at the time.

Poey's disciple, Juan Vilaró (1838-1904), was concerned with the actual position of amphibians within classification, also a problem of distribution and affinities. He, as many other naturalists, considered amphibians a "transition" between fish and reptiles. This does not necessarily imply an interpretation in terms of evolution, since "transition" or "intermediate" were terms used long before Darwin or even Lamarck. Vilaró, however, became a convinced evolutionist in later years. When discussing the position of amphibia among the classes of vertebrates he was probably following Blainville's lead in reinstating the zoological "series" as the most valid ordering of taxa of the same category.[23]

In 1873, the young naturalist Manuel Antonio Aguilera (1846-1906) was elected a member of the Royal Academy of Medical, Physical and Natural Sciences of Havana.[24] His inaugural speech dealt with biological classifications. In it he examines the different types of taxonomic ordering and defends serial (lineal) distribution against Cuvier's

---

[19] Pedro M. Pruna, *The Formation of the Hierarchical Classification of Plants and Animals* [summary of doctoral dissertation] (Moscow, Institute of History of Natural Sciences and Technology, 1980), pp. 8-13 (in Russian).

[20] On old and new terminology in taxonomy, see Alec L. Panchen, *Classification, Evolution and the Nature of Biology* (Cambridge, Cambridge University Press, 1992).

[21] Felipe Poey, *Memorias sobre la historia natural de la Isla de Cuba* (Havana, 1851), I, 85

[22] *Ibid.*, p. 357. I have not been able to identify this "Mr. Adams." He was apparently a conchologist since Poey points out that he read of his "system" in "Contrib. L. Conch., p. 190."

[23] Juan Vilaró, "Notas de Juan Vilaró," *Repertorio Físico Natural de la Isla de Cuba*, 2 (1868), 122.

[24] On this institution, see: Pedro M. Pruna, "National Science in a Colonial Context. The Royal Academy of Sciences of Havana," *Isis*, 85 (1994), 412-426.

attacks. Cuvier, he asserts, was against serial ordering because he did not distinguish between "absolute" and "relative" perfection: all beings are equally perfect, from a relative point of view; while absolute perfection has to do with "the complexity of organs as a result of the division of physiological labor, which produces more perfect organs." Clearly, organisms can only be placed within a series based on "absolute perfection." This series is a line with branches, similar to a tree: "*Natura arbori simillima*."[25]

Several years later, in 1884, José Planellas spoke on botanical classifications in his inaugural speech as a professor of Havana University. Planellas again favors the serial distribution of higher plant taxa and even proposes a "genealogical tree which might represent the morphological evolution of the plant kingdom." But he also gives us a hint regarding the beliefs of his colleagues at the School of Sciences when he says he wishes to "place myself, albeit the last, among those that here profess the theory of natural evolution."[26]

From Planella's speech it is rather evident that several professors of Havana University were evolutionists. Foremost among them was Felipe Poey himself, who had finally espoused serial distribution and the theory of evolution in his classes ("succession within the series" and "evolution" were practically synonyms with him).[27] Juan Vilaró, who answered Planellas' speech, clearly stated at the time that he was among the believers in evolutionism.[28] But perhaps the most active Darwinist within the group was Carlos de la Torre (1858-1950), Poey's favorite pupil.

Carlos de la Torre was born and raised in the city of Matanzas (about 40 miles east of Havana) and was initially attracted to natural history by Francisco Ximeno, a local naturalist, who also recommended him to Poey. In 1880, while still living in Matanzas, Torre published "A Short Exposition of Darwinism," in which he referred to some of Darwin's ideas and gave the titles of several of Haeckel's main works. Torre stated there that the theory of evolution, which emerged as a response to such problems as the existence of intermediate forms and of varieties, had actually done away with creationism. Nevertheless, the general tone of the article is explanatory, but also defensive—it seeks to shield Darwinism from "moral" criticism.[29]

In 1883, Torre defended his doctoral dissertation at the University of Madrid. It dealt with "The Geographical Distribution of the Terrestrial Malacological Fauna of the Island of Cuba." Among the causes influencing the distribution of living beings he mentions "natural selection in the struggle for existence," but Torre limits himself mainly to zoogeographical description. He does point out, however, that the areas occupied by different genera of land mollusks reflect the geographical "distribution of

---

[25] Manuel A. Aguilera, "Clasificaciones biológicas," ***Anales de la Real Academia de Ciencias Médicas, Físicas y Naturales de La Habana***, 11 (1873), 170-184.

[26] José Planellas Llanos, "Discurso leido ante el Claustro Universitario por el Dr. D. José Planellas Llanos, en el acto de su solemne recepción de Catedrático," *La Enciclopedia*, 1 (1885), 36-40, 75-81., on pp. 79 and 80.

[27] Felipe Poey, "Tabla sinóptica de los animales divididos en tipos y clases," ***Revista Enciclopédica***, 1886, p. 180.

[28] Juan Vilaró, "Discurso de contestación," *La Enciclopedia*, 1 (1885), 82-86.

[29] Carlos de la Torre, "Breve exposición del Darwinismo," *El Club de Matanzas*, 2 (1880), pp. 90-91.

their extinct ancestors, which modified by virtue of selection and adaptation, after isolation, have given rise to a large number of analogous species, representative of each genus, in every locality."[30]

Two years later, in 1885, Torre gave a speech on the advances of "natural philosophy" during the nineteenth century. He stated that E. G. Saint-Hilaire was defeated by Cuvier's personal authority (and not by reason) in their celebrated 1830 debate in Paris. Regarding Darwin's theory and its possible bearing on the origin of man, Torre—who apparently had not yet read Darwin's book on this matter—considered that "Darwin only indicates the possibility of explaining the origin of beings through the constant *selection* of favorable modifications of a small group of very simple primitive forms and remains reserved regarding the consequences which spontaneously derive from this theory, that Huxley, with his deep knowledge of Comparative Anatomy, expounds under the title of man's place in nature."[31]

In 1887 Antonio Mestre's son, Arístides Mestre (1865-1952), a physician and assistant professor at the University again took up the problem of intermediate groups, and again dealt with the amphibia, as Vilaró had years before. Mestre based his analysis on Fritz Müller's now discredited "law of recapitulation" (considerably expanded and elaborated by Ernst Haeckel). This law, Mestre asserted, confirmed the status of amphibia as a separate class within the vertebrates, closer to fish than to reptiles.[32]

Arístides Mestre defended his doctoral dissertation in natural sciences that same year. He chose a question Felipe Poey had considered among the most difficult in natural history—that of colors among animals. Poey's 1856 article on this matter intended to prove divine intervention in nature, but when it was reproduced in 1887 by a Havana journal without his previous consent, Poey sent a note to the editor of the journal stating that his ideas on the matter had changed after reading Lamarck, Darwin, Huxley, Haeckel, and Spencer.[33] Mestre's selection of a topic was apparently prompted by this new public stand of the recognized mentor of Cuban naturalists.

Mestre explained the function of colors in animals mainly in terms of mimetism and sexual selection. He derived many of his ideas from Paul Marchal, Oskar Schmidt, and Alfred Wallace, but perhaps mainly from the somewhat forgotten author of *Le Darwinisme*, Mathias Duval (1844-1907), who was widely read at the time. Mestre quotes Duval as actually saying "if the theory of transformism did not exist, it would have to be created expressly to explain the relations between the colors of animals and the surrounding environment. "

---

[30] Carlos de la Torre, "Distribución geográfica de la fauna malacológica terrestre de la Isla de Cuba," *La Enciclopedia*, 1 (1885), 291-298, 477-481, 521-525, 623-629, on pp. 294 and 524.

[31] Carlos de la Torre, "Bosquejo histórico de los progresos realizados por la filosofía natural en el presente siglo," *La Enciclopedia*, 1 (1885), 227-232.

[32] Arístides Mestre, "¿Deben los anfibios constituir un orden entre los reptiles o bien una clase intermedia entre los reptiles y los peces?," *Revista Enciclopédica*, 1 (1887), 238.

[33] Felipe Poey, "Rectificación," *La Enciclopedia*, 3 (1887), 369.

Mestre's dissertation on colors provided many examples taken from Cuban fauna. One important case was that of a butterfly, with a very peculiar distribution of colors. In his theologically inspired article of 1856, Poey had challenged anyone to offer a scientific explanation of such coloring. Mestre took up the challenge and—with the assistance of his teacher, Carlos de la Torre—offered a quite credible explanation in terms of mimetic behavior.[34]

The young doctor's ideas regarding sexual selection are summarized in a paragraph of his unpublished article "A Darwinian Woman—Mme. Clemence Royer." The obvious charm of the following quotation shows us—quite unexpectedly—the romantic side of Aristides' character:

> Every youth, every man, presents his qualities to the woman he is courting, so that she can choose the man who best combines them. And victory goes to the most apt and active of them all. What a foundation for the perfection of the races and social progress! As we know, to this rivalry we owe the georgeous plumage and lovely songs of birds, the lion's mane, and so many of the beautiful features that we see in animals. From among the roosters and peacocks that court them, the females choose those who most perfectly harmonize those sounds and colors.

The Spanish professor Francisco Vidal y Careta (1860-1923) entered the Faculty of Sciences in 1885, but postponed his inaugural speech until 1888. It dealt with the relations between plants and insects according to the geological record. It is a clear reference to co-evolution, but does not emphasize evolution itself, much less natural selection.[35] Two years later José Ignacio Torralbas (1842-1903), a well known physician and Secretary General of the Academy of Sciences, decided to obtain a degree in natural sciences and defended a dissertation on the co-evolution of insects and plants. Whereas Vidal y Careta emphasized the dependence of insects on plants, Torralbas insisted on the inverse. He also referred to the importance of natural selection in co-adaptation.[36]

One of Carlos de la Torre's students, Francisco de Francisco y Díaz, presented a dissertation on classification. He defended (mainly in the same terms as Haeckel) the viewpoint that classification has to be based on evolution, but did not go further into this matter.[37] During the defense of Francisco's dissertation, Carlos de la Torre praised the "exceptional harmony" in the teaching of natural history which characterized the Havana faculty, especially regarding theory—a clear reference to the fact that all his fellow professors were evolutionists.[38]

Carlos de la Torre was elected to the Academy of Sciences of Havana in 1889. His inaugural speech dealt with the anatomy of a relict species of freshwater fish, the Cuban gar (*Atractosteus tristoechus*). This primitive representative of bony fishes belongs to a once widely distributed group, now restricted mostly to North America. The Cuban

---

[34] Arístides Mestre, "De los colores considerados en la serie zoológica," *Revista Cubana*, 6 (1887), 403-434.

[35] Francisco Vidal y Careta, *Los insectos y las plantas* (Havana, Published by the author, 1888) .

[36] José I. Torralbas, "Los insectos y la selección natural de las plantas," *Anales de la Real Academia de Ciencias Médicas, Físicas y Naturales de la Habana*, 27 (1890), 84-114.

[37] Francisco de Francisco y Díaz, *La evolución como base de la clasificación* (Havana: P. Fernández, 1890).

[38] Arístides Mestre, "De los colores..." (see note 34, above), p. 358.

species lives only in one lake on the Zapata peninsula, in Central Cuba. In his speech, Torre showed that the so-called "cranial plaque" of this fish is not—as some eminent ichthyologists believed—the frontal bone of the cranium, but only a set of modified scales.[39] Felipe Poey went still further while answering his pupil's speech and—in a clearly evolutionary conclusion—supposed this structure to be a remnant of the external "armor" of the ancient, extinct, placoderm fishes.[40]

In 1890, Juan Vilaró was accorded the honor of delivering the opening speech of the academic year at the university. He took advantage of the opportunity to defend the evolutionary approach and, moreover, publicly proclaimed the "triumph" of evolution within the University.[41] The philosopher Enrique José Varona saw in this speech the reflection of "a new spirit, which is slowly infiltrating university studies."[42] Vilaró, however, was still under the strong influence of Lamarck, whereas other professors were closer to Darwinian selectionism.

Although Varona had pointed out that several dissertations defended at the university before Vilaró's speech where clearly in favor of evolutionism, Arístides Mestre decided to insist on this point. He thought Vilaró had not been explicit enough regarding precedence: he had clearly ignored previous references to examples of evolutionary adaptation taken from Cuban fauna mentioned by Poey (blind cave fish), Torre (distribution of mollusks), Antonio Mestre (again blind cave fish), as well as in Torralbas' and his own (Aristides Mestre's) theses. Besides, he had failed to recall Planellas' speech as an obvious precedent in the defense of an evolutionary point of view within the university. Mestre clearly states that Vilaró's ideas were still mainly Lamarckian and did not include Darwinian selectionism among its basic tenets.

2. EPILOGUE

Cuban biologists, as those in many other countries, were affected by the "crisis of Darwinism" (ca. 1890s-1930s), that is, the rejection of natural selection as the main driving force of evolution, which stemmed from some of the works of George Romanes, Herbert Spencer, and August Weismann,[43] and from the resurrection of Lamarckism. This trend, "the eclipse of Darwinism" as Julian Huxley called it,[44] was linked to the rise

---

[39] Carlos de la Torre, "Consideraciones anatómicas acerca de los manjuaríes," *Anales de la Real Academia de Ciencias Médicas, Físicas y Naturales de la Habana*, 26 (1889), 282-291.

[40] Felipe Poey, "Discurso de contestación," *ibid.*, pp. 291-293. According to a more modern viewpoint, this is probably only a case of convergence.

[41] Juan Vilaró, "Discurso pronunciado en la Real Universidad de La Habana [...] en la solemne apertura del curso 1890 a 1891," *Memorias de la Universidad de La Habana*, 1890, p. 541.

[42] Enrique José Varona, "El transformismo en la universidad," *Revista Cubana*, 12 (1890), 384-385.

[43] Some of these papers were translated into Spanish and printed in Cuban journals, among them: George Romanes, "Teoría de la descendencia, según Weismann," *Revista Cubana*, 19 (1894), 102-120; Herbert Spencer, "La insuficiencia de la selección natural," *ibid.*, pp. 200-243.

[44] See Peter Bowler, *Evolution. The History of an Idea* (Berkeley, University of California Press, 1989), ch. 9.

of genetics in the twentieth century, and prevailed until the development of population genetics and the "new synthesis" (in the 1930s and 40s) presented a more coherent model of evolution by natural selection. Arístides Mestre practically quoted the American biologist E. G. Conklin on this crisis: "Heredity is at present the central problem of that science [General Biology]; the origin of the individual is now what most attracts the attention of biologists, while the genesis of species was probably of greater interest in the past century. "[45]

Centro de Estudios de Historia de la Ciencia y la Tecnología, La Habana, Cuba

[45] Arístides Mestre, "Las leyes de la herencia y la biología aplicada," *Revista de la Facultad de Letras y Ciencias*, 27 (1918), 163-193, on p. 164. Mestre used the following direct quotation from Conklin as an epigraph to his article: "The origin of species was probably the greatest biological problem of the past century; the origin of individuals is the greatest biological subject of the present one." This conception of opposing ontogeny to phylogeny was originally that of T. H. Morgan, who tried to link embryology and genetics in his research and developed theoretical ideas on this subject in several of his most widely publicized books.

HELOISA MARIA BERTOL DOMINGUES
MAGALI ROMERO SÁ

THE INTRODUCTION OF DARWINISM IN BRAZIL

Controversies Surrounding its Reception

The reception of Charles Darwin's theory in Brazil in the decades after the publication of the *Origin of Species* was controversial in both scientific and intellectual circles. The present preliminary analysis of the theme will focus on scientific debates, since they reflected the main theoretical lines followed by Brazilian scientists of the epoch.[1] Such discussions, of course, were both part of a general system of knowledge and a reflection of the specific cultural context of the country.[2] More particularly, we will analyze the controversies associated with the introduction of evolutionary ideas in Brazil and questions related to the origin of man and races inspired by Darwin's theory of evolution. In Brazil, during the last decades of the nineteenth century, evolutionary thought and racial theory represented guiding axes for the natural sciences and strongly influenced the ideas of the intellectuals who were involved in defining Brazilian nationality. Since the process of reception of science involves a complex of interactions among scientists, their production, and society—a process which may be either institutional or political-ideological—we also intend to examine such inter-relationships.[3]

In Brazil, from mid-century on, there was an effort to insert the country in the contemporary march of "civilization." In the social area, however, this "march" faced a considerable obstacle: slavery. For those racial theorists and builders of "national" civilization in nineteenth-century Brazil, the policy was not to incorporate blacks into

[1] The authors acknowledge the support provided by the Colloquium organizers, MAST/CNPq and FAPERJ. Thanks are also due to Professor Luís de Castro Faria, anthropologist and former director of the Museu Nacional, for helpful suggestions, and to Professor Nelson Papavero of USP for bibliographical assistance.

[2] On Darwinism as a "system of knowledge invested with malleability and dynamism, which may change over time and space", see T. F. Glick, *Darwin y el Darwinismo en el Uruguay en América Latina* (Montevideo, Universidad de la República, 1989), p. 35.

[3] On the concepts of reception, style, and rationality in science, see M. Paty, *L'Analise Critique des Sciences ou Le Téatraèdre Épistemologique* (Paris, L'Harmattan, 1990), ch. 4.

Thomas F. Glick, Rosaura Ruiz, and Miguel Ángel Puig-Samper (eds.), The Reception of Darwinism in the Iberian World, 65-81.

society, since they were considered foreigners anyway. It was a different matter for Indians, for they were considered a potential alternative to slave laborers and had valuable knowledge of the Brazilian hinterland yet to be explored. The development of the natural sciences in the country ultimately contributed to the creation of a social niche for Indians who, at the time, were in the process of becoming citizens. In such a scenario of ideological debate, the reception of Social Darwinism in Brazil was situated. Darwinism in Brazil, however, was not limited to social aspects, but had repercussions in natural sciences as well.

During the second half of the nineteenth century, natural sciences in Brazil began to incorporate new areas of study which were already in evidence in other countries. New approaches to geology and anthropology were introduced which differed from traditional mineralogy and ethnography both conceptually and methodologically. It was mainly through these disciplines that evolution was first approached. Differential receptivity according to discipline explains the pattern of acceptance of, and resistance to, Darwinism. In botany and zoology Darwin's ideas were also debated, but his theory was not widely accepted.

Darwin's experience during the voyage of the *Beagle* and his correspondence with Henry Bates and Alfred Russel Wallace contributed to the development his theory.[4] Brazil was a target country for foreign naturalists, as were other countries of Latin America. Bates and Wallace traveled for several years in Brazilian Amazonia, collecting specimens and studying its singular environment. Darwin also spent four months in Brazil, in 1832, during his voyage around the world. While in Brazil, he had the opportunity of exploring a tropical forest for the first time. In his travel diary, he wrote: "Brazil, Feb. 29th.—The day has passed delightfully. Delight itself, however, is a weak term to express the feelings of a naturalist who, for the first time, has wandered by himself in a Brazilian forest." When in Rio de Janeiro, Darwin visited the Jardim Botânico, where he observed: "many plants, well known for their great utility, might be seen growing. The leaves of the camphor, pepper, cinnamon, and clove trees were delightfully aromatic; and the bread-fruit, the jaca, and the mango vied with each other in the magnificence of their foliage."[5] In spite of having visited the national Botanical Garden and having warmly exalted the country's natural richness, Darwin made no reference to Brazilian naturalists, nor did he visit the National Museum in Rio de Janeiro which, by that time, already had a natural history exhibition open to the public.

Thus Brazil not only participated in the genesis of the theory of evolution by natural selection, but also a focus of immediate support for the theory. For it was here that Fritz Müller, a German immigrant who lived in the small town of Desterro (now Florianópolis), in the southern province of Santa Catarina, produced one of the most significant early interpretations of Darwin's theory. In 1864, he became well-known in international scientific circles through the publication of a small book entitled *Für*

---

[4] R. Ferreira, *Bates, Darwin, Wallace e a Teoria da Evolução* (São Paulo, EDUSP, 1990).

[5] C. Darwin, *Journal of Researches into the Natural History and Geology of the Countries Visited during the Voyage of H.M.S. Beagle Round the World.* (London, T. Nelson, 1893).

Darwin, in which he corroborated Darwin's theory through embryological studies of crustaceans. Müller discovered the larval form called *nauplius* in higher crustaceans (Malacostraca), these larvae having been previously known only in more primitive groups. He also compared the appendages of males and females of the genus *Tanais,* thus calling attention to sexual dimorphism. In addition, he studied the air-breathing apparatus of crustaceans, having compared the structure of the hearts of Amphipoda and Isopoda. Müller also established the basis for Ernest Haeckel's biogenetic law ("ontogeny recapitulates phylogeny") by stating that "in a short period of a few weeks or months, the embryo and larval forms display before us a more or less precise picture of the transformations which occur with the species through time until they attain their present stage."[6] The repercussion of Müller's book in scientific circles in Germany and England was immediate, his work being translated into English in 1869 on the initiative of Darwin himself. Müller became Darwin's friend and collaborator, exchanging many letters with him, mainly on subjects related to associations between different kinds of animals and plants. Müller also sent Darwin specimens of one kind or another.[7]

Fritz Müller, who established himself in Brazil in 1852, produced the majority of his 248 scientific papers in virtual isolation from the Brazilian scientific and cultural center of Rio de Janeiro. It was only after he received international recognition that some national institutional funds were made available to him. In 1876, he was hired as traveling naturalist by the Museu Nacional to collect materials to expand its scientific collections. During the time he worked for the Museum (1876-1891),[8] Müller published in its journal—*Archivos do Museu Nacional*—no fewer than seventeen papers on insects, crustaceans, and the fertilization of plants, all of them related to Darwin's theory.[9]

In spite of Müller's relevant and prolific scientific production, he got far less recognition in Brazil than he got from the Germans and Darwin. According to Edgar Roquette-Pinto—director of the Museu Nacional between 1926 and 1936--the main reason for Müller's low visibility was that his papers were extremely specialized and published only in scientific journals. For Roquette-Pinto, Müller was "one of the greatest scientific monuments who flourished in South America." However, "his unbreakable will, his taste for freedom, and his philosophical principles which drove him to bless the axe-handle, explain why Müller lost his job in 1891."[10] (That is, he refused to give up field work for a post in the Museum.)

The Brazilian discussion of Darwinism began in the early 1870's. In these early debates, evolutionism and Darwinism had different meanings. It was not possibly to label as Darwinists those who were trying to explain the "origin" and "evolution" of the

---

[6] F. Müller, *Facts and Arguments for Darwin,* trans. W. S. Dallas (London, John Murray, 1896).

[7] *Apud* M. W. de Castro, *O sábio e a floresta* (Rio de Janeiro, Rocco, 1992), p. 82.

[8] F. Müller, "Notes on Some of the Climbing Plants near Desterro, in South America, in a Letter to C. Darwin, *Journal of the Linnean Society, Botany*, 9 (1865), 344-349.

[9] *Archivos do Museu Nacional*, 2 (1877), 5 articles; 3 (1878), 4 articles; 4 (1879), 5 articles; 8 (1892), 3 articles.

[10] E. Roquette-Pinto, *Glória sem Rumor* (Rio de Janeiro, Museu Nacional, 1929).

earth, or of species and races. Among those scientists who opposed Darwin's theory, some were found at the Museu Nacional. Their scientific traditions were derived from French anthropology, particularly Quatrefages, and, to a lesser degree, from catastrophist geological theories in opposition to Lyell's *Principles*.

RESISTANCE TO DARWIN'S THEORY

One of the peculiarities of the social context of the reception of Darwinism in Brazil was that the Emperor was an important player in intellectual relations with Europe generally and had a privileged relationship with the scientific world. Pedro II was the only ruler and Brazilian citizen ever to be elected to one of the eight main positions open to foreigners in the French Academy of Sciences.[11] He was elected to the Academy not for his scientific production, however, but for his interest in science and personal relationships with leading scientists of his time, some of whom were recipients of his patronage.[12] The analysis of responses to Darwin's theory in Brazil reveals a triangular relationship involving Brazilian scientists (generally educated in Europe), their European colleagues (including Darwin and his great opponent Quatrefages), and the Brazilian Emperor. Regarding Darwinism specifically—or, in this case, opposition to it—the correspondence exchanged between Pedro II and Quatrefages de Bréau is an important source of information.

In this correspondence, which extended from the early 1870's to Pedro II's death in 1891, the Brazilian Emperor always agreed with Quatrefages' anti-Darwinian ideas, revealing himself as an enthusiastic supporter of his views. Pedro II used to send, through the Museu Nacional, fossils and ethnological material to be analyzed by the French scientist.[13] In one of his letters, the Emperor states:

> La doctrine évolutive est trop decevant, quoique s'appuyant sur beaucoup de faits. . . .
>
> Le Musée continue toujours à travailler mais je vou vais qu'il étudiât d'avantage la nature dans le Brésil où ellle se presente sous les aspects si differents.[14]

In another, he remarks:

> Je vous remercie bien de votre lettre si interessante du 29 dec. L'étude de la geologie est une de celles que m'assurent le plus et j'aimeà voir les faits consttatés par une autorité

---

[11] *Comptes Rendus de l'Académie des Sciences de Paris*, 1875, 540-541: Séance du Lundi 4 mars 1875; L'Académie procède, par la voie du scrutin à la nomination d'un correspondent pour la section de Géographie et Navigation, en remplacement de feu l'amiral de Wranglell. Au premier tour de scrutin, le nombre de votants étant 57; S.M. don Pedro, empereur du Brésil, obtient 43 souffrages; M. le Géneral Sabine 7 soufrages; et M. Cialdi 2 souffrage. Il y a cinq billets blancs. S. M. D. Pedro, ayant réuni la majorité absolue de souffrages, est proclamé correspondent de l'Académie (Archives de l'Académie de Sciences de Paris).

[12] Pedro II's correspondence is deposited at the Imperial Archives in Petrópolis, Rio de Janeiro. It is well known that the Emperor used to finance scientific works of Europeans, the most being his support of Louis Pasteur.

[13] Ofício 31 de julho de 1875, Arquivo Histórico-Científico do Museu Nacional, Doc. 42, Pasta 14. On the same occasion Pedro II also sent fossils to Virchow, in Germany.

[14] Pedro II to Quatrefages, 14 December 1878 (Archives of the French Academy of Sciences, Paris).

comme vous. Je viens de lire la collection de lettres de mon ami Agassiz publiés par sa veuve. Quelle hardiesse de théorie mais comme il tâchait de ne pas se tromper sur les faits. Nous avons souvent causé la-dessus les deux fois qu'il a été à Rio. (...) La question des origines sur un sujet de discussion et je desirerais bien connaitre votre opinion actuelle sur les singes anthropomorphoides. Rien ne me repugne d'admettre comme hypothèse, cette évolution pour l'espèce humaine, mais ce sont les faits que me manquent et je crains même de dire comme Gaudry que ce sera le cynoppitecus. Hartmann vient de publier un livre sur les anthropomorphoides, mais il n'est que philosophe d'une école dont je combatrai toujours you les principes.[15]

Even after the Emperor left Brazil, he reaffirmed his hostility to evolutionism shortly before his death:

Je hâte of vous dire combien la lecture du travail of Van den Ghegn m'a interessé. Mais je continues à croire que le premier homme ne fut pas noir, ni descendant du singe. Je repéterai avec l'auteur même vaut l'ignorance avouée que l'illusions de la science.[16]

The questions here related to the descent of man and the search for explanations of racial and cultural differences. In Brazil, the methods used to demonstrate such differences were those institutionalized in French anthropology (craniometry and anthropometry) by Paul Broca and Quatrefages. These lines dominated the scene and brought about a strenuous debate in the context of Darwinian theory. These anthropological methods were introduced in the second half of the nineteenth century, when the study of man was still divided between the old ethnography and the newer practices of analyzing fossils, dating the earth, and speculating on the evolution of culture and society.

In his letters to the Emperor, Quatrefages expressed a great interest in Brazil and the production of its scientific institutions, while at the same time discussing his own theories and publications with Pedro II. In a letter written in 1873, Quatrefages, after inquiring about recent work at the Museu Nacional, comments on the publication of his own recent studies in which he introduced his notion of race based on the use of craniology to characterize human types. He concludes saying. "I do hope it may be more and more recognized that the oldest pre-historic types have still some representatives in today's populations; and that the development of human qualities and faculties is independent of the form of the skulls. The skull of the Neanderthal which some anthropologists have thought just to represent a species of brute and ferocious beasts, is found not only among the Australians, but among us, in time and in space."[17]

[15] Pedro II to Quatrefages, 6 February 1886 (Archives of the French Academy of Sciences, Paris).

[16] Pedro II to Quatrefages, dated Cannes, 17 April 1891 (Archives of the French Academy of Sciences, Paris).

[17] Quatrefages to Pedro II, 2 December 1873. In another letter, Quatrefages mentions recent work by Grandidier and Edwards, stating that it was a fatal blow to Darwin's theory with its implied simian origins. He said that Haeckel thought he had found the missing link between marsupials and apes in the lemurs, which makes the pithecoids our ancestors, no intermediate form being known. Grandidier had studied the placenta of lemurs, which Haeckel believed to have features similar to and intermediate between man and apes. However, Grandidier was able to show that, in lemurs, that Haeckel had erred, the placenta resembling those of the ungulates, the edentates and the cetaceans (Quatrefages to Pedro II, 7 January 1876, Imperial Archives, Petrópolis, Rio de Janeiro.

Brazilian Indians and Negroes could easily be accommodated in Quatrefages' scheme. Quatrefages' dialogue with the Emperor was always centered on theoretical issues. Not long after publication of Quatrefage's book on race, Pedro II sent him fossils from Brazil to be analyzed, among which were skulls discovered in the 1840's by the Danish scientist Peter Lund in a region located in the center-east of the region called Lagoa Santa, province of Minas Gerais (Pedro II also sent fossils to Virchow in Germany at the same occasion). Lund's fossils were a point of controversy between Darwinists and those who opposed evolution. For Quatrefages, these fossils did not represent American man's ancient ancestry, because they originated much more recently than the Neanderthal man, even though they did not differ much in intellect, judging by skull measurements (meaning that they were "comparable to brutal and ferocious beasts.") In 1874, Quatrefages asked the Emperor to promote expeditions to the caves which Lund had visited before. He also noted that Lund had believed to have found human bones contemporary with certain animal fossils, a conclusion he puts in doubt.[18]

The Museu Nacional naturalists, João Baptista de Lacerda and Rodrigues Peixoto, were followers of Quatrefages in anthropology. Their work was honored with the award of medals at the Anthropological Exposition of Paris in 1878.[19] Lacerda and Rodrigues Peixoto had also examined the fossils found by Lund, and the results of their findings were published in the first volume of the *Archivos do Museu Nacional* in 1876.[20] In this work, Lacerda analyzed several skulls of Indians from different parts of the country (such as São Mateus, in Ceará) and compared them with skulls from the Lund's collection. Some of these specimens were identified as being from the Botocudos tribe, which Lacerda believed to be of the most primitive type (which also included the Lagoa Santa specimens). Regarding the Ceará form, Lacerda compared it with the skull found by Lund and commented: "A skull of such constitution must correspond to a degree of intellectual inferiority very close to the anthropomorphous monkeys."[21] Lacerda did not accept any ancestral relationship between men and apes. He explained he was a follower of the anthropological tradition of "Retzius, Morton, Prichard, Wagner," as expanded by the recent studies of "Broca, Pruner-Brey, Quatrefages, Virchow, Topinard and others, whose more practical works and broader views give us today a new approach to anthropological science." In his conclusions (in which he used the same cranial measurement techniques as those of Broca and Quatrefages), Lacerda analyzed skull comparisons and insinuated a few glimpses of evolutionary principles: "The prevalence of dolichocephaly in this series brings to light a valuable argument to

---

[18] Quatrefages to Pedro II, 27 April 1874 (Imperial Archives, Petrópolis, Rio de Janeiro).

[19] The certificates and medals received by both at the Anthropological Exposition in Paris were exhibited at the Brazilian Anthropological Exposition of 1882; *Guia da Exposição* (Rio de Janeiro, Museu Nacional, 1882), pp. 45, 71.

[20] *Archivos do Museu Nacional*, 1 (1876). João Baptista de Lacerda was a physician and, at the time, was Director of the Comparative Anatomy and Anthropology Section of the Museu Nacional. He was assisted by Rodrigues Peixoto Filho. From 1880 on, Lacerda was a member of the Physiology Laboratory of the Museu, under the direction Louis Couty, a former student of Claude Bernnard.

[21] J. B. de Lacerda, "Contribuição para o estudo antropológico das raças indígenas do Brasil," *Archivos do Museu Nacional*, 1 (1876), 45-75.

prove that the type of American races is generally dolichocephalic; on the other hand, the existence in the series of some subdolichocephalous and of one mesoticephalous type seems to indicate that the primitive type of Botocudo has been modified through breeding with other races. This conclusion is so all the stronger inasmuch as mesorrhinal and leptorrhine skulls appear in the same series, thus indicating the mixing of races."[22] He adds: "Because of their reduced cranial capacity, the Botocudos should be placed closer to the New-Caledonians and Australians, those races which are most notably marked by their degree of intellectual inferiority. Their skills are very limited, and it is hard to lead them into civilization. . . . The Lagoa Santa fossil skull is one of the most valuable ones in our collection and is similar in character to the skull of the Botocudos. Its cephalic index of 69.7 indicates a higher degree of dolichocephaly than the Patagonians and Eskimos, the two most dolichocephalic races. Such a view leads us to conclude that in the course of several centuries, the Botocudos have not risen a single degree in the scale of intellectuality. Their Cloquet facial angle is 67°. They are specimens of a prehistoric race, contemporary with the fossil horse and other species already extinct."

In defense of a polygenist hypothesis, Lacerda concludes his paper saying: "If it is true that the formation of the new continent preceded the formation of the Old World as Lund would have it based on his geological observations on the Brazilian central plateau, then, as Morton says, the same beliefs, the same habits, the same rites, and even the same language should found, with small differences, in every people scattered throughout the huge American territory. The new proposal of Simonin that the American Indian is a product of America may not be out of line!" Lacerda then concludes: "it is thus fair to declare that, in regard to such questions, we have not formed an opinion, and with respect to the compass of probable hypotheses, if one were to be accepted, we would be polygenist, like Agassiz. It is possible that America was one of the centers of creation, and that later on, people immigrating from Asia or other near-by parts of the globe, mingled with the primordial race, thus producing the recent race. This is one of the greatest problems of modern science, one which perhaps the science of the future will be able to resolve.[23]

The works of Lacerda and Peixoto merited favorable comments by Quatrefages in Paris. In a meeting of the Academy of Sciences in 1883, Quatrefages asserted the importance of the Botocudos: "*En présentant à l'Académie mes recherches actuels, j'ai déjà eu à parler de ce Mémoire et j'ai été hereux de donner aux deux savants brésiliens des éloges bien mérités. . . . M Lacerda fair ressortir les différences qui distinguent ces crânes de ceux des Botocudos, différences très sensibles en effet pour l'un deux, mais bien moins pour l'autre, à en juger par les figures. Il est a rémarquer que les crânes de maraca, plus doclichocèfales que le crâne botocudo le plus allongé (index horizontal moyen, 71.39 au lieu de 73.06), se rapprochent par ce caractère du crâne fossile de*

[22] The terms "mesorrhinal" and "leptorrhine" are categorizations based on nasal index measurements (editors' note).
[23] *Ibid.*, p. 74.

Lagoa Santa découvert par Lund (69.72). La différence est seulement de 1.67."[24] Lacerda and Peixoto's findings were thus fully consistent with Parisian anthropology.

In another paper entitled "New Anthropological Studies on the Botocudos," published in volume six of the *Archivos do Museu Nacional* (entirely devoted to the Brazilian Anthropological Exhibition of 1882) Peixoto reaffirmed his theoretical orientation: "The methods we followed are those of the French school, as recommended by Broca in his *Instructions*."[25] In the same sixth volume of the *Archivos*, João Baptista de Lacerda published a paper entitled "Sambaqui Man—a Contribution to Brazilian Anthropology," which he began by describing the objectives of anthropology, the context in which he understood his own research:

> From everywhere, dedicated supporters bring valuable and copious subsidies for the solution of prehistoric problems: caves are explored, necropoli are visited, monuments which had been destroyed and worn down by time are interrogated, the characters engraved in these monuments are deciphered, and the lost traditions of remote human generations seem to recreate the warm breath of searching souls hovering over such mounds of ruins. It was man struggling with himself to reach the still obscure and mysterious sources of his origin in time and space.[26]

In fact, Lacerda made few speculations on the age of the native cultures of Brazil, having stated that

> The issue of origins is an implacable question mark which appears at every moment to disconcert the most ingenious syntheses and the most plausible hypotheses. The leading thread is lost inside the entanglement of such a labyrinth, and even today we cannot say with surety from which source came the human currents that have spread over Brazilian soil since antiquity.[27]

Nevertheless, regarding the social "evolution" of these groups, he had views of his own:

> If the unskilled builders of the *sambaquis* [kitchen middens], rough constructions lacking regular form and previous planning, had wanted to perpetuate any important fact or represent any single thought with them, such would have certainly amalgamated with other models crafted with uniformity and a certain artistic skill. In manifestations of cerebral activity of humans with respect to art or industry, there is truly an infinite gradation which ascends from the most insignificant representative of the species to its most portentous product. From the Australian and the Tasmanian, almost at the level of brutes, to the artistic brains of Michelangelo or Raphael, how many countless levels of technique, how many infinite gradations to the concept of beauty and evenness of forms. The people of the *sambaquis* certainly did not have a brain adequate for artistic production, as the Aztecs and Peruvians had; the inferiority of their brains was at such a low level that it did not allow them to think about raising monuments, the existence of which would have implied a greater degree of civilization.

---

[24] Quatrefages, "Note sur l'état des Sciences Naturelles et de l'Anthropologie au Brésil, *Comptes Rendus de l'Académie de Sciences de Paris*, 96 (1883), 308-313.

[25] F. Rodrigues Peixoto, "Novos Estudos Craniológicos sobre os Botocudos, *Archivos do Museu Nacional*, 6 (1885), 208.

[26] J. B. Lacerda, "O Homem dos Sambaquis," *Archivos do Museu Nacional*, 6 (1885), 175. *Sambaqui*, a large mound of domestic refuse marking a prehistoric settlement, is known in English as "kitchen midden."

[27] *Ibid.*, p.177.

Lacerda concludes by saying that one *sambaqui* cannot be considered an archeological monument, inasmuch as "the aboriginal races from this part of the New World left not even the slightest vestige of even an incipient civilization; they lived through the centuries in the deepest barbarism, the long and dark phase of their primitive lives lasting until today." For Lacerda, the *sambaquis* had the same origin as the *Kjokhenmoddins* of Denmark. In the final part of his essay, Lacerda discusses the measurements of bones found in *sambaquis*, and again concludes that "Everything leads us to believe that this form—the remains of which were exhumed from the *sambaquis* of Paraná and Santa Catarina—represents the lowest level on the human scale, comparable to the most primitive peoples known today including, to judge by the morphological similarities of the skull, the Botocudos."[28]

Quatrefages commented on Lacerda's work and the others included in volume six of the *Archivos do Museu Nacional* in the French Academy.[29] Besides some short comments on the work of the Canadian geologist Charles Hartt which also mention the *sambaquis*, Quatrefages concentrates on details of Lacerda's work, agreeing with his conclusions on the origin of Brazilian *sambaquis* and the Danish middens. He also agreed with Lacerda's conclusions based on his measurements of the skulls found in the *sambaquis*, adding that homogeneity in population characters, which led to the conclusion that the people of the *sambaquis* were similar to the Botocudos, bore witness to the antiquity of the ethnological traits displayed by these tribes. Quatrefages also agreed with, and praised, Peixoto's work on the Botocudos:

> Il décrit et figure douze crânes, et donne un tableau détaillé de mensurations; puis il discute l'ensemble de ces donnés et en conclut que, par les caractères craniens, les botocudos se rapprochent de la race fossile de Lagoa Santa et, par les caractères faciaux, de la race des sambaquis. L'auteur se demande si les botocudos ne seraient pas le produit du croisement de ces deux races. Cette conclusion a pour elle de probabilités et rentrerait dans celles que j'avais tirées moi-mêmes de la comparaison de diverses têtes osseuses américaines avec le crâne fossile découvert par Lund. M. Peixoto fait d'ailleurs de sages réserves, auxquelles je ne puis que m'associer.

In one of his letters to D. Pedro II, Quatrefages praised the work of the Brazilians, who he believed were now producing research of high quality. He singles out the archeological research of Ladislau Netto as of most interest to Americanists, adding that the findings of Lacerda and Peixoto were of particular interest to him, insofar as they brought insight to his own work."[30]

At that time, Ladislau de Sousa Mello Netto was director of the Museu Nacional. Netto had been a botanist when the debate on evolutionism began; only later did he begin to work on the fringes of archeology and anthropology in studies of Brazilian Indians. His position on evolution was ambivalent.[31]

---

[28] *Ibid.*, p. 202.

[29] Quatrefages, "Recherches sur les populations actuelles et préhistoriques du Brésil," *Comptes Rendus de L'Academie des Sciences de Paris*, 101 (1885), 467-470.

[30] Quatregafes to Pedro II, Paris, 11 August 1885 (Imperial Archives, Petrópolis, Rio de Janeiro).

[31] A fact also observed by L. Castro Faria, As Exposições Antropológicas do Museu Nacional (Rio, Imprensa Nacional, 1949).

In 1876, Netto published in the first volume of the *Archivos do Museu Nacional* an article entitled "Estudos sobre a evolução morfológica dos tecidos nos caules sarmentosos." In this study, he seems sympathetic to transformist ideas, although he did not identify himself as a Darwinist. Familiar with recent works on evolutionary morphology by German biologists such as, among others, Carl Nägeli (an evolutionist, but not a Darwinian), and Julius Sachs, who did support Darwin, and with Darwin's work on *The Movements and Habits of Climbing Plants* (1865),[32] Netto compared his observations with those of the above authors. His own view of evolution had more in common with Lamarck's theory, with concepts like "progressive inheritance" and "the necessity of being adapted to the environment." On the action of the environment, he has a long comment on sarmentose plants (those endowed with runners):

> sarmentose plants should have had in earlier times original proportions equal to the more vigorous and developed shrubs known today, but because they live in the most shaded and compact of present-day forests, where direct sunlight is not available, they have extended themselves by constant and continual longitudinal growth of the stem until they reach the tree tops, where they could expose their leaves and stems for the formation of sap, for which sunlight is the immediate agent.[33]

He did not agree with Darwin on the origin of the climbing plants, commenting:

> Charles Darwin believes that originally all these plants had been climbers, being transformations of the original type which today present spiraling movement only in more or less modified foliaceous and axillary organs. In accordance with the doctrine of evolution, there is no difficulty in admitting that such may indeed have been the case. However, based on my own field observations on many hundreds of types of sarmentose stems, it is mandatory to confess that I agree with the opinion of the illustrious naturalist that they are vestiges of the capacity to spiral, in view of the incomplete and imperfect ability of straight stems to turn. Nothing makes me refute this. However, rather than accept the hypothesis that such stems cannot be reduced to the helicoidal form of the others owing either to organic ineptitude or some other idiosyncratic reason, it makes more sense to suppose that only those organs which were better able to change into tendrils did so pursuant to a compensating natural law based in the precepts of evolution.[34]

In this same year of 1876, during a botany lecture read in a public course offered by the Museu Nacional, Netto criticized the Darwin and Hooker's presentation on insectivorous plants to the meeting of the British Association at Belfast in 1874. As reported by the editor of the Brazilian Horticulture Magazine, Frederico Albuquerque, Netto had said that "the two British naturalists had exaggerated their observations to

---

[32] L. S. M. Netto, "Estudos sobre a evolução morfológica dos caules sarmentosos," *Archivos do Museu Nacional*, 1 (1876), 27-39, 133-144.

[33] *Ibid*. On German biologists, see W. M. Montgomery, "Germany," in. T. F. Glick, ed., *The Comparative Reception of Darwinism*, 2nd ed. (Chicago, University of Chicago Press, 1988), pp. 81-116.

[34] Netto, "Evolução morfológica," p. 134. One may ask whether Netto, educated in France and friend of many French naturalists of the Muséum and the Botanical Society, was not merely repeating Broca's views. According to Yvette Conry, Broca thought like a structural morphologist who assigned character to a series previously determined by morphology and did not pressupose "utility." For Geoffroy Saint Hillaire and Broca, the environment provokes transformation directly; Conry, *L'Introduction du Darwinisme en France au XIXe Siécle* (Paris, Vrin, 1974), pp. 60, 61.

support their evolutionary ideas, and he even called them charlatans."[35] Netto's attack on Darwin and Hooker triggered a series of harsh critiques of Netto and his botany course by Albuquerque over the following years.[36] When, however, Netto published the syllabus of his 1878 course in the *Archivos do Museu Nacional*, he now revealed his agreement with Haeckel: "the evidence provided by the Protist Kingdom on the similarity in the formation of all cells, both animal and plant, is the best justification of the doctrine of evolution." He also mentions in the same syllabus that plants, in their struggle for life, were transformed in accordance with their potential, and that adaptability was the basis for the laws of transmutation.[37]

In 1882 Netto discussed evolution when he was honored at a special session of the Argentine Scientific Society in Buenos Aires. He had been forced to address the theory of evolution directly and comprehensively, because an Argentine newspaper announced that he was going to speak on transmutation. In a letter to Henri Baillon, in France, Netto expressed his surprise at the announcement, commenting that it was very embarrassing to him to speak of a subject he was not working on any more.[38]

In this lecture, Netto asserted that "The theory of evolution, which is still a threat to ignorance and an irritant to superstition, continues to gain ground in the camp of the refractory, at the same time as new legions of scientists profess it with growing enthusiasm, insofar as it is the most logical, the most natural, and therefore the most attractive of all those theories proposed up to now to account for the admirable epic of creation."[39] When explaining his work on botany, he more precisely demonstrated his understanding of evolution: "The *adaptation* of animals and plants to the environment in which they must live, or, better stated, to the *energy and morphological and physiological resources* which they can marshal in the struggle for life, is sometimes the cause of these profound modifications. . . . *Nutrition and climate*, which exercise such great influence on the nature of man and animals, and which can be considered the principal bases of the adaptation of each individual to the existence he is obliged to accept, have equally great value with respect to the biological conditions of plants. A *struggle for life* is a fated attribute of all living beings. The difference consists only in

---

[35] F. Albuquerque, "Curso do museo, " *Revista de Horticultura*, 1, no. 4 (1876), 63.

[36] M. R. Sá, and H. M. B. Domingues, "Os cursos Públicos do Museu Nacional," *Revista da Sociedade Brasileira de História da Ciência*, 15 (1996), 84-85.

[37] L. Netto, "Resumo do Curso de Botânica do Museu Nacional, em 1878," *Archivos do Museu Nacional*, 3 (1878), 185-199. In the same abstract, Netto speaks again of the struggle for life among plants: "plants are equipped by nature for the struggle for life, being the Alexanders and Napoleons of the plant world; the Creator gave them the energy of conquerors and did not block their way, as shown by the *Eichornia azurea*, called "lady-of-the-lakes" in Bahia, and "baroness" in Alagoas, has caused most damage to the river banks of those two provinces" (p. 198).

[38] The letter to Baillon was published by the Museu Nacional along with the lecture in the following year, *Aperçu sur la théorie de l'Évolution, Conférence faite a Buenos Aires dans la séance solenelle, célebrée en son honeur para la Societé Scientifique Argentine, le 25 octobre 1882* (Rio de Janeiro, Méssager du Brésil, 1883). Baillon (1827-1895), professor of natural history at the Medical School of Paris, was most probably Netto's teacher.

[39] L. Netto, *Observaciones sobre la Teoria de la Evolución,leídas en la Sociedad Científica Argentina* (Buenos Aires, Imprenta de La Nación, 1882), p. 4.

the kind of combat, always taking into account the circumstances and configuration of the combatants." Netto concluded by saying: "Before these object lessons provided by nature, let us attempt to break the chains that still enslave us, by means of the culture of the intellect and through the widest possible development of sociological laws!"[40]

In his Buenos Aires lecture, Netto clearly showed his interest in "social evolution." Indeed, during the 1880's, he seemed to forget about his specialization in botany and applied himself to archeology as a complement to his anthropological interests. In the same letter to Baillon, Netto regrets forsaking a secure field like botany to study Indian populations and noted that his position as Director of the Museum had led him to this decision. He added that the Museu was the only scientific institution in a position to gather and study "the remains of the last representatives of the many thousands of human beings which for centuries occupied the coast and hinterland of Brazil." He justified such a task by saying that it was necessary to save the few remnants of the descendants of those ancient masters of South America, with their traditions, their languages and their primitive ceremonies before the race disappeared completely or fused through interbreeding.[41]

Also in 1882, Netto successfully organized the Brazilian Anthropological Exposition whose aim, as revealed in the introduction to its guide, was "not only to exhibit the ethnographic artifacts and documents related to our Indians, but also to put them together in a single public repository and then, as precious relicts, exhibit them in celebration of science."[42] Volume 6 of the *Archivos do Museu Nacional* was devoted to the Anthropological Exposition, featuring articles on Brazilian Indians exclusively. Among them is an long piece by Netto covering nearly half of the volume. He begins this article by stating that the copious materials assembled for the Exposition had motivated him to publish his observations on the archeology of the Brazilian Indians, in spite of the current controversies on the subject:

> I now look at the ethnological lot of the New World and regrettably, or at least unexpectedly, must provide arguments opposed to the autochthonous American polygenist hypothesis with the respected figure of Agassiz in the forefront and which I would like to have supported. . . . If we consider that the quadrumans, as they rise in perfectibility, present areas of emergence more restricted spatially throughout time, the area of emergence of human beings must have been even more limited, at whatever point on the Earth such a great phenomenon may have occurred. The whole of the Americas is still to be studied with respect to this particular issue.[43]

Netto carried out his studies at Pacoval, a mountain on the island of Marajó, located at the mouth of the Amazon river. Analyzing its archeological remains, Netto found that symbolic inscriptions on ceramics provided elements to compare with other cultures. On this basis, he concluded that the primitive inhabitants of that place had

---

[40] *Ibid.*, pp. 8, 9, 12, 21 (emphasis ours).
[41] Letter to Baillon, in Netto, *Aperçu sur la théorie de l'Évolution*, p. iii.
[42] *Guia da Exposição Antropológica Brasileira realizada pelo Museu Nacional do Rio do Janeiro* (Rio de Janeiro, Leuzinger & Filhos, 1882).
[43] L. Netto, "Investigações sobre a Archeologia Brasileira," *Archivos do Museu Nacional*, 1882..

migrated from Asia to North America, and from there to Pacoval and had a superior intellect which may have decayed as a function of environment. He also explained that among the representations of heads of the people from Marajó displayed at the Exposition were some truly simian types. In this way, he decided to submit the material to public judgment: "the reason that made me include them in that section is based on the conventionality which I believe the "mound-builders" of Marajó used to represent their own people, giving them, by means of their physical characters, the affinities and homological analogies through which they linked themselves to certain animals, particularly to simians, which they supposed belonged to their own kin and regarded as intellectually superior. . . . they believed they were linked to monkeys." Netto made some comparisons with zoomorphic representations found in all ancient cultures and concluded that the ancient inhabitants of Pacoval had descended from a higher culture.

Netto's article was highly praised by Quatrefages, just as he had previously praised those by Lacerda and Peixoto. Quatrefages accepted Netto's hieroglyphic evidence as proof of the Oriental origin of the ancient peoples of Pacoval. He stressed, however, that such a fact could not be taken as evidence that the *ancient* inhabitants of that place had descended from an advanced civilization.[44]

Shortly after Quatrefages' comments appeared, Netto participated of the Congress of Americanists held in Berlin in 1888. During the Congress, he received a commendation from the German Emperor and was named chairman of one of the scientific sessions. In a report to the Brazilian Government, Netto commented: "I was honored to chair one of the sessions of that memorable scientific meeting and to present there researches and ideas on the palaeo-ethnological character of pre-Colombian nations of the Amazon valley."[45] In this report, Netto does not mention evolution, commenting only on his scientific methods and his contacts with foreign scientists interested in the Amazon, such as the Italian Vicenzò Grossi.

Netto's position—more evolutionist than Darwinian in the strict sense—certainly contributed to the continuing debate at the Museu Nacional. Besides Fritz Müller, there were other Darwinians there, such as João Joaquim Pizarro, then Director of the Zoology Section. In 1876, he published an article in the *Archivos* on the classification and morphology of a specimen of *Batrachychthis*, an amphibian which had been sent to the museum from Paraguay. Pizarro commended the paper to the attention of "Professors Darwin, Haeckel and Charles Martins."[46] That Pizarro was in fact a Darwinian was noted by Lacerda in his *Fastos do Museu Nacional,* where he comments that Pizarro used to scandalize ladies at public lectures at the Museum on the theory of evolution, when he emphasized the similarities between man and ape.[47]

---

[44] Quatrefages, "Recherches sur les populations actuelles et préhistoriques du Brésil," p. 470.
[45] Document published in the *Jornal do Comércio*, Rio de Janeiro, 13 December 1891 ("Sessão Gazetilha," p. 1).
[46] J. J. Pizarro, "Nota descritiva de um pequeno animal extremamente curioso e denominado 'Batrachychthis,'" *Archivos do Museu Nacional,* 5 (1876), 31-35.
[47] J. B. de Lacerda, *Fastos do Museu Nacional,* (Rio, 1905), p. 60.

DARWIN'S FOLLOWERS: HAECKELIANS AND SPENCERIANS

Public courses, like those offered at the Museu Nacional and others called "Conferências Populares da Glória" were popular cultural activities in Rio de Janeiro during the 1870s, as well as the principal platforms for debate. Indeed, most of the debates on Darwinism took place at the Conferências, initiated in 1873. First in defense of Darwin stood the physician Augusto Cezar de Miranda Azevedo, a graduate of the Medical School of Rio de Janeiro who lectured publically on the biological conceptions of Darwin from the perspective of Haeckel's monism.[48] For him, "Darwin had been in the area of zoology and botany the revolutionary who produced the same revision as Lyell had brought about in geology. Darwin had freed the natural sciences from the absurd teleological hypotheses supported by Cuvier and more recently by Louis Agassiz."[49]

Miranda Azevedo had already articulated his pro-Darwinian position in the presentation of his thesis at the Faculty of Medicine of Rio de Janeiro in 1874. On that occasion, he discussed the subject "On Darwinism: Is the steady improvement of species up to man acceptable?", in which he supported Darwin's theory.[50] Afterwards, he began to disseminate the ideas of Haeckel and Darwin, most notably in a series of seven lectures in the Glória series in 1875, titled "Darwinism—Past, Present, and Future;" "The Study and Demonstration of the Fundamental Laws of Darwinism;" "The Many Modes of Organic Reproduction;" "On Fecundation of Animal Organisms;" "Darwin's Systems;" and "Application of Evolutionary Doctrine to Man."[51]

Besides Miranda Azevedo, other speakers offered their views on Darwin from the podium at Glória. An example was Feliciano Pinheiro de Bittencourt who lectured on Darwinism in 1876: "I have spoken of Darwinism, or evolutive doctrine, trying to show that in the present state of science it is still too early to confirm that man descends from apes or is no more then an improved monkey."[52] The same year, Antonio Felício dos Santos, in a lecture on "Fashion in Relation to Hygiene," commented: "the famous Darwin, in the latest developments of his evolutionist theory, establishes the descent of man through transformations of extinct types of animals, giving us an ancestor close to the simians. In one of his latest books, *The Expression of the Emotions*, he ingeniously shows our relationship with our cousins, the recent apes. . . .While not defending this theory, I can not refuse to agree that such an argument is one of the most plausible in

---

[48] T. A. F. Collichio, *Miranda de Azevedo e o Darwinismo no Brasil* (São Paulo, Itatiaia/EDUSP, 1988), p. 41.

[49] A. C. de M. Azevedo, *Darwinismo: seu passado, seu presente, seu futuro. Conferências Populares* (Rio de Janeiro, J. Villeneuve, 1876), p. 42.

[50] Collichio, *Miranda de Azevedo*, p. 35.

[51] M. R. F. Fonseca, "As Conferências Populares da Glória: A divulgação do Saber Científico," *Revista Manguinhos*, 2, nº 3 (1995-1996), 135-166.

[52] F. P. Bitencourt, *Conferências Effectuadas na Escola da Glória* (Rio, 1882).

favor of Haeckel's hypothesis of the anthropoid simian."[53] Not all speakers at Glória defended Darwin; some called his theory subversive.

Miranda Azevedo was not only the most incisive defender of Darwin's theory at these public lectures, but he also used Darwinian ideas to justify Republican attacks on the Emperor, stating that the main cause of social problems was ignorance of Darwin's theory by those who legislate. In this sense, he asserted: "throughout the whole civilized world in present times there is a growing concern for military predominance, and what is the cause of such concern? The ignorance of Darwin's laws as they relate to the development of military laws. . . . Would it not be more advantageous to accept the consequences of Darwin's theories than to formulate stupid military laws which only serve the caprices of despotism?"[54] In this same lecture, Miranda Azevedo called attention to the positive results to be gained from the study of Darwin's laws in the education of youth and the general happiness of the people. In this way, "a country without despotism and privileged classes could be created." Such ideas, as stressed by Collichio, reflected the principles of Haeckelian Darwinism which stressed the hereditary transmission of inherited characters, including intellectual ones, and whose influence was felt in law, literature, theater, and pedagogy.

In Azevedo's thesis, one can note the belief in general laws which would later support what was called Social Darwinism. During the last years of the nineteenth century, works by intellectuals educated at the Recife Law School, such as Sylvio Romero and Tobias Barreto, began writing on social Darwinian themes, following Spencer and Haeckel.[55] At the time, the city of Recife was a center for the diffusion of German culture. Barreto actually published a German-language newspaper, in which he propagated the ideas of Haeckel and Spencer on the strength of Sylvio Romero's influence. As a follower of Spencer's brand of evolutionism, Romero wanted to produce a genealogy of Brazilian society faithful to Darwinian principles. For him, the history of a people should encompass a complete explanation of its evolutionary advance. Thus, "an evolutionary theory of the history of Brazil should elucidate the action of the environment in all its aspects . . . . study the ethnological quality of the races which formed us; delimit the biological and economic conditions of the inhabitants during early colonial times. . . . Of all the theories proposed, Spencer's is closest to the mark." In a footnote, Romero added that more such interpretations *"à la Darwin"* could be found in his books, *Brazilian Literature and the Modern Critics*; *Studies of Popular Poetry in Brazil*; and *History of Brazilian Literature*.[56] Romero

---

[53] A. F. Santos, *Da moda em relação com a hygiene. Conferências Populares* (Rio de Janeiro, 1876), p. 109.
[54] Azevedo, *Darwinismo*, pp. 41-63.
[55] Darwin's theory is discussed in many books by Tobias Barreto and Sylvio Romero. The former owned a secondary school in Recife, and the latter taught philosophy at the Colégio Pedro II and the Faculty of Law; Antonio Candido Mello e Sousa, *Formação da Literatura Brasileira*, 2 vols. (Belo Horizonte, Itatiaia, n/d), II.
[56] S. Romero, *História da Literatura Brasileira. Tomo I - Contribuições e estudos gerais para o exato conhecimento da Literatura Brasileira*, 4th ed. (Rio de Janeiro, Livraria José Olympio, 1949), p. 54.

favored miscegenation and believed that Brazilian society of the future would be white, as was demonstrated by its evolutionary history.

In an article on the theory of social evolution, Romero noted that the application of Darwin's theory to history, linguistics, law, philosophy, and sociology had begun more than twenty years before.[57] He listed the works of several writers who had applied the theory to the Brazilian case, including Tobias Barreto, Arthur Orlando, Clóvis Beviláqua, and Martins Júnior, all of whom attempted to create an ontogeny and a phylogeny of law and society. To this effect, Romero quoted Tobias Barreto: "What is observed in individuals *also happens in peoples,* whether considered in themselves or compared with one another." For Romero, however, Haeckel's idea that "each people *recapitulates all past phases* of human evolution as it develops" was an exaggeration and not wholly correct.[58] He makes clear allusions to the impracticability of the application of Haeckel's theory in sociology and disagrees explicitly with the principle stating that "every time a society moves from one region to another and a civilized group meets with others and mingles with people still in inferior cultural phases, their history retrogresses through the centuries and *starts to recapitulate all past phases of the history of humanity.*" Romero declares himself a follower of the polygenist variety of Darwinism and of biological theories (like that of G. Tarde) which accept an analogy between animals and plants. He retrieves Spencer's ideas and Schäffe's thesis, which concluded that "colonies reproduce faster and more intensively, and over a considerable period, the stages crossed by civilizations of higher culture: *it is the reproduction of phylogeny by ontogeny.*" Concluding, Romero states that races, however distinct they may appear, display analogical parallelism in the various phases of their development: "this is explicable in some cases by the identity of human nature . . . in others, by the geographical similarity of various regions capable of stimulating the production of the same ideas and tendencies; or still in others, by copying and adopting the concepts of other peoples. It is true that from the most ancient times peoples have interacted more or less in accordance with their practices and commercial facilities. . . . A people compared with itself, at the various moments of its own evolution, can here and there display similarities of a greater or lesser degree. The stages, however, do not repeat themselves, but rather unfold. That is the essence of it."

In the last years of the century, the debate on Darwinian evolutionism widened considerably. Romero's conclusions—which seemed to contradict not only some standard Darwinian interpretations of society—anthropological interpretations such as those of Quatrefages, for example, or of Ladislau Netto, for whom the environment caused the decadence of the more advanced group—were not the only ones which emerged. Besides Romero's article, the *Revista Brasileira* also ran another discussing ascent from apes to man centering on Dubois' studies of *Pithecanthropus*, which generated a debate between Herman von Ihering, the director of the Museu Paulista, and the zoologist Carl Euler. The latter, a follower of Haeckel, held that it was

[57] S. Romero, "O Haeckelismo em Sociologia," *Revista Brasileira*, 11 (1899), 200-235.
[58] *Ibid.*, p. 201 (emphasis in text).

impossible to deny that the Java findings had diminished the distance between man and ape.[59] Von Ihering—a friend of Virchow and follower of craniometry—declared that no conclusions could be reached on the *Pithecanthropus* because the most competent naturalists yet to reach a consensus. For him, the fossil was too small to be a man, too large to be an anthropoid.[60]

The *Revista Brasileira* published other articles on biology, such as one by Euler on "*Flowers and their Guests,*" in which he discusses the relationships between plant species and the insects that feed on their pollen and pollenize them,[61] and another by Domingos Freire on the biological evolution of life, entitled "*The Land of Cultivated Plants.*"[62] To what extent writings on evolution by Brazilian naturalists were based on actual experiments is hard to say. What is clear is that Brazilians marshaled the best evidence they could to support their positions in the intense debate over Darwin. A thorough analysis is required in order to reveal connections between the theory itself and its influence over scientific, social, and philosophical issues.

Museu de Astronomia e Ciências Afins -MAST/CNPq
Museu da Vida/Casa de Oswaldo Cruz/Fiocruz

[59] C. Euler, "O Pithencanthropus," *Revista Brasileira,* 9 (1897) , 33-42.

[60]H. von Ihering, "Notas e Observações - O Pithecanthropus," *Revista Brasileira,* 9 (1897), 191-192. The question of human descent from apes had also been mentioned in a newspaper article, "A Teoria Darwinista," *Jornal do Comércio,* 29 February 1892. The article, by an anonymous author, presented Robson's experiments in England on instinctual clinging to mothers by simian and human babies. According to the author, the evidence suggested man and monkey are related, or at least quite similar.

[61]C. Euler, "As Flores e seus Hóspedes, *Revista Brasileira,* 9 (1897), 93-107.

[62] A. Domingos Freire, "A Pátria das Plantas Cultivadas," *Revista Brasileira,* 9 (1897), 170-284; 216-231.

MARCOS CUETO

NATURAL HISTORY, HIGH-ALTITUDE PHYSIOLOGY AND EVOLUTIONARY IDEAS IN PERU

The past twenty-five years have witnessed the publication of a number of studies of the diffusion of Darwinism in Latin America and of its impact on political and social ideas.[1] Less attention has been paid to the influence of Darwinism on the ideas of Latin American scientists and physicians, at least in the case of the Andean countries. The lack of attention paid to these countries is in part explained by the fact that the War of the Pacific, which consumed the energies of Peru, Bolivia and Chile between 1879 y 1883, estranged this region from the rest of the world just when the Darwin debate was at its height. The occupation of Lima during the war caused the paralysis of Peruvian academic and scientific activities.

Peru, moreover, has been less visible in studies on the diffusion of Darwinism because the public debates on that subject in Lima were less spectacular than those which took place in other Latin American cities. There was no extreme political polarization over the new theory, in part because it had only one great institutional focus, namely the Universiy of San Marcos. With the exception of the confrontation between the Catholic Church and the physician Celso Bambaren, educated in París, professor of anatomy and physiology and ardent defender of both Darwin and Lamarck towards the end of the century, the situation is well described by Jorge Basadre, dean of Peruvian historians, who observed that discussions of evolution never constituted "episodes of great resonance" during this period.[2] The lack of greater attention to Darwinism is also explained by the fact that at the turn of the century Andean scientitists, apart from a few works of popularization, did not cultivate those disciplines

[1] Some of the more important works are Thomas F. Glick, "Perspectivas sobre la recepción del darwinismo en el mundo hispano," in *Actas de. II Congreso de la Sociedad Española de Historia de las Ciencias*, Mariano Hormigón, ed., 3 vols (Zaragoza, 1984), I, 49-64; *idem, Darwin en España* (Barcelona, Península, 1982); Roberto Moreno, *La polémica del darwinismo en México, siglo XIX* (Mexico City, UNAM, 1987); Pedro Pruna, *Darwinismo y sociedad en Cuba, siglo XIX* (Madrid: CSIC, 1989).

[2] Jorge Basadre, *Historia de la República del Perú* (Lima, Villanueva, 1961), VI: 2936. On Bambaren, "El Dr. Celso Bambaren (1833-1897)," *La Crónica Médica*, 29 (1912), 500.

Thomas F. Glick, Rosaura Ruiz, and Miguel Ángel Puig-Samper (eds.) The Reception of Darwinism in the Iberian World, 83-93.

such as embryology, biochemistry, or genetics which propelled Darwinism to a preeminent place in biology.

Still, it would be wrong to conclude that Darwinism amd evolution played only a marginal role in Peruvian science. To identify and describe this role one must avoid contextualizing Darwinism in the context of the supposed dilemma of victory or defeat which at times is concealed behind the notion of a "scientific revolution." Rather we must tolerate the coexistence of various theories of evolution, some mutually contradictory, and stress the search for patterning in the process of local adaptation and re-creation of scientific ideas.

The present essay has three objetives: first, to identify a line of intellectual continuity between the research of Peruvian naturalists and physicians in mid-nineteenth century and that of the beginning of the twentieth. This continuity principally refers to the relationship between high altitude and the variety of species which exist in the Andes. In second place, I want to call attention to the influence of evolutionary ideas in the work of the Peruvian physiologist Carlos Monge Medrano, who led the most active group in Peruvian biomedical research in the first decades of this century. Finally, I wish to add some reflections on how a style of scientific practice under conditions of adversity illuminates the Peruvian case.

II

Natural History as practiced, published, and taught in nineteenth-century Peru, begins with the Italian immigrant Antonio Raimondi who arrived in 1850. Raimondi, a self-taught naturalist, began to classify the geological and mineralogical collections of the laboratory of physics and natural sciences of the reorganized Faculty of Medicine of the University of San Marcos. From 1851 he was professor of natural history in that Faculty and, later, the first dean of the Faculty of Sciences. Over nineteen years he divided his time between university teaching and field trips to the interior of the country, botanizing, collecting minerals and fossils, and establishing the location and altitude of the principal geografical landmarks.

Raimondi's excursions covered around 45,000 kilometers of Peruvian territory and permitted the identification of important new mining sites, the addition of new species with possible medical and industrial applications to the fauna and flora of Peru, and the elaboration of widely used maps of the country. Raimondi's varied interests reflect the scarcity of scientific talent in nineteenth-century Peru, the encyclopedic style of Natural History and the modernizing projects of a civil elite who saw in the export of natural resources and in road construction two of the principal avenues of economic progress.

In his publications, Raimondi stressed a concept that goes back to the writings of late colonial naturalists and which lived on afterwards: the failure of European scientific works to include American materials. According to Raimondi, biology texts published in Europe lost some of their relevance in countries like Peru, because they contained no examples of New World flora. This idea was reinforced by the local belief that there was something different or unique about the Andean region which could not be fully understood by European naturalists because they were not fully acquainted with it. In his first book, published in Peru in 1857, titled *Elementos de Botánica aplicada a la medicina y a la industria*, Raimondi cited Lamarck, arranged his materials on Peruvian botany according to taxonomic and geografical criteria, and described his first excursions through the Andes. He was impressed how "the high mountains" and the climate of high altitudes, had "such a great influence on the distribution and variety of plants." One of his important observations was that, in the Andes, plants might serve to "determine the altitude over sea level."[3] This observation would be taken up later and developed further, first by Peruvian botanists and afterwards by physiologists.

In the first volume of his most important book, *El Perú* (1874), Raimondi mentions Darwin and briefly explains the polemic stimulated by his works (without specifying any), but provides no additional details on this debate nor takes any position on it. I think there are at least two reasons for his stance. First, Raimondi himself did not understand the debate. Second, he shared the perception of other Andean naturalists that what they studied was different from what was studied in Europe. This perception was not only a reflection of provincialism, but of naturalists' desire to make original and useful contributions to science and to the society in which they lived. Thus there was a close relationship between the classical parameters governing the classification and identification of new species that Raimondi used and the local political perception of Natural History as a useful activity. For Raimondi, "inasmuch as natural sciences are in their infancy in Peru. . . it was necessary to introduce them through their utility."[4]

Raimondi's ideas continued to be disseminated at San Marcos by his most important student, the physician and naturalist Miguel Colunga who for many years held the chairs of Medical Natural History, Zoology, and Botany (in which he established more precise criteria for systematic botany). Colunga wrote a text book in which he presents the "dichotomous method" of Lamarck, as well as other systems of classification.[5] It is important to note that Colunga held important posts in the Faculty of Sciences (he was dean between 1899 and 1902), and that his personal library was given to the library of the Faculty of Medicine at San Marcos upon his death in 1914.

III

At virtually the same time that Raimondi was pursuing his studies and publications, a direct contact—as it were—with the authors and ideas of English evolutionism was established by a British naturalist then exploring the Andes, and the Peruvian Amazon and coast. The man in question was Richard Spruce, friend and colleague of the co-discoverer of natural selection, Alfred Russel Wallace, the two volumes of whose *Notes of a Botanist on the Amazon and Andes* appeared in 1908, fifteen years after Spruce's death.

Spruce reached South America in 1849 and travelled through Peru and Ecuador between 1855 and 1864. His trip was motivated in part by the earlier explorations of

[3] Antonio Raimondi, *Elementos de Botánica aplicada a la Medicina*, (Lima, Calle de Compas, 1857), p. 285.
[4] *Ibid.*
[5] Miguel Colunga, *Elementos de Botánica* (Lima, 1878). On Colunga see Hermilio Valdizan, *Diccionario de la Medicina Peruana* (Lima, Larco Herrera, 1928), II: 139.

Wallace and Henry W. Bates in the Amazon.[6] Spruce spent the first years botanizing in the jungle of Pará, in Brazil, financing his tour by sending specimens and collections to England, especially to Kew Gardens in London. Then he relocated at Tarapoto in the Peruvian Amazon for two years. He discovered many new species and established the botanical fundamentals of the genus *Hevea*, the source of Amazonian rubber. Towards the end his stay in South America, he collaborated with Clements R. Markham in provisioning British plantations with cinchona seeds. Spruce believed his mission to be that of other European naturalists of the period: "to transform the wild forest into a fertile garden."[7]

It is interesting to hear the echo of Darwinism from a naturalist working alone in the jungles of the Amazon. According to a letter written by Spruce sometime in the 1850s, he thought (possibly echoing Lyell) that "[those] same laws and forces now in existence have been so throughout all eternity."[8] Nevertheless, his letters and and most ambitious articles, those with evolutionary observations, were considered deficient by English naturalists, including Darwin and Wallace, who thought he attributed too much emphasis to the fixing of characters by the action of the environment.[9] Although Spruce seems not to have been in contact with Raimondi or any other Peruvian naturalist, some of his manuscripts reveal, if not local influence, then at least an effort to adapt evolutionary ideas to the indigenous population. These ideas are not so different from those sustained by Peruvian physicians and naturalists at the turn of the century.

One of the most interesting and least known of Spruce's writings is a manuscript in the archives of the Linnean Society in London titled "Notes on the possible aclimatization of Europeans in tropical South America," six handwritten pages written in response to some questions from Wallace. It is undated but most likely was written towards the end of his life, after his return to England, that is, around 1870.[10] Although the principal motivation of this document is to emphasize the economic advantages to possible immigrants, there are several additional interesting points.

According to Spruce, white Europeans can adapt well to life in the Amazonian valleys and the Peruvian coast provided they take necessary measures with respect to shelter, health, and diet. He notes that in the coastal regions, neither the fertility nor the health of European mothers or fathers had been diminished. He was more cautious regarding the effects of settlement in the Andes, attributing the scant European presence there to the cold, the rarified atmosphere, and the altitude of the region. Near the end of the document, Spruce observes that in spite of the fact that South American Indians must all belong to the same race, there was a marked difference between native

[6] On Spruce, see ***Richard Spruce (1817-1893) Botanist and Explorer***, M. R. D. Seaward and S. M. D. Fitzgerald, eds. (Londres, Botanical Gardens of Kew, 1996).

[7] "Rough draft letter written to agent of Ecuador, G. P. Prochett in reply to enquiries as to the feasibility of forming a colony of Europeans in Forest of Canelos, Los Baños," Dec. 1857, in Richard Spruce Papers, Archives de Kew Gardens, Londres, [hereinafter cited as Spruce Papers, Kew].

[8] Spruce to W. Wilson of Warrington, May 28, 1870 (Spruce Papers, Kew).

[9] R. E. Schultes "An Unpublished Letter by Richard Spruce on the Theory of Evolution," ***Biological Journal of the Linnean Society***, 10 (1978), 159-161; *idem*, "Richard Spruce and the Potential for European Settlement of the Amazon: An Unpublished Letter," ***Biological Journal of the Linnean Society***, 77 (1978), 131-139.

[10] Richard Spruce, "Notes on the possible aclimatazation of Europeans in tropical South America" (no date. Richard Spruce Papers, Linnean Society, London).

inhabitants living at sea level and those from high altitudes. He attributes this difference to varying patterns of disease and, above all, to the difference of climate and altitude. Based on his observations he reaches the conclusion that the Indians of the Peruvian mountains constitute a racial "variety," distinct from that of the coastal Indians, both having descended from some common North American ancestor. The result, according to Spruce, was one human variety acclimatized to bear the extreme heat of the coast, while the other was able to bear the opposite conditions. Near the end, he responds ambiguously to the question: Would not the same have happened (that is, the emergence of two varieties of a single race) in the white man? He concludes that it would have, were the necessary time alloted, "because one can [deleted: "cannot"] doubt that the inherent adaptibility is the same in both [Indians and whites], or (if not) that the white man possesses it in a higher degree."

IV

At the end of the nineteenth century, the ideas and contradictions of Raimondi and Spruce on the evolution of life in the Andes and on the impact of altitude, continued to weigh on Peruvian physicians and naturalists. Moreover, there appeared Eugenics propagandists like Carlos Enrique Paz Soldán, whose ideas are summarized in Nancy Stepan's recent book, and also defenders of social Darwinism, described by García Jordán, who observes how in Peruvian society of the post-War period and turn of the century, filled with hopes and disappointments, Comptean positivists blamed national decadence on the indigenous Andean "race" and urged attracting European immigration.[11]

In this way a mood favorable to natural selection was created and metaphors or ideas inspired in evolution and natural selection soon appeared in the University. For example, the physician Carlos Bambaren, nephew of Celso Bambaren, the earliest defender of Darwinism in Peru, wrote a series of articles on vitalism, the new science of genetics, the rediscovery of Mendel, Galton's biometry, and evolutionism in general in the main medical journal of the times.[12] In one of these articles he opposes the mechanical application of Mendelian inheritance to human races and predicts that, once human physiology is studied, "*Mendelian* genetics will be converted into *Lamarckian* genetics."[13]

Around the beginning of the century, Carlos I. Lissón, a mining engineer and doctor in sciences from San Marcos, who taught geology and paleontology, began his paleontological research. From the start of the century, Lissón completed a series of studies of Peruvian fossils and in 1913 published his *Edad de los Fósiles peruanos,* a

---

[11] Nancy Leys Stepan, *The Hour of Eugenics* (Ithaca, Cornell University Press, 1995); Pilar García Jordán, "Reflexiones sobre el darwinismo social, inmigración y colonización, mitos de los grupos modernizadores peruanos (1821-1919)," *Boletín Francés de Estudios Andinos*, 21 (1992), 961-977.

[12] For example, Carlos A. Bambaren, "Una teoría general de la vida," *La Crónica Médica*, 28 (1911), 352-354; "Las doctrinas vitalistas y el neovitalismo del profesor Gasset de Montpellier," *La Crónica Médica*, 29 (1912), 454-461; "El transformismo, la teoría de las mutaciones de Hugo de Vries," *La Crónica Médica*, 29 (1912): 512-527, "La Herencia en biología y en patología generales," *La Crónica Médica*, 30 (1913), 183-185.

[13] *Idem*, "Una ciencia nueva, la genética," *La Crónica Médica*, 29 (1912), 386-388, on p. 388.

book subsequently reprinted a number of times. This work included a paleontological map, along with studies of syncrony among fossil fauna and flora.[14] At the same time, there appeared regional botanical studies impelled by local nationalism and by the desire to establish some congruity between indigenous names and uses and the scientific names of plants. Fortunato Herrera's studies on the flora of Cuzco are a case in point.[15] Herrera was professor of botany and plant biogeography in the Faculty of Sciences at the University of Cuzco. Like Raimondi, he wanted to define flora in function of the geographical altitude of its range. Nevertheless, as the years went by, both Lissón and Herrera were co-opted by the University administration and the utilitarian dimension of their professions. Thus Lissón became the promotor of the Geological Society of Perú, which had broad contacts with the mining industry, and Herrera devoted more time to university administration and the directorship of various museums.

The relationship between evolutionary ideas and high altitude was taken up by Augusto Weberbauer, a German botanist who began to work in Peru at the beginning of the twentieth century. He had studied at the University of Breslau, where he graduated in 1894. As an associate of the museum of that university he visited Peru several times and, in 1911 published, first in German and afterwards in Spanish, *El Mundo Vegetal de los Andes Peruanos*. In 1922 he completed a doctorate in the Faculty of Sciences of San Marcos. His thesis, "Studies on the relations between the anatomical structures of leaves and altitude over sea level," was about the influence of climatic and geological changes on Peruvian flora.[16]

Starting in 1925 Weberbauer taught systematic botany in the Faculty of Sciences of the University of San Marcos, and from 1935 was head of the botany seminar and director of the University Botanical Garden, a dependency of the Faculty of Medicine. It is important to stress that Weberbauer was a researcher of great prestige both within the university and without. Medical students were obliged to take classes in the Faculty of Sciences as part of their pre-clinical trainaing, and Weberbauer's influence on them was important. His botanical activities led to an agreement with the Field Museum of Natural History in Chicago which backed his excursions in different parts of Peru in exchange for his sending part of the specimens he collected.

Weberbauer renewed the debate over the relation between variability in Peruvian botany and the altitude of the Andes. According to Weberbauer, Andean plants located at a higher altitude had specific characters marked by the environment. Among these were small leaves, pods and petioles, which appeared to be buried in the earth or flattened against it. Likewise, he found that roots were disproportionately enlarged with respect to the aerial part, and buds were retracted to take advantage of the higher

---

[14] Carlos I. Lissón, *Edad de los fósiles peruanos* (Lima, La Opinión Nacional, 1913; reprinted in 1917, 1924 and 1942. On Lissón, see "Carlos I. Lissón," *Boletín de la Sociedad Geográfica de Lima*, 65, no. 1 (1948).

[15] The principal works of Fortunato Herrera are *Nombres indígenas y técnicos de algunas especies botánicas espontáneas en el Departamento del Cuzco* (Cuzco, El Trabajo, 1915); *Contribución a la flora del Departamento del Cuzco* (Cuzco, El Trabajo, 1912); *Chloris Cuzcoensis* (Cuzco, Rozas, 1926); *Estudios sobre la Flora del Departamento del Cuzco* (Lima, Sanmartí, 1930).

[16] It was published as the book *Estudios concernientes a las relaciones entre la estructura anatómica de las hojas y la altura sobre el nivel del mar* (Lima, Americana, 1922).

temperature of the earth and the water. That is to say, everything appeared to be conditioned to survive in high altitudes, cold, and atmospheric aridity.

V

The importance assigned to high altitude in the evolution of life in the Andes had a great influence on physicians, most especially on Carlos Monge Medrano, the pioneer of high-altitude physiology in Peru. As was the case in other countries, interest in evolutionary ideas was more accentuated among medical doctors, especially professors of physiology and anatomy. The reason for this is well-known: Faculties of Medicine offered a critical mass of students plus resources, prestige, protection, and the possibility of preserving a balance between theory and the demands of utility.

Monge studied medicine at San Marcos, with an additional year at the School of Tropical Medicine in London. On his return to Peru he left aside somewhat the clinical studies had had initially interested him. According to Monge, in England he had learned that physicians from countries like Peru, with so many infectious and tropical diseases, ought to be explorers and naturalists. His interests are reflected in his unsuccessful candidacy for the chair of Medical Natural History in 1914.[17] Finally, he became professor of clinical medicine, from which position he worked out a singular confluence between physiology and biology.

In 1927 he organized the first medical expedition to Oroya and Morococha, in the central mountains of Peru. This and other expeditions were designed to demonstrate the capacity of Andean natives to adapt to an environment where oxygen was scarce. He wanted to refute the conclusions of Joseph Barcroft, a Cambridge University physiologist who years before had concluded that the environment of those same mountains had diminished the physical and mental capacity of Andeans.

Elsewhere I have examined the social, cultural, and institutional factors that shaped the research programs of Monge and the Institute of Andean Biology as an example of scientific excellence at the periphery.[18] Here, I want to focus a bit more on the relationship between Monge's ideas and general concepts of evolution and natural selection. Although neither Darwin nor Lamarck are cited in his early works, it is possible to perceive neo-Lamarckian influence, as understood at that time, that is, an emphasis on the fixing and inheritance of acquired characterists and the role of the environment in those processes. But that was not to be his definitive position, inasmuch as in the 1940s Monge cites Darwin in references to the infertility of species taken to high altitudes, mentioned in *The Variation of Animals and Plants under Domestication* (London, 1868). He increasingly invoked the term natural selection to support his arguments.[19]

---

[17] Carlos Monge Medrano, *Programa de Historia Natural Médica* (Lima, Sanmarti, 1914).

[18] Marcos Cueto, *Excelencia Científica en la Periferia: Actividades científicas e investigación biomédica en el Perú, 1900-1950* (Lima, Tarea, 1989).

[19] Carlos Monge Medrano, *Acclimatization in the Andes: Historical Confirmations of "Climatic Aggression" in the Development of Andean Man* (Baltimore, Johns Hopkins University Press, 1948). Not until the 1930s did the idea of natural selection win broad support: Michel Ruse, "Darwinism Fleurit!" *Isis*, 88 (1997), 111-117.

In Monge's view, "life was nothing more than the expression of hereditary potential," and the variety of biological species in the Andes was conditioned by a process of "natural selection" at high altitudes.[20] Most of his observations were directed at human beings and perhaps can be ascribed to an indigenist desire to vindicate the Andean population. Nowhere did he explicitly identify himself either as a Darwinian or an anti-Darwinian, nor did he ever provide any theoretical justification for his experimental results. But he did combine evolutionist ideas of European origin with locally stimulated nationalist ideas to propose that "the high-altitude man was a climato-physiological variety of the human race."[21]

For Monge, Andean populations, in order to adapt to the environment in which they had developed, had acquired certain characteristics which were transmitted to its progèny. In the process of migrating to high altitudes those species survived which best adapted to environmental conditions of drought, cosmic and ultraviolet radiation, and—most importantly—to hypoxia, that is the scarcity of oxygen typical of high altitudes. Furthermore, Monge thought that the species which had survived at high altitudes were the result of changes appropriate to the conditions created by such altitudes. Nevertheless, Monge never made clear whether he was referring to individuals or to populations, inasmuch as he considered that "aclimatization [to high altitude]" could be attained by individuals.[22] Perhaps this kind of reasoning lay behind Monge's participation in at least one Latin American eugenics meeting in the 1930s, along with other physicians and biologists espousing different versions of eugenic theory, whether nationalist, contradictory, or neo-Lamarckian.[23]

Some of the empirical factors most frequently mentioned by Monge were anatomical features of some species found in the Andes, an environmental barrier which signaled the presence of illnesses typically found at high altitudes such as mountain disease (*mal de montaña*) and the infertility displayed by some sea-level species when transported to high altitudes. Monge held that such infertility was proof that only some species had been able to survive in that environment.[24] He made these arguments around the same time that the Institute of Andean Biology, which Monge headed, was engaged in collaboration with the Peruvian government and the National Agrarian Society on a series of successful experiments on the use of artificial selection to adapt to high altitudes varieties of sheep which had historically been considered infertile or of low fertility there. Thanks to this research, Monge quickly made the connection between artificial selection and natural selection. Monge and his associates drew similar conclusions from anatomical observations obtained in physiological

[20] For example, the term appears in "Chronic Mountain Sickness," *Physiological Reviews*, 23 (1943), 166-148.

[21] For example, in Carlos Monge Medrano, "Biología Andina," *Revista de la Universidad de Arequipa*, 20 (1948).

[22] Carlos Monge Medrano, "High Altitude Disease," *Archives of Internal Medicine*, 59 (1937), 32.

[23] Carlos Monge Medrano, "Biología Andina, la raza andina, características biológicas del hombre de los Andes," in *Segunda Conferencia Panamericana de Eugenesia* (Buenos Aires, Fracoli y Binde, 1934), p. 73.

[24] Monge's studies on infertiliy are summarized in Carlos Monge Medrano, Mauricio San Martín Jorge Atkins and José Castañon, "Aclimatación del ganado ovino en las grandes alturas. Fertilidad e infertilidad reversible durante la fase adaptativa," *Anales de la Facultad de Medicina*, 28 (1945), 15-31.

studies of exercise which suggested the greater metabolic efficiency of Andeans when compared to inhabitants of lowlands.

Neither Monge nor his associates attempted to verify the biological implications of their physiological postulates experimentally. They were not even aware of this weakness, inasmuch as, from 1928 on, they assumed "that the secret of acclimatization must lie in changes at biochemical and genetic levels."[25] They likewise supposed that, in addition to certain anatomical features, human beings acclimated to high altitudes must have biochemical characteristics which are not lost when such persons change their residence to sea level.[26]

Although Monge never attempted to establish a precise phylogeny of Andean man, or any other species, he believed that congenital adaptation to high altitudes had originated in prehistoric times. He used historical chronicles and other documents to rework the history of Peru as a struggle of organisms to adapt to the climatic agressivity of high altitude as the principal factor explaining the outcome of various events such as the expansion of the Tahuantinsuyo, the Spanish conquest, the Independence movement, and the country's successes and failures generally. According to Monge, one of the main virtues of the Inca Empire was its ability to understand the problems created by high altitude, both for those populations who moved into the mountains as well as for those who migrated to sea level, which explains why the Incas supervised and controlled migrating populations and the movements of armies. In the same manner he believed that this appreciation of ecological factors had been lost during both colonial and republican periods, which in great part explains Peru's decadence.[27] Obviously, one of the connotations of these ideas was that future development depended in part on the capacity of administrators to accept the advice of high-altitude physiolgists.

In rexamining Peruvian history, Monge revealed one of his presuppositions—popular at the time but greatly criticized nowadays—namely that the population of the Andes was, and had always been, static.[28] That is to say, that it had never had much contact, through migration or miscegenation, with the lowland population. This supposition was shared by many physicians, geographers, and intellectuals of those days, all of whom contributed to the establishment of the idea that Peru was clearly divided into three geographical and demographic regions: the coast, the mountains, and the jungle. At present, this construction is no longer taken as a certainty, and most

---

[25] The 1928 argument is cited and repeated in Carlos Monge, Jorge Mejía, Victor Prti and Arturo Salas, "Sobre algunos puntos de la bioquímica en la sangre considerada como un sistema físico químico en las alturas habitadas en el Perú," *Anales de la Facultad de Medicina*, 21 (1938), 237-262 and Carlos Monge Medrano, "Posibles mecanismos bioquímicos adaptativos a la vida en las Alturas," *Anales de la Academia de Ciencias Físicas, Exactas y Naturales* (1939). Likewise in Carlos Monge Medrano "Hombre y Ambiente. El concepto de aclimatación," *Anales de la Facultad de Medicina*, 38 (1955), 1-18.

[26] Carlos Monge Medrano, "Physiological Anthropology of the Dwellers in America's High Plateaus," en *Proceedings of the 29th International Congress of Americanists* (Chicago, University of Chicago Press, 1952), pp. 361-373. One of Monge's students, Mauricio San Martín, repeats the same arguments in "El Hombre y ambiente, genética y antropología," *Anales de la Facultad de Medicina*, 38 (1955), 35-42.

[27] Monge's historical studies are summarized in *Acclimatization in the Andes*.

[28] On the physiology of exercise, see for example, Carlos Monge Medrano, *et al.*, "El rendimiento cardiovascular al esfuerzo en el hombre de los Andes," *Anales de la Facultad de Medicina*, 17 (1935), 29-42.

Peruvian high-altitude physiologists, biochemists and geneticists admire Monge as a pioneer, while rejecting certain of his conclusions.

VI

The Peruvian case suggests that there was some continuity in ideas on evolution between nineteenth-century naturalists and physicians of the early twentieth century. Evolutionary ideas were absorbed into a local tradition of Natural History research and exploration that assigned an important place to the effects of high altitude. It also suggests some characteristics and limitations of the style of scientific practice under conditions of adversity. Some of the components of this style include small working groups, scarcity of resources, and scant cultural prestige. The style is characterized operationally by the concentration of energy and subjects of research, the use of utilitarian and nationalistic arguments to justify science, the use of economical and accessible technologies, and the construction of horizontal, national, and international networks of communication. These factors are aimed at attenuating the discontinuity of scientific energies typical of poor nations, at enhancing the survival of scientific activity, and eventually the attainment of acaedmic excellence in selected fields. Elsewhere I have described in detail Monge's creative use of technologies and international networks, as well as his nationalist arguments. Here I want emphasize the themes of concentration and utilitarianism as favorable, but in the final analysis, limiting factors in the reception and development of evolutionary ideas in Peru.

One characteristic which distinguishes the adaptation of the ideas of evolution and natural selection in Peru is that of concentration. It has been considered crucial to have a scientific community of standard size, because only a small fraction of it will produce original work. In counties where researchers work with scant resources, unstable institutions, and low salaries, succesful institutions must concentrate resources and personnel, at least intially. Such concentration acted like an umbrella which, in contrast to a model of decentralization and the development of parallel chairs, protected researchers, concentrated the development of multiple disciplins around the study of a single problem, amd lent uniformity to the training of students. Thematic concentration also permitted scant disposable resources to be focused on the resolution of a few problems that won international recognition in a short space of time. The Institute of Andean Biology developed various experimental programs in physiology and then began to work in other areas of biology and medicine that ran parallel to the fairly isolated research of the holders of established chairs at the University of San Marcos. But concentration could also be a limiting factor. Partly because of their concentration on phyisiology, anatomy and clinical medicine related to high altitude, other disciplines like biochemistry, embryology, and genetics were not cultivated.

A second favorable, though sometimes problematical, characteristic is the permanent tension with the utilitarianism or with the rhetoric of the usefulness of scientific work that both Raimondi and Monge practiced. The survival of a scientific undertaking under conditions of adversity requires that its backers promise some degree of utility. Peruvian naturalists and physiologists had to compete for the scarce resources dispensed by the State which could redirect them towards priorities perceived as more urgent. The negative aspect of utilitarism can be seen when local research problems received higher priority than more general theoretical pursuits.

The influence of evolutionary ideas in Peruvian Natural History and physiology suggest that the various contradictions, ambivalences, and dissonances coexisted and were perpetuated in part because they were functional in the local context and because they were congruent with the style of practicing science under adverse conditions. Does this mean that we ought to stop studying scientific activity in Latin American countries only as a function of an ideal theoretical model, developed in Europe, to which local scientists either did or did not accommodate? Perhaps.[29]

Instituto de Estudios Peruanos/ Universidad Peruana Cayetano Heredia

29 This essay was translated by Thomas F. Glick.

FRANCISCO PELAYO

REPERCUSSIONS OF EVOLUTIONISM IN THE SPANISH NATURAL HISTORY SOCIETY

1. INTRODUCTION

In 1871, during the markedly liberal environment created by the Sexennial Revolution (1868-1874), the Spanish Natural History Society (SEHN) was founded.[1] It was established at the instigation of an active group of naturalists whose primary objectives were to promote the study of Natural History in Spain, discuss the most relevant problems relating to this discipline, and have a vehicle through which to publish their work. This early initiative was highly successful, with the number of members in the first three decades fluctuating between 290-300 and, in some years, exceeding four hundred. University professors and teachers from the Faculties of Science, Medicine and Pharmacy; researchers and curators from the Natural Science Museum and the Botanical Gardens in Madrid; teachers from Secondary Education Institutes, Seminaries, Schools and Colleges; engineers with different specialities, students, Natural History enthusiasts, *etc*., comprised the long list of naturalists who joined the SEHN.

During its initial phase, which more or less lasted from the time the society was set up to the start of the Twentieth Century, the SEHN not only met its objectives as regards the publication of catalogues and monographs on Spanish earth sciences, flora and fauna, but also—to a large extent—dealt with issues which arose in Spain within the field of Natural Sciences by using its periodicals as a medium of expression.

The open nature of the society meant that it was possible for naturalists, who represented various ideologies, to participate in it. Some of these differences were reflected in the antithetical positions taken by the SEHN naturalists in scientific debates at their meetings. One issue that the SEHN was able to agree upon unanimously, however,

[1] This study was funded as part of DGICYT grant PB 95-0095.

Thomas F. Glick, Rosaura Ruiz, and Miguel Ángel Puig-Samper (eds.), The Reception of Darwinism in the Iberian World, 95-110.

was the decision to memorialize Darwin and his work in a session held on May 3, 1882, a few days after the English naturalist's death.

Apart from the single exception of Mariano Paz Graells (1809-1898), professor of Zoology and Comparative Anatomy at the Central University, all the most important Spanish naturalists of the latter decades of the Nineteenth Century belonged to the society, a large percentage of whom actively participated in the Darwin controversy. This would imply that the SEHN is an excellent vehicle for assessing the impact of evolutionism on Spanish science towards the end of the Nineteenth Century.

DARWINISM AND SPANISH PALEONTOLOGY (1873-1876)

Although paleontology was undoubtedly the discipline on which Darwinian theses had the greatest impact, it is true to say that the theory of evolution was discussed in depth in the Society's publications, in such a way that rather than *debates*, at least from the viewpoint of this science, one should talk about *stands* on Darwinism.

Generally speaking and perhaps very much simplifying the question, in the SEHN's initial years, two groups of geologists and palaeontologists were prominent. First was a conservative group, whose leading figures were Juan Vilanova y Piera (1821-1893), Federico Botella y Hornos (1822-1899), José Solano y Eulate (1841-1912), and Josep Landerer i Climent (1841-1922). The former two had studied in France, where they had acquired classical training in geology and paleontology by attending courses given by Elie de Beaumont (1798-1874) at the École des Mines in Paris, amongst others. These had a profound effect on them, especially Botella, who was the leading spokesman in Spain for De Beaumont's personal catastrophist ideas on the formation of mountains.[2] For his part Solano was a follower of Vilanova and his successor in the chair at Central University in 1877[3], while Landerer was a very active palaeontologist, anti-Darwinist and a great believer, as were the other three, in the harmony between natural sciences and religion.

On the other side stood a group that was less homogenous, but more liberal and which was made up of naturalists close to the Institución Libre de Enseñanza, such as José Macpherson (1839-1902)[4], Salvador Calderón (1853-1911), and Francisco Quiroga

---

[2] On Botella see J. M. López Azcona, "Mineros destacados del Siglo XIX: Federico Manuel María de Botella y Hornos," *Boletín Geológico y Minero*, 100-3 (1989), 162-173.

[3] Cf. A. F. Gredilla, "Noticia necrológica del Excmo. Sr. D. José María Solano y Eulate...," *Boletín de la Real Sociedad Española de Historia Natural*, 13 (1913), 104-116 and Archivo General de la Administración de Alcalá de Henares (AGA), Sección Educación y Ciencia, Leg, 5402-6: José Solano Eulate. Expediente de Cátedra y Leg. 1423-4: José Solano y Eulate. Expediente Personal.

[4] On Macpherson, consult J. Rodríguez Mourelo, "D. José Macpherson. Noticia necrológica," *Boletín de la Sociedad Española de Historia Natural*, 2 (1902), 342-356; E. Hernández Pacheco, "El geólogo gaditano D. José de Macpherson y su influjo en la ciencia" *Asociación Española para el Progreso de las Ciencias. Congreso de Cádiz*, 1927, I, 75-92 and E. Alastrue, *La personalidad y la obra de Macpherson (1839-1902)*, Discurso de apertura de curso de la Universidad de Sevilla (Seville, 1968).

Rodríguez (1853-1894)[5], not forgetting the institutionist Antonio Machado Núñez (1815-1897)[6], one of the strongest defenders of Darwinism in Spain.

2. CREATIONISM AND ANTI-DARWINISM IN THE SEHN

The first references to Darwinism in the society are critiques allegedly based on data provided by paleontology. The two members sharing these views were Juan Vilanova and Josep Landerer. Vilanova was a character of great authority and a scientific heavy-weight in Spain, besides being very well known among European geologists, palaeontologists and prehistorians for his faithful attendance at international conferences.[7] In 1852 he was appointed professor of Geology and Paleontology at Central University and, in an initial phase spanning twenty years, Vilanova taught both disciplines at the university level. Subsequently, in 1873, during the First Republic when the subject was divided into two,[8] he opted for the paleontology chair.

Vilanova's paleontology, mainly influenced by French ideas, was characterized by an attempt to reconcile the Biblical story of creation with paleontological data. Vilanova maintained that throughout time different creations had taken place, the most visible manifestation of which was the independence of flora and fauna fossils belonging to the different geological periods. For this reason, he believed in the fixed nature of species; they had a limited existence because they carried a "germ of death" which brought about extinction. Vilanova, consistent with these ideas, proved to be anti-Darwinian in outlook.

Josep Joaquim Landerer graduated in Sciences from the University of Valencia. His link with naturalists and geologists was largely due to Vilanova, who supported his membership both at the SEHN and Societé Géologique de France.[9] He thought along the same lines as Vilanova, that is, he was a creationist palaeontologist, a determinist, and therefore an anti-Darwinian.

---

[5] His obituary was written by S. Calderón, "Francisco Rodríguez Quiroga," *Anales de la Sociedad Española de Historia Natural*, 23 (1894), 150. By "institutionist," I mean a person in the intellectual orbit of the Institución Libre de Enseñanza.

[6] Cf. S. Calderón, "D. Antonio Machado y Núñez," *Anales de la Sociedad Española de Historia Natural*, 26 (1897), 136-142. Also see the teaching record Machado y Núñez in the General Archive of the Administration at Alcalá de Henares (AGA), Expediente 896-59.

[7] The most complete compilation of material on the life and works of Juan Vilanova is R. Gozalo Gutiérrez, ed., *Homenaje a Juan Vilanova y Piera* (Valencia, Universitat de València, 1993).

[8] *Gaceta de Madrid*, September 24, 1873, nº 267, pp. 1829-1830, Decree of 23rd September 1873 signed by the President of the Republic, Emilio Castelar: "Exposición de la Junta de Profesores del Museo de Ciencias Naturales."

[9] On the life and works of Landerer see R. Gozalo Gutiérrez and V. Navarro Brotóns, "Josep Joaquim Lànderer i Climent," *Ciencia i Tecnica als Països Catalans: una aproximació biográfica* (Barcelona, Fundació Catalana per a la Recerca, 1995), pp. 459-492, and the same authors, "Josep Joaquim Lànderer (1841-1922): entre creacionismo y transformismo," *Geogaceta*, 19 (1996), pp. 185-186.

The first article in which Landerer presents critical approaches on Darwinism is "Explanation of the Chart of Early Times."[10] In this work Landerer, a man of strong Catholic convictions, was concerned, just as Vilanova had been, with reconciling the book of Genesis with scientific data. From the start he maintained a clear stand in favor of a creationist theory of the origin of life: "It is not worth looking for secondary causes to explain the appearance of fauna, because they do not exist. For the most logical and scientific explanation of the origin of life, one should revert to the primary cause, the will of the all powerful Supreme Being."[11]

Landerer's rejection of the Darwinists stemmed from the fact that, in his opinion, they subordinated the fundamental principles of science to preconceived ideas, claiming to explain "the succession of organized beings brought about by the transformation of species, believing that the incessant action of external factors throughout innumerable centuries, affected the organism, transformed it, and from evolution to evolution was able to run through the immense journey of man's cryptogram. Darwinists, being overly inventive, make all living beings of this kind descend from the primordial cell—which they always define *ex cathedra*—this homogenous corpuscle equipped with the endless ability to develop from one organism into another."[12]

For Landerer, Darwinism was utterly absurd, as he declared that without openly violating the laws of nature, it was impossible to imagine, amongst other things, that Cretaceous teleost fish could descend from ganoids—which they preceded in time—or discover any trace of radical anatomical transformation in man, since the time in which he had lived in caves.[13] There existed further paleontological data, according to Landerer, which contradicted Darwinian theories. He therefore maintained that if the transformation of the species were true, those species belonging to the same race had to come from one another in chronological order; however, there are certain species of fish—anenchelon, enchodus, dapedius—which from their first appearance, were represented by various species at the same time.[14]

When explaining the development of life on Earth, Landerer supported progressivism based on multiple and successive creations. Multiple, because life had not started, as he explained to the Darwinists, with a single primordial cell, but with older deposits belonging to the Palaeozoic seas of the Silurian, where there existed a variety of simple organisms, such as zoophytes, molluscs and crustaceans, and at the same time in the layers immediately above, contemporaneous with vertebrates of the fish variety. Successive, because subsequently, during the Secondary period, there had lived reptiles of a greater organic complexity than previous ones. In the Tertiary, mammals appeared, pachyderms first, then rodents and ruminants, and lastly carnivores. And "only when all the successive creations of organized beings had been completed, when the earth had gradually cooled and increased in size by a series of rapid revolutions or slow changes, did we reach the stage we are at today, when man walked upon the last Eden on the planet, taking

[10] J. Landerer, "Explicación del cuadro sinóptico de los tiempos primitivos," *Anales de la Sociedad Española de Historia Natural*, 2 (1873), 341-378.

[11] *Ibid.*, p. 347

[12] *Ibid.*, p. 347.

[13] *Ibid.*, pp. 347-348.

[14] *Ibid.*, p. 348.

possession of this universe he dominates, thanks to the divine breath given to him by the Creator."

Landerer agreed with Vilanova's hypotheses of the independence of flora and fauna fossils characterized by the different geological periods, and the presence of a "germ of death" in species which determined their extinction. In a subsequent article also published in the SEHN's *Anales*,[15] Landerer stated that refuting the theory of evolution meant adding nothing particularly new to what had already been said by such eminent naturalists as Joachim Barrande (1799-1892),[16] François Jules Pictet, Juan Vilanova and J. L. Armand de Quatrefages (1810-1892). Landerer was not prepared to make even a minimum concession to Darwinism, and for that reason he did not hesitate to criticize Pictet, who, he said, had been carried away by conjectures by admitting to a certain limited transformism.[17]

THE PALEONTOLOGICAL PROBLEM OF *EOZOON CANADENSE.*

A controversial question which greatly affected the SEHN was the much debated issue of the organic nature and age of *Eozoon canadense*. The first sample of this alleged organism had been found in 1858 in Precambrian rock in East Canada, and in theory the director of Geological Survey of Canada, William Logan (1798-1875), thought that it might be organic remains. Some years later, Logan observed similar forms in limestone near Ottawa, which were identified by John William Dawson (1820-1899), rector of McGill University in Montreal, as foraminifera, a finding later confirmed by William Carpenter (1813-1885), the leading authority in this field. The first refutation of the organic nature of the *Eozoon* occurred in 1886 and was made by William King (1809-1886) and Thomas Rowney (1817-1894), mineralogists from Queen's College, Galway (Ireland), who maintained that the supposed organic remains were in reality of crystalline origin. This in turn gave rise to considerable controversy which would last many years.[18]

The importance of the *Eozoon* for the debate on Darwinism lay in the fact that, if it were organic, it would support Darwin's theory of an earlier appearance of life on Earth, as well as an increase in inorganic complexity from a "single basic form." From the 4th and successive editions of the *Origin of Species*, Darwin mentioned that in 1859 he had already

[15] J. Landerer, "El piso tenénico o urgo-áptico y su fauna," *Anales de la Sociedad Española de Historia Natural*, 3 (1874), 345-373.
[16] J. Barrande was a French geologist, a devout Catholic and an anti-Darwinist, who had considerable influence in Spain. He proposed the existence of "basic fauna" in the Silurian period, the first records of life, made up of trilobites and other invertebrates, such as brachiopods, crinoids, etc. On Barrande, see C. de la Valée Poussin, "Joachim Barrande et sa carrière scientifique," *Revue des Questions Scientifiques*, 16 (1884), 5-71.
[17] On F. J. Pictet's stand on transformism, see G. Laurent, *Paléontologie et évolution en France: 1800-1860* (Paris, C.T.H.S., 1987), pp. 452-458.
[18] Cf. C. O'Brien, "*Eozoon Canadense*, the Dawn Animal of Canada," *Isis*, 61 (1970), 206-223.

suggested the existence of living beings prior to the Cambrian period, which is why the discovery of the *Eozoon* served to confirm his words.[19]

The first reference to the *Eozoon canadense* in the SEHN occurred during the session of March 4, 1874, when Vilanova spoke on the subject.[20] Vilanova justified his public lecture at the Anglo-American section of the Paris Exhibition in 1867, by alluding to the great importance given to these remains by scientific opinion, as it could be used as a base for "more or less daring and brilliant theories." The sample brought from the USA which appeared in the geological collections of the Jardin des Plantes had been extracted from a pre-Silurian horizon, which signified that the *Eozoon canadense* had to be considered "the dawn/dawning of life."

For Vilanova, an initial problem stemmed from the fact that all the samples he had seen had been discovered in rocks of an age which was difficult to determine, inasmuch as they were found in all types of fossilferous terrain, from the Laurentian or pre-Silurian system in North America to the Tertiary terrain of Tuscany and the island of Elba. However, in actual fact, of all the samples found, the only deposit which was of any importance for the evolutionist argument was the Laurentian, due to its key position in the history of the Earth.[21]

The fibrous structure of the sample gave rise, in Vilanova's opinion, to its apparent organic composition, which is why it was not surprising that those people who were inclined to believe that fossil organisms existed in the oldest geological strata, were influenced by their imagination and saw in the *Eozoon* "the start of the animal series from which the actual human microcosm had to come."[22]

Analyzing the *Eozoon* from the point of view of paleontology, Vilanova noted ironically that, if it were thought that this foraminifera had been the first representative of organisms, as the evolutionists claimed, thanks to "natural selection and the struggle for life" which no one disputed, then it must possess all the miraculous qualities the Darwinists freely attributed to it; meaning that with all these positive factors the *Eozoon* would have multiplied and produced other more perfect foraminifera flooding the primordial and Silurian seas. Besides, if longevity was (as it is now) closely related to the size of the species, it would have to be assumed that merely because of its considerable mass, the *Eozoon* would have enjoyed a long life.[23]

However, Vilanova believed that paleontological data did not confirm these alleged theories at all. Outside Canada there were many areas in which examples of this fossil forminifera had been discovered, and even in the United States its vertical extension in the geological horizon in which the alleged invertebrate was found was very limited. This is why its existence, contrary to the hypotheses put forward, had been very limited both in terms of time and space. Vilanova stated that in North America there had been found no

[19] C. Darwin, *Origen de las especies por medio de la selección natural,* tr. from 6th English edition by Enrique Godinez (1877), reprint ed. (Madrid, Akal, 1985), p. 378.
[20] J. Vilanova, "La estructura de las rocas sepentínicas y el *Eozoon canadense," Anales de la Sociedad Española de Historia Natural,* 3 (1874), 261-266.
[21] *Ibid.,* p. 262.
[22] *Ibid.,* p. 263.
[23] *Ibid.,* pp. 264-265.

direct-line descendants from the *Eozoon* or proto-organisms, except in much later geological terrain.[24]

These and other discrepancies, such as the enormous size of the *Eozoon* in relation to that of real foraminifera, clearly indicated that the paleontological data did not coincide with "the good intentions of evolutionist theory and consequently, far from being masses of Laurentian serpentine and calcium silicate resulting from the organic secretion of the *Eozoon canadense*, it is more likely an illusion based on the actual structure of the serpentine and amphibolic rock, more than enough to convince those people who have wanted to believe in foraminifera as the start of the organic series."[25]

That same year, 1874, Vilanova published another article in which he discussed the organic origin of the *Eozoon*.[26] He repeated the same arguments he had presented at the Society's session and finished by saying, "These and many other aspects which we have omitted through lack of time, are the reasons behind our disbelief that the *Eozoon* is truly an organic being, as the evolutionists would like to have us think because of what they have said about it and all the fuss they have made, for lack of better data; which is precisely what serious paleontology does not or ever will want to give them."[27]

A second article on the *Eozoon* read at the session of December 1, 1875, was published in the *Anales* two years later.[28] Its author was the Marquis of Ribera, about whom virtually no biographical and academic data exist, apart from the fact that he belonged to the Council of State, the German Geological Society, and was a member of the SEHN from 1872, occupying the presidency in 1876.[29]

Unlike Vilanova, the Marquis of Ribera agreed that the *Eozoon* might be a fossil organism. To support his argument, Ribera used several paragraphs from the reports of Logan, Dawson, and Carpenter who had studied the remains of the alleged fossil. Ribera pointed out that these studies confirmed the age of the rocks where the *Eozoon* had been discovered.[30] In Ribera's opinion, the difficulty in obtaining good samples from the rock where the *Eozoon* was found was one of the problems preventing foreign palaeontologists from making the observations necessary to correctly identify the samples. Ribera went on to say that, given the fact that only seventeen years had passed since its discovery, it was completely normal that the organic nature of the *Eozoon* was still being debated in

[24] *Ibid.*, p. 265.
[25] *Ibid.*, pp. 265-266.
[26] J. Vilanova, "El Darwinismo ante la Paleontología," *Revista de la Universidad de Madrid*, 3 (1874), pp. 383-403
[27] *Ibid.*, p. 407.
[28] Marqués de la Ribera, "El *Eozoon canadense*," *Anales de la Sociedad Española de Historia Natural*, 5 (1876), pp. 27-43.
[29] Cf. *Actas de la Sociedad Española de Historia Natural*, 1 (1901): The news of the death of Marquis de Ribera, communicated to the Society at the session of January 9, 1901 by the president, Blas Lázaro Ibiza, provided no information on the deceased.
[30] *Ibid.*, p. 29.

scientific circles, even more so considering that very few naturalists had had the opportunity of observing it under a microscope.[31]

Those favoring the existence of the *Eozoon*, whom Ribera supported, declared that what they saw through the microscope was not mere fantasy and that the organism actually existed. Dawson and Carpenter had therefore been able to determine the zoological species to which the *Eozoon* belonged, based on both the structure of the rock containing the organic remains and the similarity of the remains with other beings from subsequent periods and with present marine species. They compared these with the different examples of the *Eozoon* and were able to study and classify them, mainly using Carpenter's systematic work, *Introduction to the Study of the Foraminifera* (London, 1862).[32]

With respect to the inconsistencies cited by Vilanova regarding the difference in size between the *Eozoon* and other rhizopods of the foraminifera variety, Ribera replied by referring to Dawson, who assured that there was no reason to maintain that foraminifera had to be small, because samples of colossal dimensions of this kind had been discovered in lower Silurian strata. Ribera recognized that he supported the belief in the existence of the fossil. Carpenter had shown him slides under his microscope proving the organic origin of these fossils, as well as their similarity to the foraminifera, the group to which the British palaeontologist attributed them.[33]

VILANOVA'S CRITIQUE OF GAUDRY'S EVOLUTIONISM

Vilanova continued to express his anti-evolutionist ideas at the Society's meetings, most particularly in a report presented during the session of August 1875 entitled "Gaudry's *Protriton petrolei* and the Theory of Evolution."[34] He justified reading this paper by stating that, in his opinion, the Society could not shut itself up and discuss only discoveries made in Spain and its overseas colonies. Rather it must open itself, as it had done on other occasions, to subjects of general interest in Natural History, although they exceeded the strict limits of the by-laws.

Taking advantage of this proposal for a more open attitude, Vilanova commented on a study by the French evolutionist Albert Gaudry (1827-1908), professor of paleontology at the Muséum d'Historie Naturelle, in Paris, in which he referred to the discovery of the fossil remains of amphibians in Palaeozoic terrain, which seemed to confirm the evolutionist theory. The argument put forward by Gaudry was that the present types of batrachians seemed to be very recent, because only a few of their fossils had been found in Tertiary strata, which was why it was considered very strange that a creature of so simple a structure within the vertebrate group had appeared so late in the course of time. This fact was used as a serious objection to the theory of the progressive development of organisms.[35]

The discoveries of these batrachian fossils in the upper Palaeozoic terrain of Muse and Autun (France) were, for Gaudry: "What a Darwinist expected from old terrain. Because

---

[31] *Ibid.*, p. 31.
[32] *Ibid.*, pp. 31-32.
[33] *Ibid.*, pp. 40-41.
[34] It was published in the *Actas de la Sociedad Española de Historia Natural*, 4 (1875), 83-88.
[35] *Ibid.*, p. 83.

of the shortness of the tail, the body and limbs looked like those of a salamander, while the head was like a frog's. This helps reduce the distance which today separates the urodeles from the anurans, providing a link between these two groups of amphibians."[36]

After introducing the anatomical arguments that the French palaeontologist had expounded in his work, Vilanova asked whether this information, together with the abundant paleontological material on Palaeozoic terrain, confirmed or opposed evolutionary theory. He began by saying that if the *Protriton*, so-called because it was the precursor of the salamander, were actually an amphibian, one would have to explain its sudden disappearance in the Palaeozoic period, and the appearance of examples of its kind in the Tertiary period. From that, Vilanova said: "If evolution is considered to be limited to class, I do not know how the interruption of the species can be explained, and it could almost be described as a nullification of natural selection and struggle for life during the Triassic, Jurassic and Cretaceous periods, the materials from which show no remains of these vertebrates. If, on the other hand, evolutionist theory is not defined within the limits of class, rather by the type of vertebrate, it is still not clear how reptiles far more perfect that the *Protriton* had come to appear in previous geological periods, in which case the sequence of transformation is reversed, i.e. from perfect to imperfect."[37]

To illustrate the latter argument, Vilanova used the *Archaeogosaurus* from the Carboniferous period. Besides, continued Vilanova, apart from ophidians and chelonians the rest of the group of reptiles appeared suddenly and simultaneously at the end of the Palaeozoic, and reached maximum development in the Jurassic. With this information, he concluded, it was impossible to consider the existence of organic evolution, or that the *Protriton*, separated by such a huge space in time, belonged to a class like amphibians, which continue today almost as they were in the Tertiary period. Vilanova concluded with these words: "Please forgive me Gaudry, I do admire your tireless work and endless enthusiasm for paleontology, but I feel your desire to match the facts to a theory you hold dear is rather too exaggerated."[38]

Vilanova's interest in Gaudry's *Protriton petrolei* would not end in this report. In sessions held on October 4 and November 8 of the following year, 1876, he presented a fossil he had acquired during a trip to Autun and its surrounding area and, reiterating his anti-evolutionist convictions, turned around the arguments Gaudry had advanced with this batrachian in support of transformism. He asserted that this Permian fossil, which could represent the embryonic state of the reptiles, had preceded the emergence of the large secondary saurians, some of which were organically highly complex. The fact that amphibians appeared again in the Tertiary after these great reptiles indicated a contradiction in evolutionist theory.[39]

Creationist and anti-Darwinian paleontology of the conservative group of the SEHN reached its culmination in the "Paleontology Program" which Vilanova presented at the

[36] *Ibid.*, p. 84.
[37] *Ibid.*, p. 88.
[38] *Loc. Cit.*
[39] *Actas de la Sociedad Española de Historia Natural*, 5 (1876), 80, 85-86.

session of May 3, 1876.[40] The reason for the presentation was to present the curriculum of the new subject of paleontology, as Vilanova was planning to teach it at Central University starting in 1877. One of the problems that this science posed, in Vilanova's opinion, concerned the unity of plan which had presided over both the creation and subsequent development of organisms. This gave rise to a very important issue, which was how and when had inorganic matter been transformed into organic matter: was it "with the divine breath of the Creator as we believe, or by the single action of general laws of matter when encountering favorable conditions for this great event to take place, as others suppose."[41] Vilanova, consistent with his lifelong beliefs, pronounced in favor of the supernatural explanation. One of the theoretical aspects of paleontology was to examine and discuss the laws which had governed the development of life on Earth. This part was very important as it encompassed highly relevant issues, such as whether organisms as a whole consisted of single or multiple series, and whether species were predetermined and immutable or variable to the point where they could evolve into another. On this latter point Vilanova again claimed to be anti-Darwinian, since he did not accept that the appearance of organisms had followed a gradual series of slow and successive developments as the transformists maintained.

EVOLUTIONISM AMONG THE NATURALISTS OF THE INSTITUCIÓN LIBRE DE ENSEÑANZA

The other paleontological and geological trend observed in the SEHN was that represented by the group of members linked to the Institución Libre de Enseñanza (ILE), who accepted Darwinism, although not uncritically.[42] In this group's work, references are found to organic evolution as a general law of Nature. It is therefore hardly surprising that these naturalists, apart from some criticisms, welcomed Darwin's theory. This was the case with Augusto González de Linares (1845-1904)[43], Enrique Serrano Fatigati (1840-1918), Blas Lázaro Ibiza (1858-1921), and the aforementioned Salvador Calderón Arana.

González de Linares, one of the first Spaniards to lecture on Darwinian theory, was professor of Natural History at the University of Santiago de Compostela when he was removed from his chair by the education minister, Orovio, as a result of the First

---

[40] "Programa de Paleontología," *Actas de la Sociedad Española de Historia Natural*, 5, (1876), 50-64. This "program" was the inaugural lecture of Vilanova's paleontology class at the Gabinete de Historia Natural in Madrid on October 3, 1878, later published as a 29 page leaflet entitled *Lecciones de Paleontología* (Madrid, 1878).

[41] "Programa de Paleontología," p. 51.

[42] On Krausism, see V. Cacho Viu, *La Institución Libre de Enseñanza* (Madrid, Rialp, 1962); A. Jiménez-Landi, *La Institución Libre de Enseñanza y su ambiente. Los orígenes* (Madrid, Taurus, 1973), and J. López Morillas, *El krausismo español*, 2nd ed. (Madrid, Fondo de Cultura Económica, 1980). The Krausist concept of nature has been outlined by D. Núñez, *La mentalidad positiva en España. Desarrollo y crisis* (Madrid, Túcar, 1975) and J. Sala Catalá, *Ideología y ciencia biológica en España entre 1860 y 1881* (Madrid, CSIC, 1987), pp 45-65.

[43] On González de Linares see S. Calderón "Noticia necrológica de D. Augusto González de Linares," *Memorias de la Sociedad Española de Historia Natural*, 2 (1904), 437-453 and B. Madariaga, *Augusto González de Linares y el estudio del mar* (Santander, Institute of Marine Studies, 1972).

University Crisis, so called, in 1875. As a result, he moved to Madrid to teach at the Institución Libre. During these initial years, 1877-78, González de Linares commented favorably on Darwinian theory, although he criticized several aspects of it from a Krausist perspective, especially its mechanicism, which he considered a mistake.[44] Blas Lázaro Ibiza, who taught at the ILE between 1880 and 1885 had a similar position. In a debate on Darwinism at the National Youth Circle in 1881, he appeared to favour evolutionist theses, but distanced himself from the materialistic views of Haeckel.[45]

In another institutionist, like Enrique Serrano (1840-1918) one can see certain ideas on the existence of a progression and transformation of Nature, which he justified by considering both processes to be essential for the evolution of organisms. Even recognizing, as anti-Darwinian palaeontologists had done, that organic types of a highly complex nature had appeared in earlier geological periods, Serrano Fatigati still espoused general, sequential progress in the animal order wherein the existence of each of the species was a necessary condition for the existence of all of those which subsequently appeared.[46] Serrano also criticized the creationism of the catastrophists, emphasizing that the process dominating the succession of the species was a slow and gradual evolution. At the same time as he himself moved away from a creationist stance, Serrano Fatigati ended up accepting both the transformist hypothesis—which explained the adaptation of organisms to the variation of environmental conditions via a slow, ordered progression—and spontaneous generation.[47]

Later on Serrano Fatigati stated that individuals, species, genera, and kingdoms passed through different phases which changed via an uninterrupted and continual series, just as (he said) Darwinism claimed.[48] He emphasized that this theory maintained that all the organic groups had arisen from simple forms which, by dividing up into varieties, ended up forming new species that, from the point of view of order, were more complex. In this sense, it was clear to him that environmental conditions influenced the evolution of organic forms, a fact confirmed by paleontological data, as Gaudry had shown.[49]

Based specifically on a quotation from Gaudry, Serrano Fatigati took the opportunity to introduce into his concept of transformism some distinctions which he thought distanced himself from materialism. Gaudry had written that "the Supreme Being has not separately created the successive species of the geological periods, but rather has produced some

[44] González de Linares considered Nature to be a unitary organism, the diverse manifestations of which were expressed in the form of transformations.

[45] See *El Imparcial*, April 17, 1881. On Blas Lázaro Ibiza as an evolutionist, see A. González Bueno, "Actitud de Blas Lázaro e Ibiza (1858-1921) ante la corriente evolucionista," in *Actas II Congreso de la Sociedad Española de Historia de las Ciencias* (Zaragoza, 1984), I, 419-427 and *idem*, "Blas Lázaro Ibiza (1858-1921)" in *Dictionnaire du Darwinisme et de l'Evolution*, P. Tort, ed. (Paris, PUF, 1996), II, 2598-2599.

[46] E. Serrano Fatigati, "El progreso de la materia," *Revista de la Universidad de Madrid*, 2 (1873), p. 83.

[47] *Ibid.*, p. 85.

[48] E. Serrano Fatigati, "La evolución de la Naturaleza" *Revista de la Universidad de Madrid*, 4 (1874), pp. 292-317, 496-512.

[49] *Ibid.*, pp. 502-503 n. 1.

from others," which in his opinion distinguished materialism from transformism "without noting in it a great scientific truth entirely independent of that system, and we could even say contradictory, if one heeds the way in which it is presented nowadays."[50]

SALVADOR CALDERON AND THE PALEONTOLOGY OF VERTEBRATES

Salvador Calderón, another instituionist, had been one of the founders of the Ateneo Propagador de las Ciencias Naturales, a scientific and cultural association which eventually merged with the SEHN.[51] After winning the chair of Natural History the Secondary Education Institute in Las Palmas in 1874, he too lost his position in 1877 when he refused to respect the Orovio decree. This also resulted in his not being allowed to compete for the chair of Geology in the Central University, left vacant when Vilanova's chair was divided in two.[52] After these political outrages, Calderón became associate professor at the Institución Libre de Enseñanza.

One of his first works published in the *Anales* of the SEHN, in 1876,[53] was titled "Enumeration of the Fossil Vertebrates of Spain."[54] In it, Calderón justified his interest in vertebrate paleontology because it clarified such important questions as past geographical connections between Spain and Africa or the disappearance of species.[55] However, Calderón also indicated that this field was also important, beyond its geological and paleontological interest, as it illuminated the problem of "specific centers" of the ancestors of present-day fauna.[56] Calderón found that the results of Spanish vertebrate paleontology did not contradict "the principle of the process of continual organic perfection" established by eminent naturalists who had observed progression from the appearance of fish, continuing through that of amphibians, reptiles and birds, to the marsupials and "ordinary" mammals. Richard Owen (1804-1892), who in Calderón's opinion could not be considered a transformist, had found a perfect balance between the degree of organic complexity of the four types of mammals and the chronological order in which they had appeared on the Earth.[57]

---

[50] *Ibid.*, p. 311 n. 2.

[51] On S. Calderón see E. Hernández Pacheco, "El profesor D. Salvador Calderón y Arana y su labor científica," ***Boletín de la Sociedad Española de Historia Natural***, 11 (1911), pp. 405-445. Also AGA, Alcalá de Henares, Education and Science Section, Legajo 5.404-63: file on the Chair of Salvador Calderón y Arana, and Legajo 236-4, personal file of Salvador Calderón y Arana.

[52] See AGA, Alcalá de Henares, Education and Science Section, Legajo 5402-6: file on competition for the Geology chair.

[53] S. Calderón, "Enumeración de los vertebrados fósiles de España," ***Anales de la Sociedad Española de Historia Natural***, 5 (1876), pp. 413-443. It would be translated and published as "On the Fossil Vertebrata Hitherto Discovered in Spain," ***Quarterly Journal of the Geological Society of London***, 33 (1877), 124-133.

[54] The evolutionist stance of Calderón has been discussed by F. Pelayo, "Salvador Calderón Arana (1851-1911)," in Tort, ed., ***Dictionniare du Darwinisme***, I, 488-489.

[55] Calderón, "Enumeración de los vertebrados," p. 414.

[56] *Ibid.*, p. 415.

[57] *Ibid.*, p. 416.

CALDERON AND NEO-LAMARCKISM

In the early 1880s, in discussing an article by the French palaeontologist François Fontannes (1839-1887), Calderón took up the problem of the variation in fauna over the course of geological periods.[58] He did not agree with those naturalists who thought the successive fossil faunas to be "a product of migration," which were due to changes in environmental conditions. To Calderón's mind, the naturalist who appealed to migration to explain links between paleontological fauna and those of present-day regions "would also have to accept the gradual, successive and radical change in animal forms, and be obliged to revert to transformism, which is what the theory of migration tries to avoid."[59] From his point of view, one either accepted the old and discredited theory of the periodic extinction of fauna and replacement by others, or else recognize that species had been perfecting themselves or evolving during the geological periods. Calderón concluded that "the creation of species through the triumph of certain varieties and the disappearance of others in the struggle for life is a principle established by Darwin, although not correctly applied to paleontology."[60]

A little later, Calderón evaluated the Neo-Lamarkian hypotheses of Edward Drinker Cope (1840-1897).[61] His earliest reference to Cope came in one of the Society's sessions in mid-1889, in which he stated that he had received publications sent by Cope, whom he considered to be that naturalist, Darwin and Wallace aside, who had best promoted the doctrine of evolution.[62] Calderón pointed out that the two main causes that Cope cited as modifying the structure of the skeleton and dentition were use and disuse. From the numerous data in his possession, Cope deduced a fundamental law which was that identical causes produce identical effects. However, one had to take into account that this process was not instantaneous, but rather each form originated from those which had preceded it in time, and homologies had to be established within each type. In this way, the fins of fish were essentially different organs from the limbs of the ichthyosaurs or the whale. Therefore, "the same causes working on these organs, although similar, were never morphologically identical."[63] Calderón concluded by emphasizing that although the principle of a mechanical need as a cause of such developments in organisms was not new

---

[58] S. Calderón, "Una idea sobre la renovación geológica de las faunas," *Boletín de la Institución Libre de Enseñanza*, 8 (1884), 231-232. Calderón mentioned Fontannes's article "Sur une des causes de la variation dans les temps des faunes malacologiques, à propos de la filiation des *Pecten restitutensis et latissimus*," *Bulletin de la Societé Géologique de France*, 13 (1884), 357-364.

[59] Calderón, "Renovación geológica," p. 231.

[60] *Ibid.*, p. 231.

[61] On Cope's Neo-Lamarckism, see M. Gaudant and J. Gaudant, *Les théories classiques de l'évolution* (Paris, Duond, 1971), pp. 110-116, and P. Bowler, *El eclipse del darwinismo* (Barcelona, Labor, 1985), pp. 136-159.

[62] *Actas de la Sociedad Española de Historia Natural*, 18 (1889), pp. 81-84.

[63] *Ibid.*, p. 84.

in Comparative Anatomy, no one had previously managed to study it in such depth as Cope.

The following year, Calderón discussed the application of Cope's ideas in "Observations on the Dentition of Rodents."[64] Here Calderón states that the old principle of use proposed by the North American School of paleontology was not sufficient to explain the origin of the dental system of rodents out of the more homogenous systems of earlier geological periods. It was necessary, he said, for use to act upon species which had specific characteristics, which could have appeared as "mere abnormal individual cases" in various groups of mammals, making up constant varieties in which some deformation had expressed itself.[65] Calderón derived this point from Fontannes, who had explained how some similar, contemporaneous forms, when subjected to the same influences, had become extinct at the same time as others had evolved.[66]

3. DARWINIANS AND ANTI-DARWINIANS IN THE SEHN

In addition to initial criticisms of Darwin's thesis by the palaeontologists Landerer and Vilanova, evolutionism was rejected in several regional systematic works, such as those of the malacologist Gerónimo Macho Velado (1826-1899)[67] and the botanist Estanislau Vayreda y Vila (1848-1801).[68] Both maintained the same view as the above-mentioned palaeontologists, in that they supported creationism, the harmony of science and religion and consequently adopted an anti-Darwinian stance. Macho Velado, for example, asserted in his report that "Darwinian fatalism is only a hypothesis, which is not modern in science, and for which there is no positive proof."[69]

On May 3, 1882, a SEHN session took place in which, at F. Botella's suggestion, the members met to honor Darwin who had recently died. Botella emphasized the admirable work of the English naturalist, both in terms of his scientific discoveries and for having renewed Lamarck's theory. Máximo Laguna (1826-1902) presided over the meeting and emphasized Darwin's scientific merits, irrespective of the opinion held on his theories.[70]

That same year, in a lecture at Ateneo Científico y Literario in Madrid, M. Laguna, whose attitude with respect to evolution was ambiguous at best, discussed the role of both *Eozoon canadense* and *Bathibius Haeckelii* (a sample of the latter retrieved in the North Atlantic had initially been identified as a micro-organism) in the formation of the first organisms, from the point of view of Darwinism. In his opinion, evolutionary theory could

---

[64] S. Calderón, "Observaciones sobre la dentición de los roedores," *Anales de la Sociedad Española de Historia Natural*, 19 (1890), pp. 279-297.
[65] *Ibid.*, p. 291.
[66] *Loc. cit.*
[67] G. Macho Velado, "Moluscos de agua dulce de Galicia," *Anales de la Sociedad Española de Historia Natural*, 7 (1878), 235-248.
[68] E. Vayreda y Vila, "Plantas notables por su utilidad o rareza que crecen espontáneamente en Cataluña..." *Boletín de la Sociedad Española de Historia Natural*, 9 (1880), 53-130.
[69] *Ibid.*, p. 238.
[70] Cf. D. Fernández Galliano, "Darwin y la Sociedad Española de Historia Natural," *Boletín de la Sociedad Española de Historia Natural*, 80 (1984), 36-40.

neither be proved nor rejected, because neither sample could securely de identified as a fossil organism, and older, simpler beings might still remain undiscovered.[71]

Such negative or ambivalent attitudes with regard to Darwinism did not represent the general feeling among members of the SEHN. From Society's beginning, the leading Spanish naturalists accepted and supported Darwin's theory of evolution, namely Machado, Ignacio Bolívar (1850-1944)[72], Eduardo Boscá (1843-1924), Víctor López Seoane (1834-1900), José Gogorza (1859-1926), etc. The defense of Darwin by members such as Rafael García Álvarez (1828-1894), Odón de Buen (1863-1945) or Gregorio Chil y Naranjo (1831-1901) brought them into conflict with the Catholic hierarchy, who condemned their scientific works. For that reason, it is not surprising that some members, including Machado, Antonio Vila Nadal (1861-1933), and J. Gogorza attempted to reconcile transformism and religion. Gogorza, professor of Organography at the Central University, published a leaflet entitled "Refutations of an anti-Darwinist," in which one of his accusations was that the theory of evolution did not recognize God. To his mind this was wrong, "as neither Darwin nor any other serious Darwinist can exclude the idea of God, and if any members have become carried away by their own ideas on issues surpassing the limits of science, then these Darwinists have done more damage to the theory of Natural Selection that have the anti-Darwinists."[73]

However, the notable presence of a nucleus of pro-Darwin naturalists at the core of the SEHN was scarcely reflected in the works published in the *Anales*. There were few reports on research carried out according to Darwinian principles in the initial decades.[74] Subjects such as biogeography, adaptation, or heredity were hardly touched upon in the systematic works presented at the SEHN. In this respect, one can cite the taxonomic studies carried out by I. Bolívar (1876)[75] and his follower Manuel Cazurro (1865-1935) (1888)[76], as well as the reports published by O. de Buen (1883), J. M. Castellarnau (1848-1943) (1888)[77] and J. Gogorza (1891).[78]

Odón de Buen was undoubtedly the most committed and active Spanish naturalist who openly espoused republican, liberal, and Darwinian ideals, a stance which cost him both

[71] M. Laguna, "¿Qué son las plantas?" *Revista Contemporánea*, 41 (1882), 385-405.

[72] Cf. A. Gomis Blanco, "Ignacio Bolívar y Urrutia (1850-1944)," in Tort, ed., *Dictionnaire du Darwinisme*, I, 363.

[73] J. Gogorza, *Refutaciones a un antidarwinista* (Salamanca, 1897), p. 14.

[74] In this respect, see Sala Catalá (note 42, above) and S. Casado de Otaola, *Los primeros pasos de la ecología en España* (Madrid, Residencia de Estudiantes/M.A.P.A., 1997), esp. pp. 75-154.

[75] I. Bolívar, "Sinopsis de los ortópteros de España y Portugal," *Anales de la Sociedad Española de Historia Natural*, 5 (1876), 79-130, 259-372; (1877), 249-348, and 7 (1878), 63-129.

[76] M. Cazurro Ruiz, "Enumeracion de los ortópteros de España y Portugal," *Anales de la Sociedad Española de Historia Natural*, 17 (1888), 435-513.

[77] J. Castellarnau, "Unidad del plan generativo en el reino vegetal," *Anales de la Sociedad Española de Historia natural*, 18 (1888), 31-74.

[78] J. Gogorza, "Influencia del agua dulce en los animales marinos," *Anales de la Sociedad Española de Historia Natural*, 20 (1891), 221-270.

ecclesiastical condemnation and administrative exile.[79] His support of Darwinian theory was well defined in the initial approach of his work published in the SEHN titled "Geographical-botanical Notes on the Central Region of the Iberian Peninsula."[80]

To conclude this essay, I would like to use a quote from Odón de Buen himself in his *Tratado de Geología*—the first edition of which had been condemned by Church—in which he sets out perfectly a defense of the liberal principles held by Spanish Darwinian naturalists: "We should point out, lastly, the main agent of all intellectual progress, whose influence on the development of Geology has definitely been more effective than in any other area of science: This agent is Freedom. Without freedom of research, or freedom of instruction, or freedom of the press, science would not have advanced by leaps and bounds."[81]

University of Valencia

[79] Cf. J. Josa, "Odón del Buen y del Cos (1863-1945)," in Tort, ed., *Dictionnaire du Darwinisme*, I, 457-459.

[80] O. de Buen, "Apuntes geográfico-botánicos sobre la zona central de la península ibérica," *Anales de la Sociedad Española de Historia Natural*, 12 (1883), 421-440. Odón de Buen started his report in the following way: "Geographical-botanical studies have taken on truly tremendous proportions. Begun by Humboldt, they received a considerable boost in natural history from Darwin's theory. Searching for facts to support evolutionist theories first, and slowing down later with its triumph, different scientists studied the links between plants, the soil they inhabited and the climate in which they lived, interpreting in the true sense vegetable associations as a means of defense against external influences. The migrations of plants via the expansion of continents was compared and the relationship between vegetation of different countries seen..." (p. 421).

[81] O. de Buen, *Tratado elemental de Geología (General y Particular de España)*, 2nd ed. (Barcelona, 1896), p. 460.

SUSANA PINAR

DARWINISM AND BOTANY

The Acceptance of Darwinian Concepts in Nineteenth-Century Spanish Botanical Studies

The aim of this paper is to assess the impact of Darwin's theories on some of the botanists and other professionals connected with the plant kingdom such as forestry experts, although, in the case of the latter, their work was frequently focused on subjects like entomology, mineralogy, or geography. We will try to show the lines of thought followed by both groups and their endorsement, whether whole-hearted or lukewarm, of the principles of adaptation, natural selection, struggle for existence as well as other currents they espoused. Our analysis will be mostly descriptive, as this is only a preliminary study not intended to encompass all the bibliography available on a subject which has scarcely been studied from this perspective.

A quick summary of the prevailing trends reveals botanists' concern for systematics which, after experiencing a surge of support as a result of the shift towards natural classification systems, then declined, due partly to the fact that any new development required a thorough analysis of the constitution of genera and families, a type of research that formed the bulk of those studies published between 1860 and 1908. Classifications of higher categories did not experience significant changes until the introduction of "descent theory," which shed light on the relations of affinity observed between living organisms, the main spur to natural systems of classification and to the development of disciplines such as phytogeography. This topic was closely linked to the renewed debate over the concept of species which, when brought into doubt, contributed to the decline of taxonomy.

Aware of the decay of systematics and of botany in general, Simon de Rojas Clemente claimed in his *Ensayo sobre las variedades de la vid* that "while the natural history of plants... might be no more than a methodical list of their characters, perhaps accompanied by some citations . . . of the name or names which some give them, and of some indication of the sites where they have been seen and of the season when they flower or bear fruit, no one can completely vindicate the futility, aridity and monotony

Thomas F. Glick, Rosaura Ruiz, and Miguel Ángel Puig-Samper (eds.), The Reception of Darwinism in the Iberian World, 111-126.

which cause many talented men to shun its study or obliges them to view it with disdain. It is now time that we aspire to expand it and make it respectable, extending our researches to the latitude, altitude, exposure, weather, topography, and climate in which each plant lives, to its morphology and uses, and in sum, to whatever relationships each might have with the other beings and phenomena of the universe. Then indeed will botany lead us to great and important results."[1] Certainly, the interest of authors of the last quarter of the nineteenth century was centered on that kind of endeavor.

As a whole, concepts like evolution through gradual change, adaptation to the environment, and the struggle for survival that natural selection entailed were well received, at least at an early stage. They were, after all, facts obvious to those already working on the mechanisms of competition and species invasion in the plant kingdom, as well as for those studying the effects of environmental factors like light, heat and so on.

Although religious arguments were also brandished to refute evolutionist theories, they were not as virulent as in other disciplines, where they were employed to deal with topics regarding man's origins and the origins of morality. A second phase, in the period spanning the end of the nineteenth and early twentieth centuries, was marked—as elsewhere—by the rejection of the principle of natural selection by most naturalists and a surge of support for vitalist ideas, embodied here by García Maceira and Castellarnau. Thus, though the discourses articulated by botanists and foresters differed as to level of religiosity, both subscribed to the idea of a vital energy, the ultimate stimulus responsible for the changes experienced by living creatures and for making existence possible. That position clashed, understandably, with that maintained by reductionists who interpreted vital phenomena as mere physicochemical processes.

BOTANISTS AND NATURALISTS

It seems appropriate to begin with Miguel Colmeiro (1816-1901). Appointed professor of organography and plant physiology at the Natural Science Museum of Madrid in 1857, he later held the chair of phytography and, from 1868 to his death, the directorship of Madrid's Botanical Garden, replacing Mariano de la Paz Graells, who was removed from that post after the Revolution of 1868. A few years later, together with a group of Spanish naturalists that included Ignacio Bolivar and Joaquín González Hidalgo, he founded the Spanish Society of Natural History. He was its first president and published the bulk of his botanical writings (written between 1872 and 1899) in its *Anales*. He also participated, along with Aureliano Maestre de San Juan, in the foundation of the Histological Society of Madrid in February 1874.

Apart from his numerous taxonomical writings and compilations, he wrote his *Estabilidad de las especies en el reino vegetal* [Stability of Species in the Plant Kingdom] in 1860 as his lecture of reception in the Academy of Sciences. Although moderate in tone, this was the first work critical of evolution written in Spain after the publication of *On the Origins of Species* (1859). In his assessment of this text, Sala

---

[1] See M. Colmeiro and E. Boutelou, ***Discursos leidos ante la Real Academia de Ciencias Exactas, Físicas y Naturales, contestación de Esteban Boutelou*** (Madrid, E. Aguado, 1877), p. 45.

Catala (1987) concluded that it was a product of the movement of opposition launched in 1860 by Louis Agassiz, who stressed the instability of variations and underestimated the influence of the environment in the formation of new organisms.[2] Colmeiro's analysis of man-made modifications, as in cultivated plants, is similar to, but as weighty as the argumments wielded by Agassiz.

In his lecture, Colmeiro divided the alterations incurred by plant species into four types: 1) highly ephemeral changes caused by cultivation conditions; 2) monstrosities, individual in character and alien to the normal channels through which species develope; 3) genuine varieties, permanent thanks to the different mechanisms of multiplication employed (tubers, cuttings, etc.), but not the result of sexual reproduction and 4) 'breeds or subspecies', depending on their designation, with a high degree of permanence, preserved through sexual reproduction. To achieve that, however, they had to be isolated in a way that, in Colmeiro's own words "inhibits the fertilizing action of pollen from other individuals, an organic disposition that is not opposed to the production of seeds, the persistence of the circumstances capable of maintaining the forms obtained, and a long enough time to permit them to consolidate or not disappear as a result of atavism,"[3] conditions that Colmeiro regarded as impossible to perpetuate through spontaneous growth, without man's intervention. The text does not show, nonetheless, the existence of a firmly-elaborated stance, at least insofar as the development of plants in times prior to the emergence of man is concerned: "the action of a long time under diverse conditions, could have influenced man, in a previous epoch, to have had more influence in the derivation of new plant forms, and in their preservation or destruction, but what then took place is not at all applicable to the state in which our planet is found, after its many, serious physical and geographical changes."[4] Thus, he did not explain the reason for such a limitation, although the idea he conveys is clearly associated with that of the 'revolutions of the globe.' It must be added that Colmeiro set the boundaries of his statement on species within a historical chronology, using evidence unearthed in Egyptian, Greek and Roman archaeological sites, as well as studies of nomenclature.[5] The archaeological argument had been, and continued to be used, just for that purpose. The paleontologist Justo Egozcue (1893) identified it as one of the devices employed by those obsessed by the notion of fixity, together with the remark that numerous species of similar configuration had been found on the base of tertiary strata, thus recalling that those divisions had been established by Lyell in consideration of the percentage of fossil organisms identical to the living ones that had been found there.

---

[2] J. Sala Catalá, *Ideología y ciencia biológica en España entre 1860 y 1881. La difusión de un paradigma* (Madrid, CSIC, 1987; Cuadernos Galileo de Historia de la Ciencia, no. 8), pp. 44-45.

[3] M. Colmeiro, *Estabilidad de las especies en el reino vegetal* (Madrid, 1860), pp. 16-17. See also pp. 12-15.

[4] *Ibid.*, p. 18.

[5] *Ibid.*, pp. 6-12.

Returning to Colmeiro, in the same year that he published *Observaciones y reflexiones hechas sobre los movimientos de las hojas y flores de algunas plantas con motivo del eclipse del Sol del 18 de julio de 1860*, he verified the scant influence exerted by the environment on the movement of some plants. A few years later he criticized studies of carnivorous and insectivorous plants in an article published in the *La Revista Europea* entitled "Las plantas aprehensoras de insectos" (1876), which focused on the 'digestion theory,' first expounded by Burnett (1829) and tested on the genus *Sarracenia* and subsequently by Curtis (1834) and Candy (1868) on the *Dionacea*. In this article, he stressed the importance of studies carried out since 1874, both that by Darwin on the genera *Drosera* and *Utricularia*, and by Hooker on *Sarracenia* and *Nepenthes*. Although he admitted the existence of plants capable of catching insects, in greater number in fact than those defined as insectivorous (as the latter category included also those that because of the viscosity of their leaves, stalks, etc. were equally capable of capturing them), he did not accept the phenomenon of digestion, nor its importance, given the small amount of plants where such phenomena occurred. Digestion, he said, "is not sufficiently demonstrated, and even when it might be with respect to those few plants that can be called carnivorous, the digestion of animal substances accidentally deposited or apprehended by whatever means could never be considered a necessary and important medium of nutrition for them, however admirable and worthy of study they might be."[6]

There were also a number of articles that questioned the boundaries between the animal and vegetable kingdoms as conventionally understood. Among such articles published in the *Revista Europea* we find one on carnivorous plants by J. D. Hooker in 1875, and the following year, T. H. Huxley on the frontier between the animal and plant kingdoms. Among Spanish authors interested in these phenomena were Salvador Calderón, a disciple of Francisco Giner de los Rios and a renowned Krausist, who read a paper entitled "Consideraciones sobre la alimentación de los vegetales en relación con los descubrimientos de las plantas carnívoras" (1876) to the Spanish Natural History Society. In it he discussed the generalization of the process of digestion of organic substances by plants, but in a sense opposite to Colmeiro's argument. Calderón identified three processes by which plants (as well as animals) obtain organic material: 1) necrophagy, mainly practiced by fungi and bacteria on the decomposing bodies of other creatures; 2) plasmophagy, parasitical feeding at the expense of a host; and 3) biophagy, entailing first the capture and then the ingestion of a living creature. He takes the process even further, pointing out that it can be observed in all plants at the microscopic level thanks to the phenomenon described by Tyndall, namely the condensation of germs on any damp surface, a device plants could resort to thanks to the condensation of water on their leaves and stems. From this, Calderón concluded that "the proof of the idea we have been analyzing completes the circle of the struggle for life. . . . it transcends dynamic and physiological problems of the greatest interest, refuting distinctions thought up to characterize the two organic kingdoms (no longer sustainable) which attribute to plants a role of mere subordination, and establishes as a common property the law that all living beings take mineral food directly from the

[6] M. Colmeiro, " Las plantas aprehensoras de insectos," *Revista Europea*, 119 (1875), 522-523.

mineral world, organic food from the organic world."[7] The notion, adhered to by Krausists, that all matter shared common vital processes of chemical analysis and synthesis had been previously noted by Salvador Calderón and Enrique Serrano y Fatigati in their *Estudios de Filosofia Natural* (1870). In this work nutrition was regarded as the fundamental mechanism of living creatures, because it generated a trophic chain of relations among them, that is "an ordered scale of complexity in faunas and floras,"[8] where the cell was considered the vital unit due to its capacity to consume organic material. In the same line were other influential works by Serrano on "Insectivorous Plants in Spain" (1878) and "Biological Physics: Nutrition and Light" (1879), as well as those by another Krausist biologist, Augusto González de Linares, such as "Haeckel's Morphology: Antecedents and Critiques" (1877), "The Plant Cell: A Contradiction that Permeates Contemporary Botany" (1878) and "On the General Form of the Higher Plants" (1879). González de Linares tried to end the dualism between the organic and inorganic worlds, offering the view that "everything was organic, not material," endowed with the capacity to evolve simply because it was active, so that change was the only law that life observes.[9]

In his response to Colmeiro's 1877 reception lecture in the Academy of Sciences, titled "Sketch of the Origin and Progress of Botany and Notice of the Authors who Have Contributed Most to its Advancement," Esteban Boutelou seems to favor the theory of evolution or, at least, displays competent knowledge of its main concepts. Beyond outlining the history of plant systematics, he alludes to recent achievements in the fields of plant physiology, morphology, reproduction, and so forth. As regards systematics, he remarks that no significant changes had occurred after the introduction of so-called natural classification systems, although much progress had been made in knowledge of specific families or genera. Henceforth, the only predictable change might come via new 'physical-natural theories' of the plant kingdom, especially the 'genealogical' doctrine or that of 'descent:' that is, the study of varieties, ontogeny and phylogeny, or paleontological evolution of the group, all of which would contribute to the explanation of the principles of 'natural selection' and the '*struggle for survival*' and would clarify the concept of species held currently by evolutionists, who had abandoned the conventional meaning up to this time. The new concept of species amounted to "perpetually variable individuals, that is, according to my limited understanding of the concept, a group of individuals who share a certain number of characters over a long time."[10] On the taxonomical level, he described the shift away from the family method towards what he labeled "natural classification in the form of a genealogical tree" where plants were grouped according to their descent and consanguinity. Such classification

[7] S. Calderón, "Consideraciones sobre la alimentación de los vegetales en relación con los descubrimientos de las plantas carnívoras," *Anales de la Sociedad Española de Historia Natural*, 1876, pp. 98-105.
[8] Sala Catalá, *Ideología y ciencia biológica*, pp. 57-58.
[9] *Ibid.*, p. 109.
[10] Colmeiro and Boutelou, *Discursos*, p. 4.

schemes could be tested, in spite of the difficulty in designing such tests and of the inadequacies of the paleontological register.

Similarly, he devoted some paragraphs to lay out the new orientation of plant geography, whose main issue centered on whether species had a nucleus of origin and, if that was so, was it single or multiple. Colmeiro's view was that, as result of the law of variability, a single hearth from which a species could later spread was most plausible. But the location of the original hearth was, he thought, an impossible task since plants did not necessarily occupy now the same place where they had emerged and could be found in very different ones because of changes in weather conditions, soil, and so forth. On the paleo-phytogeographical level, he presented the latest discoveries, claiming that "we can observe in the evolution of the plant kingdom the two laws of differentiation and perfectibility as a result of selection in the struggle for existence."[11]

Colmeiro's best student adopted the opposite view regarding evolutionist theories. According to González Bueno (1982), Blas Lázaro e Ibiza (1858-1921) has been considered one of the scholars who introduced Darwinism into Spain since, despite not publishing his first significant lecture on the subject until 1900, in which the concept of adaptation to the environment was substantiated with facts. Before this time, however, Spanish botanists had already become acquainted with and expounded that concept. Here, our main concern is with his efforts to disseminate the new concepts of species and plant evolution, as well as to underline the marginalization endured by evolutionists within the Academy of Sciences up to 1897, a fact that doubtless had an effect on Lázaro e Ibiza's lecture.[12] The concept of species and its limits was dealt with in the first edition of his *Compendio de la Flora Espanola*. In it he considers as the most plausible hypothesis that of the formation of new species through slow modifications through a considerable number of generations, together with the concept of 'natural selection,' conceiving the latter as that selection resulting from the struggle for existence among plants competing for various environmental resources.[13] The first scientific work in which, according to Sala Catalá, the notions of struggle for survival and the principle of natural selection were formulated clearly and without reservations was that written by Odon de Buen in 1883, thus inaugurating biogeography in Spain.[14] The concepts themselves, however, had already been mentioned by Esteban Boutelou.[15]

As Lázaro e Ibiza noted in his lecture to the Academy of Sciences on the occasion of his election in 1900, it was necessary to leave faith aside in order to "treat the matter with exquisite restraint, so it does not annoy those with contrary opinions." Unbridled passions on this issue, so pervasive in popular writings on Darwinism in Spain reduced the Darwinist debate to a clash of ideologies, thus neglecting its actual reception by the scientific community, where it spread less dramatically than on the popular level.

[11] *Ibid.*, p. 43.

[12] Sala Catalá, *Ideología y ciencia biológica*, p. 88.

[13] B. Lázaro e Ibiza, *Compendio de la Flora Espanola*, 2 vols. (Madrid, Hernando, 1896), pp. 29-30. The same comments are also found in later editions.

[14] O. de Buen y del Cos, "Apuntes geográficos-botánicos sobre la zona central de la Peninsula Ibérica," *Anales de la Sociedad Española de Historia Natural*, 12 (1883), 421-440. See, on the same topic, Sala Catalá, *Ideología y ciencia biológica*, p. 86.

[15] In his reply to Colmeiro, n. 1 *supra*.

Lázaro e Ibiza described the diffusion of Darwinism thus: "The theory has spread and ramified so profusely, that there no longer remains any interesting problem in the natural sciences that has no connection with and which does not feel its influence to a greater or lesser degree."[16]

In the rest of his lecture, he mentions several strategies adopted by plants in order to survive and compete in different environments, enabling him to "demonstrate that it is not merely hypothetical to assert that the struggle exists, but rather we confront a reality backed by sufficient proof."[17] He goes on to stress the importance of adaptation to the environment, referring to the laws of inheritance as the reason for the spatial and temporal unity of species and their historical continuity, noting that those species with greater capacity to adapt had more chances to perpetuate themselves, something that, put in other words, amounted to confirming once again the concept of 'natural selection':

> in spontaneous vegetation, just as in cultivated, there is revealed a tendency towards adaptation, opposed and evened out always by the law of inheritance. Thanks to this antithesis, the vegetation of a given country does not change suddenly, and species can be considered as inmutable within our chronological units.
>
> This tendency towards differentiation is, from the biological point of view, a strong and healthy virtue, and it bears within it a preservative tendency which makes the continuity of life forms possible. Without it, species already created and established in a world whose variations have been so great and profound, the history of life on our planet would not have been able to continue and would have been extinguished when the external conditions to which, in other times it had adapted, disappeared. Far from such a state of affairs, plant organisms, wisely endowed with this power, defend themselves from the evolution of the planet by carrying through on their own, in parallel fashion, adjusting to the environment, so long as the latter do not become incompatible with all forms of life.
>
> Therefore, considering the circumstances in which we see this evolution in out times, we must consider the possibility of differentiation and adaptation to the most varied conditions as a very powerful weapon, for those species which possess this faculty to the greatest degree have the greatest probability of perpetuating themselves.[18]

Some years later, the Association for the Progress of Science, of which he was vice-president, commissioned a lecture entitled "The Concept of Plant Populations and their Continuous Variability," which opened the section devoted to Natural Sciences at its annual meeting in Madrid. As in the previous lecture, Lázaro used a public event to introduce new theories that had recently emerged in Europe, specifically the concept of "syndynamics," as coined by Warming and Cowles.[19] Another lecture article that fits neatly with those of Lázaro e Ibiza was presented to the Academy of Sciences by the

---

[16] This quotation and the previous one are found in Lázaro e Ibiza, *Armas defensivas empleadas por los vegetales en la lucha por la vida* (Madrid, L. Aguado, 1900), p. 9.
[17] *Ibid.*, p. 11.
[18] *Ibid.*, pp. 60-61.
[19] A. González Bueno, "Actitud de Blas Lázaro e Ibiza (1858-1921) ante la corriente evolucionista," in *Actas del II Congreso Nacional de Historia de la Ciencia. Zaragoza*, ed., M. Hormigón, 3 vols (Zaragoza, 1982), I: 424-425.

forestry expert Carlos Castel y Clemente under the title "Value of the Agents that Determine the Distribution of the Plants on the Earth" (1899) focusing, as its title indicates, on the influence of environmental factors on the distribution of plants, emphasizing the importance of the struggle for existence. Máximo Laguna replied to that lecture with one dealing with the characteristics and distribution of martime flora.

EVOLUTION AND FORESTRY

The modern discipline of scientific forestry was introduced in Spain by Agustín Pascual González (1818-1884) who studied at the Therandt Forest Academy near Dresden, Germany.[20] When he returned to Spain, he became a professor at the first School of Forestry Engineers, established in Villaviciosa de Odón (Madrid) in 1848. This School had been established thanks to the effort of a group of 'enlightened men' attached to the Botanical Garden of Madrid and to the local Economic Society, including the lawyer Bernardo de la Torre Rojas, and Agustín Pascual, who had been a student of the mathematician José María Vallejo and of the botanist Antonio Sandalio de Arias. The principles of scientific forestry centered around the rational exploitation and preservation of forests, the control of forest land by the state to protect it from private rapacity, and a unified geographical vision that considered vegetation along with climactic, hydrographic, topographical, and pedological factors.

As part of his effort to introduce German science, Pascual had to translate new concepts into Castilian. A case in point was a series of articles about the origin of the term "forest," published in 1868 in the *Revista Forestal, Económica y Agrícola*, in which he synthesized his thoughts on the development of forestry science. He referred to the transition from woods to forest or what for Pascual was equivalent, from communal property to private ownership, a process that led to the concept of "forest" as an object and the beginning of scientific forestry. In its early stages, the forest vegetation was subject to a fixed, immutable and absolute regulation, during what may be called the 'dogmatic phase.' This phase gave way to a regressive movement with a marked skeptical undercurrent that ignored variety, disregarded local circumstances of climate, soil, and vegetation, and ultimately led to the disrepute of forest production. Finally, a harmonious and rationally-inspired movement emerged whereby the superiority of the synthetic method was established. To this were later added studies on the principle of rent, establishing the relationship between potential and growth, with the result that the notion of "forest" replaced that of the medieval "woodland."[21]

Once the definition of the term 'forestry' and the unified concept of forest were consolidated, the relationship between natural forms and their environment, characteristic of the integrative vision of plants and environmental conditions of forest science, emerged. As one Spanish forester expressed it: "when spontaneous growth from different localities are compared, notable differences are found. . . .And yet, in midst of this apparent confusion are discovered marvelous harmonies, which demonstrate that plants have not been randomly sown on the face of the globe. Rather

[20] L. Olazábal, "D. Agustín Pascual," *Revista de Montes*, 192 (1885), 33-50.
[21] A. Pascual, "Sobre el vocablo: Forestal," *Revista Forestal*, 1 (1868), 715.

each one has its special habitats because its life is subject to invariable laws, against which only rarely is it possible to struggle advantageously. Thus is it observed that vegetation is always linked to accidents of the terrain, which have such great influence in the constitution of the climate, as a result of which it acquires a particular character related to the localities which it occupies."[22] This harmony extended also to production and exploitation techniques, as well as to administrative and judicial dimensions.[23]

In his recent book *Ciencia y Política de los Montes Españoles*, Gómez Mendoza depicts nineteenth-century Spanish forestry as originally based on Romanticism and Schelling's *Naturphilosophie* and, later on, in the harmonious rationalism of Krausism, positivist evolutionism, and Neo-Kantian philosophy.[24] Here he mentions Pascual who, in 1878, put forward his views on the modern philosophical bases of forestry science, tracing its origins back to Kant, passing through successive phases---analytical, synthetic and harmonious—finally reaching the last phase thanks to the application of induction and analogy to a mass of accumulated empirical data. This increasing accumulation of new facts—though it gave impetus to the development of the field—prevented, because of its fragmentary nature, its unification under a single doctrinal corpus. This process crystallized in two main theories, representimg two aspects of the same thing. The first was the "law of evolution" which postulated the sequential order of phenomena, its range of application including all domains. Pascual did not accept Darwinism integrally; he pointed out that serious objections could be raised against it in relation to its negative effect on the notion of reality. Spencer or John Stuart Mill, for whom evolutionism led inevitably to skepticism, had already noted this, but because of the dissemination of evolutionary doctrine into the domain of morality: "for the evolutionists moral life is regulated by uniform regularity and is directed by a determinist force, always inflexible, always rigid, and never susceptible, in the final analysis, either to merit nor demerit. For Darwin the moral sense in man is the highest level of what, in animals, is social instinct, and the idea of justice is explained by the force, ever active, of gradual transformations: heredity, habit, etc. . . . The blind force of determinism blocks out freedom in the world of morality and does not conform to the almost idolatrous cult that evolutionists accord the freedom of thought."[25] He considered this doctrine accurate, however, insofar as far as the gradual development of evolution was concerned.

The second were the new philosophies that held that science only dealt with secondary causes, thus relativizing knowledge, a notion Pascual rejected because, for him, being and knowledge were equivalent and it was necessary to unify induction and

---

[22] Antón y Villacampa, "Arboricultura. Ideas generales," *Revista Forestal*, 1 (1868), 252.
[23] J. Gómez Mendoza, *Ciencia y Política de los Montes Españoles (1848-1936)* (Madrid, Icona, 1992), pp. 76-77.
[24] *Ibid.*, pp. 78-79.
[25] A. Pascual, "Necrológica. Don Miguel Bosch y Julia," *Revista de Montes*, 3 (1879), 441-467, on pp. 449-450.

deduction.[26] This reminds us of Schelling's 'absolute,' but it also suggests Krausist leanings to the extent that, as has been noted, its key theme was unity within diversity, an idea implicit in the very foundations of the forest as a unit, as mentioned above.

Pascual traced the first reference to Darwin in the *Revista Forestal* to 1872, a reference in its fifth issue to an article written by M. Martin in the *Revue des Deux Mondes* (Dec. 15, 1871), titled "Sobre la creación del mundo orgánico según los naturalistas ingleses y alemanes de la nueva escuela" [On the creation of the organic world according to English and German naturalists of the new school]. It praised, in contrast with the deductive/a priori systems of Schelling, Steffens, Kielmeyer and Carus, the solid set of observations produced by modern naturalists, especially Darwin. [27] Ten years later the *Revista de Montes* reprinted an obituary of Darwin by the great *regeneracionista* Joaquín Costa (1846-1911) from the *Boletín de la Institución Libre de Ensenañza.*

As Pascual accurately noted, the new theories seduced many "because of the apparent clarity of the principle, its simplicity and it universal application, conditions which accounted for its dissemination in some Forestry schools just as the time when instruction in Forestry Science was beginning in Spain."[28] We will now assess the extent of this "seduction," briefly outlining the degree of acceptance of Darwinism within the community of forestry experts, confining our study to articles and authors appearing in the *Revista de Montes*, the official journal of Spanish forestry.

First of all, a certain level of indecision can be observed among the forestry experts. Antonio García Maceira was the leading figure of the anti-Darwinian current. A forestry expert specializing in entomology, he wrote an article in 1875 on "El hombre terciario y la hipótesis del poligenismo [Tertiary man and the polygenist hypothesis]" in which he dismissed both concepts, criticizing the theories of Favre and Hamy. Around the same time A. Parada y Barreto published "Ligera reseña del origen e historia del mundo" [Light review of the origin and history of the world], with a clear creationist bias. A bit later in "Una nueva teoría sobre la vida" in the *Revista de Montes* (1888), García Maceira criticized C. von Naegeli's book *Mechanisch-physiologische Theorie der Abstammungslehre* (1884), that is his theory of the "idioplasm," understood as that part of the protoplasm responsible for the genetic configuration of the organism—which, by the way, he did not identify with the nucleus—relying on the comments of the German authors Jurgens and Dressel. According to García Maceira, this theory was superior to Darwin's as it included within the organism the cause of its own alteration: "it cannot be denied that Naegeli's point of view in his treatise on descent is much solider on many points than that of Darwin and his supporters. While Darwin, by way of explaining successive change, emphasizes natural selection in the struggle for existence, shaping organisms in the play of external influences, Naegeli always advocates the internal cause of continuous improvement."[29] García Maceira, however, rejected Naegeli's theory too, because he thought it was mechanicist and accused Naegeli of ignoring

[26] *Ibid.*
[27] "Crónica," *Revista de Montes*, 5 (1872), 106-109.
[28] Pascual, "Necrológica. Don Miguel Bosch y Julia," p. 447.
[29] A. García Maceira, "Una nueva teoría sobre la vida," *Revista de Montes*, 12 (1888), 442.

studies of the cell by Reinke, Strassburger, Schmid, and Froman. These authors defined protoplasm as the result of the aggregation of blastema's crystalline molecules which, mixed in soluble or insoluble form, formed a mucous and half-fluid mass of an albuminous nature whose activity was due to differing states of excitation exhibited by different groups of molecules found in each strand or mycelium. In sum, Naëgeli's theory obscured, rather than illuminated, the process of life.

On the occasion of the Catholic Congress celebrated in Madrid in 1898, García Maceira wrote a leaflet entitled *Exposición y exámen del Darwinismo* (1898), employing arguments borrowed from Van-Tieghem, E. Fabre, M. Girard and, especially, Wingand, to attack the theory. In a biographical sketch, however, García Maceira branded Wingand as a materialist because "when he destroyed the idea of an external transformation of types, proposing instead a regular, internal evolution, he omitted from it any metaphysical complement, and forged a mechanical theory of life, just like his opponent, Darwin."[30] He goes on to quote Kant to claim that no simple mechanism of nature was sufficient to explain its organic products.

Responding to this article, Juan José Muñoz de Madariaga, commended García Maceira's efforts to rectify scientific errors and repress the deviations from religious beliefs spawned by the thesis of Darwin and his followers. In his leaflet, García Maceira had rejected evolution through natural selection, variability and inheritance and included a chapter entitled "internal generation" in which he railed against Kolliker's school, Naegeli's protoplasmic and mechanical theory, and Hartmann's doctrine, all because of their atheism.[31] Muñoz also reviewed the Jesuit Juan José de Urraburu's markedly Catholic book, *Origen de los seres vivientes segun sus diversas especies y exámen del transformismo*, translated from a Latin original, which he highly recommended.

García Maceira supported the idea of gradual evolution, but only as the result of the flow of vital energy according to a pre-arranged plan, which presented an internal tendency towards modification in which selection through the struggle for existence provided no advantage. He concluded, therefore, that the "law of natural selection" was not a means of transformation but the expression of a set of results.[32] The concept of vital energy appears repeatedly in several of his writings, but it is best exemplified in an article on biological materialists:

> The Biological materialists, Darwinians and Lamarckians, both of whose systems seek harmony, greatly exaggerate the influence of the environment as the sole cause of modifications not transmissible through heredity and always of a physiological and superficial nature, insufficient to explain the systematic transformation of types. As Lamarck accorded extraordinary importance to habit, resulting from the use or disuse of organs, the [materialists] take that idea, adding to it Darwin's natural selection and the

---

[30] A. García Maceira, "Zoologos ilustres. Wingand," *Revista de Montes*, 23 (1899), 338-340, on p. 338.

[31] J. J. Muñoz, "Bibliografía: Exposición y exámen del darwinismo, *Revista de Montes*, 22 (1898), 171-172.

[32] García Maceira, "Zoologos ilustres. Lamarck," *Revista de Montes*, 23 (1899), 44-47.

struggle for existence . . . . The modification of organisms by external influences always presumes (and on this point the materialist biologists are silent) a preexisting attitude and internal tendency towards modification, without which the organism would die in an unfavorable environment, instead of accommodating itself. Yet they deftly escape from this truth by presuming that functional assimilation always gives as an immediate result the adaptation of organisms to their environment. . . . In spite of their efforts, the materialist biologists have not been able to destroy the concept of vital energy.

Similarly, in his biography of Edmond Perrier—whom he criticizes scathingly for his materialism—García Maceira insists that science can describe life but not explain it: "the concept of life is not explained by science. Those who subscribe to the spiritual conception of life do not seek to destroy any advance of science thereby. They simply say that what appears through observation, that which is certain, cannot be denied; that vital energy and vital phenomena are the masters of physical forces, not their slaves. They say that there is both mechanism and agency in life."[33]

Turning to Lanessan, García Maceira notes that the French author uses arguments on symbiosis, co-operation, and mimetism to make the struggle for existence more palatable, emphasizing the protection and defense of the weak, and that in the social life of animals the weak are not always victims of the strong. He could not forgive Lanessan, however, for extending his argument to human society: "Given such an erroneous principle, it is incredible to read Lanessan's terrifying assertion: 'All men must be granted the full exercise of all their rights. Family property, which constitutes the most powerful impediment to progress, must be suppressed, because of the perturbations it induces in the process of sexual competition." Lanessan is one of those notable, but confused, naturalists, who does not respect the boundaries between animal and human society, and who attempts to resolve the loftiest questions with the ignominious criterion of a gross empiricism. . . . This class of naturalists—born and raised in the bosom of Darwinism, by necessity fierce revolutionaries—must be stopped at the threshold of the moral world, where we will tell them: Go back!"[34]

García Maceira also criticized Darwin for "having established an equality or identity between an ideal relationship and a genealogical or embryogenic one[35] or, in other words, "it is false that affinity/resemblance entails identical order." He also criticized Haeckel for extending evolutionary theory into ethics and politics, fields where Darwin had been more cautious.[36]

Among those who assumed an ambivalent stance we may mention Máximo Laguna y Villanueva who, as time passed, grew gradually closer to evolutionary theory. In 1877 he claimed that the theory contained "certain bold hypotheses"[37] and supported the

---

[33] García Maceira, "Zoologos ilustres. Perrier," *Revista de Montes*, 23 (1899), 300-302.

[34] García Maceira. "Zoólogos ilustres. Lanessan," *Revista de Montes*, 23 (1899), 274-277, on p. 177. Lanessan's works were first translated into Spanish by the physician and naturalist Romualdo González Fragoso, a mycologist and staunch Darwinian: *Estudio sobre la teoria de Darwin* (1884) and *La lucha por la existencia y la unión para la lucha* (1884). He also translated Hanstein's *El protoplasma considerado como base de la vida de los animales y vegetales* and Haeckel's *El reino de los protistas*.

[35] García Maceira, "Zoólogos ilustres: Wingand" (n. 30 *supra*)., p. 339.

[36] García Maceira, "Zoologos ilustres. Haeckel," *Revista de Montes*, 23 (1899), 392-396.

[37] M. Laguna, *Progresos verificados en el conocimiento de la reproducción de los vegetales y en especial de la fecundación de las plantas fanerogamas* (Madrid, E. Aguado, 1877), p. 18.

existence of a tendency towards the increasing perfection of organisms as they climb the "tree of life," whose apex was man, with his brain, soul and world of ideas.[38] This notion recalls once again the transformation experienced by the *Scala naturae* when it had to encompass relations between living creatures. Laguna also stresses harmony among different organisms.

Apart from the texts previously mentioned, the *Revista de Montes* published several reports as independent leaflets, some of them devoted to anthropology, a discipline wherein the controversy was more acrimonious, such as *El porvenir de las Razas*, a lecture by Rafael Álvarez Sereix at the Madrid Geographical Society in 1895, inspired by his translation of *El problema de la vida*, a book by J. F. Albert de Pouget, marquis of Nadaillac, in 1893.

With regard to the advocates of evolutionism, let us begin with José Secall e Inda (1853-1918), one of the editors of the *Revista de Montes*. He thought of himself as an evolutionist, in contrast with his own teacher Pedro de Avila. That is at least what is suggested in the dedication of his work "Catálogo metódico de las plantas lenosas silvestres o asilvestradas que se observan en San Lorenzo de El Escorial y sus alrededores" to Avila in 1888: "your disciple in principles, always hearing your lectures and resolving my doubts, still a great abysm separates us in those questions that today divide the field of science. You, a tenacious partisan of the permanence of species, I an unabashed evolutionist, each day more persuaded of the greatness of that theory and the subjectivity of the idea of species."[39] His researches were mostly botanical. He advocated the teaching of evolutionism at Spanish universities and sternly criticized those scholars in fields other than natural sciences for imparting lessons on subjects they had not mastered, such as evolutionism: "Men learned in moral and political sciences and in knowledge of our great poets and writers, but lacking in the most elemental knowledge of Natural History, they are given to speak nonsense with the greatest calm and stolidity about subjects which they believe they understand with the same facility with which they master a political speech or with which, comfortably reclining in box seats, they listen to a popular operetta."[40]

JOSÉ MARIA CASTELLARNAU

José María Castellarnau worked on many new botanical topics of theoretical interest, as well as on techniques of microscopy. He wrote the first study of evolutionary morphology published in the *Anales de la Real Sociedad de Historia Natural* in 1880.[41]

---

[38] *Ibid.*, pp. 95-96.

[39] J. Secall, *Catálogo metódico de las plantas lenosas silvestres y asilvestradas que se observan en San Lorenzo de El Escorial y sus alrededores* (Madrid, Moreno y Rojas, 1888), p. 3.

[40] This quotation in Secall, "Falta de Cultura, " *Revista de Montes*, 24 (1901), pp. 5-8.

[41] J. M. Castellarnau, "Estudio micrográfico del tallo del Pinsapo (Abies pinsapo Boiss.)," *Anales de la Sociedad Española de Historia Natural*, 9 (1880), 401-464. See Sala Catalá, *Ideología y ciencia biológica*, p. 105.

He defended the notion of evolutionary change through gradual and permanent adaptation, perpetuated by reproductive mechanisms, and rejected both fixity and successive creations: "as long as the Linnean concept of species and that of successive creations have been dominant in the Natural Sciences, the full value of laws that control life has not been appreciated. The autonomous and independent species proclaimed by Linnaeus and supported by Cuvier and Agassiz, if not opposed to the unity of the organic world, at least has not provided any demonstration of its existence. One of the advantages of the theory of evolution is that it provides that unity. . . . Life is just one thing, the same for animals as for plants, and since it appeared in the early ages of earth it has developed without cessation, on the impulse of favorable conditions, continuously modifying its way of being. . . . Plants, just like animals, are nothing more than manifestations of that life, which has reached us in the thousand different forms, transitory and changeable, which constitute species. Thus, the organisms which nowadays populate the surface of the earth are only the latest forms in the long series which have continuously and without interruption dated back to the first manifestations of life. Inheritance and adaptation, that is to say, the faculty of transmission of inherited and acquired conditions, by means of reproduction, and the ability to adapt oneself to the environment so as to make existence easier, explain how with the aide of time, the immense differences that are observed between distant points of the same series have been formed. These relationships are not always visible on first view and have been explained by different causes. . . . The first, that no series, whether plant or animal, exists complete, but rather present themselves interrupted and with great lacunae. The second. . . . consists in the imperfect knowledge which still nowadays we have of different plants and animals. . . . But these gaps are not so great, nor our knowledge so deficient, that we would not be allowed to trace the pattern of the plant kingdom in its broadest design, putting in evidence the intimate connections that exist between their primordial groups. . . ever since the theory of selection [sic] and descent was proclaimed. . . . as the happiest and clearest of all the hypotheses that ever predominated in the Natural Sciences, making clear that the organic plan of the Phanerogams and the Cryptogams is the same."[42]

In 1914 Castellarnau wrote a study of the "General Morphology of Plants According to Biogenetic Laws" and in the 1920s, he delivered two addresses to the Academy of Sciences on biological reductionism and the purpose of living creatures: "Pueden explicarse químicamente los fenomenos esenciales de la vida?" [Can the basic phenomena of life be explained chemically?] (1922) and "De la finalidad de los fenómenos en los seres vivos" [On the finality of phenomena in living beings] (1928). The most remarkable aspect of both articles is their relative withdrawal from Darwin's theory, although he did explicitly embrace it again six years later, when he wrote: "of *The Origin of Species*. . . today only the 'principle of evolution' remains on firm ground—a principle which in reality is not alone that of Darwin, although his due is universally acknowledged sentimentally—inasmuch as we do not have conclusive proofs that demonstrate it. Biologists do not at present agree that fluctuating variations

---

[42] Castellarnau, "Unidad del plan generativo en el reino vegetal," *Anales de Historia Natural*, 17 (1887), 31-33.

are transmitted by heredity, nor that new species are formed through the operation of selection, whether natural or artificial."[43] Conceding that he was not an authority on the subject, he relied on the opinions of William Bateson who had a saltatory view of evolution and on those of T. H. Morgan, of whom he said that, despite accepting evolution, he had rejected 'natural selection' for it was not capable of producing new species, just of increasing the number of individuals of the selected type.[44] He also cited works by Hertwig and Loeb, both opponents of natural selection.

Castellarnau was also opposed to biological reductionism. In his writings life takes on the category of a real and independent entity, unexplainable by any set of physio-chemical reactions. Although he tries to distance himself from the vitalist school of Claude Bernard, he quotes him in support of vitalism in the course of a brief history of vitalist theories in the nineteenth century.[45] Here he subscribes to J. B. S. Haldane's statement that "one must investigate life as it is and not as physicists and chemists would like it to be, because now the hour has arrived for Biology to recover its freedom, and to develop by itself as an active, experimental science which has ahead of it a whole world to conquer. The phenomena of life. . . .must not be interpreted by the same standard by which we today interpret those of inorganic matter, becaue life is a basic reality and not a simple artifact."[46]

Later on, Castellarnau returned to the purpose of living creatures, distinguishing two types of functions, one physiological, assimilated into the need to adapt to the environment (thus, the purpose of an organ is the object of its function), and another teleological, whose roots can be traced back to Kant: "to judge a thing's organization as an end of Nature, is completely different from supposing the existence of that same thing as a design of Nature; because for the latter one must know, in addition to the concept of a possible end, the end that Nature proposes for it, and this implies the existence of some suprasensory relationship between Nature and Being which is beyond teleological knowledge."[47] Although there was, apparently, an initial physical motive in any biological phenomenon, its genuine cause lay in the very 'reason to live,' which expressed itself in various ways depending on the habits of each species, thus reflecting again the idea of 'life' as an entity independent of creatures, a force which shapes their biological reality. This vital idea emerged from the earth itself and continued through the reproduction and evolution of living creatures in a process of evolution that stuck to a plan virtually contained in the cellular origin of each individual and that transmitted itself to all the creatures' cells through inheritance. For him, this was but the present

---

[43] Castellarnau, *Pueden explicarse químicamente los fenomenos esenciales de la vida?* (Madrid, Real Academia de Ciencias Exactas, 1922), p. 26.

[44] *Ibid.*, p. 27.

[45] *Ibid.*, pp. 14-26.

[46] *Ibid.*, p. 38. The quotation is from J. B. S. Haldane, *Organism and Environment as Illustrated by the Physiology of Breathing* (New Haven, Yale University Press, 1917), p. 89.

[47] Castellarnau, *De la finalidad de los fenómenos en los seres vivos* (Madrid, Real Academia de Ciencias Exactas, 1923), p. 9.

remembrance of previous events so that each evolutionary step was necessary in order to lead into the next.

As I do not wish to go beyond the early twentieth century, I will conclude by stressing the importance of the issue of purpose in the evolutionary discussions of those years. Thus, Castellarnau's speeches were not the only ones at the Academy of Sciences to deal with this point. Pedro M. González Quijano's *Azar y Determinismo* (1925) can also be consulted, alongside Leonardo de Torres Quevedo's reply to it.

Centro de Estudios Históricos, Madrid

MIGUEL ÁNGEL PUIG-SAMPER

DARWINISM IN SPANISH PHYSICAL ANTHROPOLOGY

PRELIMINARY CONSIDERATIONS

This paper investigates the reception of Darwinism, or more generally the reception of evolutionary theory, among Spanish anthropologists in the nineteenth century and the first decades of the twentieth.[1] Before entering into this subject, however, one must first locate the problem in its Spanish context and demonstrate whether the conclusions and stereotypes evident in the principle works on Darwinism in Spain hold true for the case of anthropology.

In general, studies on the impact of Darwin in Spain have been predominantly sociological in character. The Darwin polemic provided an interesting research problem in the development of an "externalist" approach to the history of Spanish science, following the now-classic scheme proposed by Thomas F. Glick in 1969.[2] Glick's formulation permitted one to trace everything from the quality and interests of the scientists of the receiving country to political and ideological implications, as well as the paths and protagonists of the dissemination of ideas (translations, institutions, etc.). Following a similar line some years later, although insisting more on the connections between the Darwin polemic in Spain and the development of Positivism, Diego Núñez provided the first general work, which also contains a valuable selection of texts.[3] Shortly thereafter, Glick offered an overview of the Darwin polemic in Spain in his work *Darwin en España*.[4] That book was published in 1982, the centennial of Darwin's

[1] This work forms part of the research project DGICYT no. PB94-0060. John-Marshall Klein (University of Texas at Austin) provided the translation.

[2] T. F. Glick, "La recepción del darwinismo en España en dimensión comparativa," *Asclepio*, 21 (1969), 210-211. This scheme applied in more general form appears in the volume edited by T. F. Glick, *The Comparative Reception of Darwinism* (Austin, The University of Texas Press, 1972).

[3] D. Núñez, *El darwinismo en España* (Madrid, Castalia, 1977). The connections with the Positivist movement can be seen in his previous work, *La mentalidad positiva en España: desarrollo y crisis* (Madrid, Túcar, 1975).

[4] T. F. Glick, *Darwin en España, Barcelona* (Barcelona, Península, 1982).

Thomas F. Glick, Rosaura Ruiz, and Miguel Ángel Puig-Samper (eds.), The Reception of Darwinism in the Iberian World, 127-141.

death, an occasion which inspired a variety of institutions and publications to dedicate special studies to the scientist and his theories.[5]

Among the conclusions drawn by this externalist focus, perhaps the most prominent was the notion of the Spanish scientific community's weakness during the era of the Darwin debate, weakness characterized by the isolation of Spanish scientists and by Spain's backwardness in experimental science with respect to other European countries.[6] Some authors see this lag as the consequence of the absence of a genuine bourgeois revolution in the country.[7] In similar fashion, it has now become a virtual cliché that the Darwinism debate in Spain took place largely outside of the academic world as a broad ideological, political and religious confrontation. It is generally accepted that it was with the "Glorious Revolution" of 1868 that the debate between Darwinians and anti-Darwinians began. Naturally, this debate took place against a political backdrop of liberal versus conservative confrontations and, more broadly, the multiple confrontations between progressive and reactionary ideologies, as well as between materialists and Catholics.

The conclusion regarding the extra-academic locus of the debate, without being false, needs to be modified because the few works on different scientific communities in various areas and institutions indicate that there was indeed a scientific debate centered on Darwin's ideas and about evolution in general. It is one of the present article's objectives to demonstrate this point further.[8] The second conclusion appears clear in general terms, although there was not always a mechanical correspondence between scientific attitudes and political ideologies. Indeed, some cases reveal multiple contradictions, as Alvaro Girón has demonstrated recently in a study of the relation between Spanish anarchism and the evolutionary thesis.[9]

One would also have to modify Glick's notion, later generalized, that it was basically in the medical community that Spanish scientists debated evolution in the nineteenth century. A preliminary evaluation of the case of anthropology permits us to distinguish two periods tied professionally to Medicine and later on to the Natural

---

[5]Among these, we would highlight the monographic issue of the Barcelona journal *Anthropos* (October 1982) on "El darwinismo en España en el primer centenario de la muerte de Ch. Darwin (1882-1982)". The issue offers a chronology of Darwinism in Spain and interesting contributions by Thomas F. Glick , Diego Núñez and José Florit, among others. In addition, the reception of Darwinism in Spain, Portugal and the Iberian world was the central theme of the II Congreso de la Sociedad Española de Historia de las Ciencias (Jaca, 1982): M. Hormigón, ed., *Actas del II Congreso de la Sociedad Española de Historia de las Ciencias*, 3 vols. (Zaragoza, SEHC, 1984).

[6] This idea is discussed and qualified by J. Coello in, "Los científicos españoles del XIX y el darwinismo," *Mundo Científico*, no. 14 (May 1982), 536.

[7] Núñez, *Darwinismo en España*, p. 14; and J. L. Peset, S. Garma, and J. S. Pérez Garzón, *Ciencias y enseñanza en la revolución burguesa* (Madrid, Siglo XXI, 1978).

[8] For examples, see the following works presented at the II Congreso de la Sociedad Española de Historia de la Ciencias, A. Galera, M. A. Puig-Samper and F. Pelayo, "El darwinismo en la Sociedad Antropólogica Española" (*Actas*, I, 389-402); A. González Bueno, "Actitud de Blas Lázaro Ibiza (1858-1921) ante la corriente evolucionista" (pp. 419-427); F. Pelayo, "La Paleontología. Un argumento para rebatir al darwinismo en el intento de armonizar Ciencias Naturales y Religión" (pp. 475-488); L. Sequeiros, "Impacto del darwinismo en la Paleontología española. Juan Vilanova y Piera (1821-1893)" (pp. 523-538).

[9] A. Girón, *Evolucionismo y Anarquismo en España (1882-1914)* (Madrid, CSIC, 1996).

Sciences, as will be shown later on.[10] For the twentieth century, the lack of research is evident and perhaps for this reason only the doctors and biologists of the Spanish school of neurohistology have received any attention. Among these, Cajal, Achúcarro, and Río-Hortega head the list, without doubt due to their impact on the world scientific scene. Physiologists and pathologists like Marañón and Novoa Santos or geneticists like Zulueta and Nonídez have received some attention as well. However, botanists, zoologists, geologists, paleontologists, and anthropologists have been largely overlooked, giving rise to an overworked stereotype of the solitary Spanish scientist who emerges almost by spontaneous generation, artificially removed from the context which surrounded him.[11]

Something similar occurs with the idea that the operational reception of Darwinism in Spain occurred very late and was out of step with its theoretical or rhetorical reception. More internally-oriented studies will have to be produced before we can reach definitive conclusions to complete the pioneering studies of José Sala Catalá.[12]

EVOLUTIONISM IN THE SOCIEDAD ANTROPOLÓGICA ESPAÑOLA

It appears an incontrovertible fact that the Darwinism polemic in anthropology was not chronologically distinct from the rest of the academic disciplines, since the first articles to discuss Darwin's theories did not appear until the decade of the 1870s. This fact supports José María López Piñero's assertion that Spanish academic life prior to the Revolution of 1868 was fairly languid and characterized by the individual efforts of members of what he calls the "intermediate generations" and by the lack of scientific associationism.[13]

This did not mean, however, that interest in anthropological topics was nonexistent in the prior to 1868. Without entering into even more remote antecedents, one can point to the 1834 creation of an Anthopology Section in the recently founded Real Academia de Ciencias Naturales in Madrid, with the participation of Joaquín Hysern—future president of the Sociedad Antropológica Española—and Francisco Fabra Soldevila, author of *Filosofia de la legislación natural fundada en la Antropología o en el conocimiento de la naturaleza del hombre y sus relaciones con los demás* (Madrid, 1838).[14]

The following decades saw the production of some isolated works like that of José Varela de Montes, *Ensayo de Antropología* (Madrid, 1844-45), a precursor of Spanish

[10] M. A. Puig-Samper and A. Galera, *La Antropología española del siglo XIX* (Madrid, CSIC, 1983).

[11] An approach to this context is found in Luis Alfredo Baratas, *Introducción y desarrollo de la Biología experimental en España entre 1868 and 1936* (Madrid, CSIC, 1997).

[12] J. Sala Catalá, "El evolucionismo en la práctica de los biólogos españoles del siglo XIX (1860-1907)," *Asclepio*, 33 (1981), 81-125; and *Ideología y Ciencia Biológica en España entre 1860 y 1881* (Madrid, CSIC, 1987).

[13] J. M. López Piñero, "La literatura científica en la España contemporánea," *Historia general de las literaturas hispánicas* (Barcelona, Vergara, 1968), VI: 679-684.

[14] A. Gomis Blanco, "Hace 150 años se fundó la Academia de Ciencias," *Llull*, 7 (1984), 93-99.

medical anthropology,[15] and important lectures by three medical doctors with degrees in Natural Sciences all of whom were associated later on with the Sociedad Antropológica Española. The first of these was Rafael Martínez Molina, founder of the Instituto Biológico, the title of whose doctoral dissertation was *El hombre considerado en sus relaciones y bajo la influencia de los agentes naturales* (1853). The second presentation came from the pen of Sandalio Pereda, one of the founders of the Sociedad Antropológica Española, who titled his doctoral work *Unidad específica de las razas humanas* (1858). This work criticized Lamarckian evolutionism using creationist and fixist positions:

> Lamarck, in denying the existence of species with constant organic characters, basing himself principally on specific mutual affinities, did not distinguish the real fact—the primitive form, permanent from the Creation—from the accidental and contingent one which, under the influence of physical agents, determines varieties and races.[16]

In his discussion Pereda used paleontology to disprove biological evolution, as would a number of Darwin's detractors shortly thereafter:

> Paleontology, in its great and important progress, daily demonstrates the independence of fossil flora and fauna, and the constancy of forms in the respective species. The law of the gradual perfection of beings is at least exaggerated, if not false.

With reference to man, Pereda adopted a stance that would become a constant among the Spanish anti-Darwinists ("fear of monkeys," as Julio Caro Baroja wryly termed it): "To believe in such changes amounts to accepting the ridiculous transformation of anthropomorphic monkeys into humans."[17] The doctor's critique of variations under artificial conditions is quite interesting, as this is precisely one of the elements used by Darwin to elaborate his idea of selection. Pereda argued:

> A naturalist does not go to flower gardens or agricultural schools to examine primitive organic forms. He must find them in the field, in the habitat of animals and plants, both of which are independent of the action that breeding and cultivation exercise over accessory organic modifications, which are the origin of races and varieties.[18]

The third lecture was that of Manuel María J. de Galdo, another of the founders of the Sociedad Antropológica, addressed the Academia Médico-Quirúrgica Matritense on "The Importance of Anthropology in Medicine," in which he stressed the significance of Broca's concept of nature. As I have indicated elsewhere, one of the factors underlying the creation of the Sociedad Antropológica Española in 1865 was the decisive influence that Paul Broca, founder of the Sociéte d´Anthropologie de Paris, exercised on Pedro Gónzalez de Velasco, an anatomist who had previously pushed for the renovation of morphological studies in Spain. The Spanish Anthropological Society was the fourth such created in Europe, after those of Paris (1859), London (1863), and Moscow

[15] E. Ronzón, *Antropología y antropologías. Ideas para una historia crítica de la antropología española. El siglo XIX* (Oviedo, Pentalfa, 1991).

[16] S. de Pereda y Martínez, *Unidad específica de las razas humanas* (Madrid, Eusebio Aguado, 1858), p. 4.

[17] *Ibid.*, p. 5; J. Caro Baroja, "El miedo al mono o la causa directa de la Cuestión Universitaria en 1875," *En el Centenario de la Institución Libre de Enseñanza* (Madrid, Tecnos, 1977), pp. 23-41.

[18] Pereda, *Unidad específica*, p. 7

(1863).[19] In addition to Velasco, the founding board of the Spanish Society included F. Delgado Jugo, a pioneer of Spanish ophthalmology; Manuel María José de Galdo, professor of Natural History and director of the Academia Médico-Quirúrgica in Madrid; Sandalio de Pereda, also a professor of Natural History and a member of the Academies of Science and Medicine; Matías Nieto Serrano, a neo-Kantian physician; Ramón Torres Muñoz de Luna, professor of Chemistry and former student of Justus von Liebig; and Juan Vilanova, professor of geology and paleontology at the University of Madrid.[20]

The focus of nineteenth-century Spanish anthropology can be appreciated by three facts: first, of fifty-eight founding members of the Spanish Anthropological Society, 40 were medical doctors; second, the influence of French Positivist anthropology; and, third, the halting introduction of Darwin's works in Spain.[21] It is significant that the society proposed as the first point of its agenda the "classification of the races and varieties of the human species and debate over it's origin." A speech by education minister Manuel Orovio also proved significant. Orovio was responsible for causing the infamous University Crises which resulted in the ouster of progressive professors.[22] He spoke at the Society's inaugural meeting to urge the need to study the whole human being, both moral and spiritual, while keeping one's sight on heaven, font of inspiration and source of the illumination necessary for uncovering the truth.[23]

Early scientific records of the Anthropological Society have not been found. Even so, the Society's discussions evidently caused profound unease among Spain's most reactionary and clerical elements. One can infer this from a speech by Delgado Jugo who noted that the Society's members had been branded "heterodox freethinkers" who discussed subjects openly and outside of "the iron circle of dogma." Delgado Jugo clearly felt that this reaction was unwarranted given his prudence in tabling until later such thorny topics as the principles of German philosophy pertaining to man and in committing the Society's first efforts to the study of the primitive inhabitants of the peninsula and the origin of language.[24]

In contrast to the standard view of the role of the Revolution of 1868 in the creation of scientific societies and institutions, the Sociedad Antropológica Española temporarily ceased to function during the revolutionary period and remained inert practically up to

---

[19] M. A. Puig-Samper, "El doctor González de Velasco y la Antropología española en el siglo XIX," *Asclepio*, 34 (1982), 327-337.

[20] Biographical data can be found in J. María López Piñero, T. F. Glick, V. Navarro and E. Portela, eds., *Diccionario histórico de la ciencia moderna en España*, 2 vols. (Barcelona, Península, 1983). See also C. Ortiz and L. A. Sánchez Gómez, eds., *Diccionario histórico de la Antropología española* (Madrid, CSIC, 1994).

[21] A. M. Verde Casanova, "La primera sociedad antrpológica española," *Actas del I Congreso Español de Antropología* (Barcelona, Universidad de Barcelona, 1980), II: 17-38.

[22] A. Jiménez-Landi, *La Institución Libre de Enseñanza* (Madrid, Taurus, 1973).

[23] *El Siglo Médico*, 1865, p. 383.

[24] F. A. Delgado Jugo, *Discurso leido en la inauguración de las sesiones de la Sociedad Antropológica Española, verificada el domingo 21 de febrero de 1869* (Madrid, T. Fortanet, 1869).

the restoration of the monarchy. The explanation for this stems from the massive participation of many of its members in political, cultural, and educational reform activities during the revolutionary years. The Society was composed of a mix of Krausists, progressives and democrats. Some of the most prominent members included politicians like Manuel Becerra, Nicolás Salmerón, Nicolás María Rivero, Segismundo Moret, and Manuel María José de Galdo—made mayor of Madrid by the Revolution. University reformers included the rector Fernando de Castro and Pedro González de Velasco.[25]

Despite these circumstances, some of the most prominent scientists associated with anthropology like Galdo, Velasco, Pereda, Hysern, Martínez Molina, and Vilanova founded the Sociedad Española de Historia Natural in 1871 along with other leading naturalists such as Ignacio Bolívar, Marcos Jiménez de la Espada, Miguel Colmeiro, Francisco Martínez y Sáez, and others.[26] This Natural History Society published in its *Anales* (first period, 1872-1891) works by eminent evolutionary biologists like Augusto González de Linares, Antonio Machado y Núñez, Odón de Buen, Salvador Calderón, Luis Simarro, and Manuel Antón. Some of these authors were significant members of the Institución Libre de Enseñanza, a body which would shortly thereafter name Darwin as an honorary fellow.

In 1874, the Sociedad Antropológica Española was reinstalled with a governing board formed by Joaquín Hysern, Rafael Ariza, Pedro González de Velasco, Francisco María Tubino, Manuel Calderón, Juan Vilanova, Manuel María J. de Galdo, and Angel Calderón. The Society also then created the *Revista de Antropología*, in which appeared a number of discussions of evolutionism in a more tranquil tone than that reflected in the mainstream press of the period. This journal complemented the medical journal directed by Velasco, *El Anfiteatro Anatómico Español,* which also published articles on Darwin. The first volume of *El Anfiteatro*, created in 1873, contained some anthropological comments signed by Carlos María Ferrer, including one dedicated to the "Refutation of Some Utopias." In his early writings, Ferrer used anatomical arguments to flatly reject the idea that human bipedalism arose as a consequence of habit and education. Subsequently, he entered fully into the debate about the possible simian origin of man. Ferrer asserted that the idea that man descended from the apes represented an erroneous conclusion based on Lamarck's theories and, to an even greater extent, on those of Darwin. Ferrer considered Darwin a "naturalist of great merit" who had arrived at an explanatory theory of the origin of the species by successive transformations, based on mechanisms like the struggle for life and natural selection. Ferrer argued this process should be rejected as incompatible with physiological laws. Even so, Ferrer demonstrated that, following Darwin, it was impossible to make man descend from apes, given that Darwinist theory explained the phenomenon of speciation with a diverging phylogenetic tree in which backward steps were impossible. Consequently, humans could never have arisen from quadrumana.

[25] María V. López-Cordón, *La revolución de 1868 y la I República* (Madrid, Siglo XXI, 1976); J. J. Gil Cremades, *Krausistas y Liberales* (Madrid, Seminarios y Ediciones, 1975); J. López Morillas, *El krausismo español*, 2nd ed. (Madrid, FCE, 1980).

[26] J. L. Martínez Sanz, *Medio siglo de ciencia española: la Sociedad Española de Historia Natural* (Madrid, Universidad Complutense, 1982).

This impossibility had been demonstrated as well by the anatomical studies of Richard Owen, Duvernoy and Gratiolet, and even Carl Vogt, who admitted the possibility of a common trunk but never the direct passage from ape to man.[27]

The anti-Darwinists in the Sociedad Antropológica included its own president, the homeopathic physician Joaquín Hysern who in 1874 wrote the first article in the *Revista de Antropología,* "On the Native Unity of the Human Species, or the Relationship by Universal Consanguinity among all the Races of the Human Species, Dispersed throughout All the Regions of the Earth."[28] Hysern showed himself in this article to be a firm creationist, very critical of materialism and sensualism, a convinced monogenist, obsessed with the anatomical differentiation of man with respect to the other primates. Even so, he admitted in a qualified manner the mechanisms of natural selection and the struggle for life in the animal kingdom, but always made an exception for man by invoking an idea of mutual protection to oppose the notions of selection and survival.

Among all the anti-Darwinians in the Sociedad Antropológica, the most scientifically enlightened was Juan Vilanova y Piera. Professor of Geology and Paleontology at the Universidad Central, he was well-versed in Darwinian theory and tried always to harmonize scientific findings with religion. He opposed evolutionary ides from his very first articles in 1866 in the *Revista de Sanidad Militar y General de Ciencias Médicas* up to his compendious article, "Origen, antigüedad y naturaleza del hombre," which appeared in the *Revista de Antropología* in 1874.[29] Vilanova disputed the simple explanation of the appearance of life insinuated by Darwin because this led, according to Vilanova, directly to admitting spontaneous generation, an idea by then abandoned by all the principal European thinkers. In addition, Darwin's formulation would cast doubt on the slow, successive transformations favored by the Darwinians and advanced by Lamarck. Vilanova likewise declared himself a believer in the principle of the fixity of species, limiting biological variability to the narrow limits of races and varieties for whose origin he felt free to use such Darwinian concepts as the struggle for life and natural selection, without however daring to accept any change in species. Instead, he clung to his religious beliefs and the lack of paleontological data.

Even so, Vilanova always maintained a position intermediary between those whom he considered evolutionary materialists and the religious integrists. One can see this in

[27] Carlos María Ferrer, "Refutación de algunas utopías," *El Anfiteatro Anatómico Español*, 1 (1873), 230, 242-243.

[28] "De la unidad nativa del género humano, o del parentesco por consanguinidad universal entre todas las razas de la especie humana, diseminadas por todas las regiones de la tierra,"*Revista de Antropología*, 1 (1874), 9-17, 81-95, 161-170.

[29] J. Vilanova y Piera, "Origen del hombre," *Revista de Sanidad Militar y General de Ciencias Médicas*, 3 (1866), 678-681; "Antigüedad de la especie humana," *ibid*., 3 and 4 (1866-67) (thirteen articles); "Origen y antigüedad del hombre," *Boletín Revista de la Universidad de Madrid*, 1 (1869), 233-247, 449-462, 641-663; *Origen, naturaleza y antigüedad del hombre* (Madrid, Compañía de Impresores y Libreros del Reino, 1872); "Origen, antigüedad y naturaleza del hombre," *Revista de Antropología*, 1 (1874), 53-64, 125-136, 185-203.

his debate with Manuel de la Revilla in the *Revista Europea* (1874-1880).[30] Vilanova also discussed "Prehistoric Times" at a January 17, 1882 lecture held at the Ateneo Científico y Literario of Madrid as part of a course on Universal History given jointly with José Moreno Nieto and Eduardo Saavedra. In his presentation, Vilanova dismissed the link that some established between prehistoric science and evolutionism. In his opinion, when one applied the strictest Positivist standards (and as such was circumscribed by the limits of observation), the human remains found up till then belonged to the human species and never to the supposed species intermediary between man and the greater anthropoids as the evolutionists claimed. He criticized the invention of Haeckel's *alalus* or "voiceless man" as fantastic, but also reproached ultra-orthodox Catholics for failing to recognize the antiquity of man and his works, so clearly demonstrated by the discoveries of Boucher de Perthes.[31]

Still another anti-Darwinian physician was E. Moreno Caballero, a Valencian collaborator of Velasco in the *Anfiteatro Anatómico Español*, who wrote a series of articles between 1877 and 1880 titled "Genealogía del hombre." This period marked the high point of the Darwin polemic in Valencia, where in 1878 Darwin was defended in debates at the Ateneo Científico by such eminent figures as Amalio Gimeno, professor of Medicine also associated with Velasco, Celso Arévalo, professor of Natural History, and the psychiatrist José María Escuder.

Moreno Caballero read Darwin's texts in their first Spanish translations: *El origen del hombre* (Barcelona, 1879) and *El origen de las especies* (Madrid, 1877). He began his critique of Darwinism with an attempt to demonstrate the falsehood of Malthus's thesis based on mathematical calculations of population growth. Moreno Caballero concluded that the law of subsistence was regulated exclusively by the division of labor which always brought with it an excess of production, so that the struggle for existence was not necessary. He contrasted as well the concepts of life and transformation, ridiculing Haeckel´s Monera, and criticized the different definitions of natural selection given by Darwin. Moreover, he considered the hypothesis of the survival of the fittest a truism.[32]

Andrés del Busto advanced a curious attempt to harmonize opposing positions. Del Busto was a professor of Medicine in Madrid, a founding member of the Anthropological Society, an active member of the Histological Society and a tireless advocate of experimental medicine in reformist journals like *España Médica, Iberia Médica,* and *El Anfiteatro Anatómico Español.* He presented his idiosyncratic approach in a speech entitled "El código de la naturaleza. Estudios acerca de las leyes de la materia y de la vida," delivered on the occasion of his acceptance into the *Academia de Medicina* in 1877. In it, he attempted to adapt the evolutionary hypothesis to the idea of

---

[30] L. A. Sánchez Gómez, "La antropología española del úlitmo tercio del siglo XIX a través de las revistas culturales de la época," *Revista de Dialectología y Tradiciones Populares*, 41 (1986), 211-236.

[31] *Curso de Historia Universal. Conferencia pronunciadas por los señores don José Moreno Nieto, don Juan Vilanova y don Eduardo Saavedra durante el curso de 1882* (Madrid, Ateneo Científico y Literario de Madrid, , 1883).

[32] E. Moreno Caballero, "Genealogía del hombre," *El Anfiteatro Anatómico Español*, 5 (1877), 336-337; 6 (1878), 147-148, 231-232; 8 (1880) , 34-36.

an initial divine creation, denying any continuity between inorganic and organic matter but accepting the evolution of species from proto-organisms:

> Nature does nothing by leaps. There are spaces so narrow between flora and fauna that differences can scarcely be established. There are organisms with such ambiguous characters, that much work is required to include them with surety in one kingdom or the other. It is as if they passed timidly through the gates of life without knowing where to settle down. Those were the true grandparents of species, as they left the pristine shadows, whose first germs were created by divine will from the dust of rocks or from the slime of the seas. From there they came, perpetuating their species . . . . These beings are the beginning of animal life, whose ladder is extended today to around 300,000 species, the least numerous being the most complex or perfect. . . . The sea is an immense battlefield where organisms appear by the million, and by the million they perish in the incessant and progressive war of the species.[33]

Rafael Ariza Espejo was one of the purest representatives of Darwinism in the Sociedad Antropológica Española. This Sevillian physician was the founder of otorhinolarygology in Spain and a collaborator of Pedro González de Velasco in the Escuela Libre de Medicina,[34] along with other doctors like Carlos María Cortezo, Federico Rubio, and Luis Simarro. These had already proclaimed their support of Positivism in the Ateneo de Madrid along with the Cuban journalist José Perojo, one of the principal promoters of the spread of evolutionary thought in Spain through the *Revista Contemporánea* (1875-1907), which he edited. Simarro also worked with Ariza in the Sociedad Histológica Española, a point of entry for the evolutionist approach in Spanish histology, a process which would later bear fruit in the work of Cajal and the Spanish neurohistological school.[35]

In 1874, Ariza published an article in the *Revista de Antropología* entitled "Diferencias especificas de las razas humanas" in which he accepted Darwinian evolutionism. According to him, it invalidated classic definitions of species and introduced broad new notions that were useful to science. In this sense, Ariza accepts without discussion the struggle for existence, which he relates to the concept of adaptation, as well as the idea of useful changes and the Darwinian mechanism of natural selection. He also considered probable the existence of intermediate species from which man and the anthropoids could have arisen, his only hesitation being the practical difficulty of differentiating species from varieties.[36]

Francisco María Tubino y Oliva, general secretary of the Sociedad Antropológica and a pioneer Spanish prehistorian, was perhaps the most radical defender of

---

[33] A. del Busto, *El código de la naturaleza. Estudios acerca de las leyes de la materia y de la vida.* (Discursos pronunciados en la Real Academia para la recepción pública del académico electo Dr. D. Andrés del Busto, el día 3 de junio de 1877) (Madrid, Rojas, 1877).

[34] *Escritos médicos del doctor don Rafael Ariza y Espejo*, 3 vols. (Madrid, E. Teodoro, 1888).

[35] J. J. Campos and R. Llavona, eds., "Los orígenes de la psicología científica en España: el doctor Simarro," *Investigaciones Psicológicas*, 4 (Madrid, Universidad Complutense, 1987).

[36] R. Ariza, "Diferencias específicas de las razas humanas," *Revista de Antropología*, 1 (1874), 18-31, 96-109, 171-184.

evolutionism at this time. Tubino had gained notoriety through his participation in the debates on Positivism in the Ateneo as well as through his evolutionist articles in the *Revista Contemporánea*, joined by fellow Darwinians like Perojo and Pedro Estasén, and in the *Revista de España* (1868-1894), in the company of evolutionists such as Eduardo Echegaray and Enrique Serrano Fatigati. One Tubino's clearest expositions is his article "Darwin y Haeckel. Antecedentes de la teoria de Darwin" [37] There he explains the ideas of Maillet, Robinet, Buffon, Lamarck, Geoffroy Saint Hilaire, Bory de Saint Vicent, etc. before reaching Darwin and Haeckel. Naturally, he represents Darwin one of the most eminent scientists of the age, criticizing him only for his refusal to apply his theory to man in the *Origin of Species*.[38] Haeckel, on the other hand, struck Tubino as more consistent in his evolutionary doctrine. Indeed, it was Tubino who proposed Haeckel as an honorary member of the Sociedad Antropológica Española in March 1874, thus establishing himself as one of Haeckel´s strongest defenders in Spain along with the Valencian anatomy professor, Peregrín Casanova. Tubino, along with Joaquín Hysern, also stimulated the creation in Havana in 1877 of the Sociedad Antropológica de la Isla de Cuba, established as a correspondent of the Spanish society in Madrid. Tubino worked through the latter's Havana members Luis H. Delmas, Juan Santos Fernández, and Gabriel Pichardo. Without entering into details which have already been studied by Pedro M. Pruna and Armando García, it should be noted that the Havana institution served as the point of entry for evolutionism in Cuba.[39]

The Anthropological Society also enjoyed a close relation with the Canary Islands through the anthropologist Gregorio Chil y Naranjo, who was also in contact with the Museo Antropológico founded by Pedro González de Velasco in 1875. Chil was a member of the Societé d´Anthropologie de Paris and the main promoter of the Museo Canario, a Canarian anthropological and historical museum inaugurated in 1880. Co-founder with Juan Bethencourt of the Gabinete Científico de Santa Cruz de Tenerife (1877), Chil introduced Darwinism in the Canary Islands. His most important work were the *Estudios históricos, climatológicos y patológicos de las Islas Canarias*, published in Las Palmas in three volumes between 1876 and 1891. In these, following of Lyell, Darwin, Haeckel, Huxley and Broca, Chil tried to give an evolutionary explanation for the origin of the islands' aborigines and even their geology. This earned him the public condemnation of the Catholic Church and ecclesiastical censorship of his works.[40]

---

[37] *Revista de Antrología*, 1 (1874), pp. 238-256, 356-385, 401-428, 481-496.

[38] F. María Tubino, "Recientes publicaciones sobre la ciencia prehistórica," *Boletín Revista de la Universidad de Madrid*, 2 (1870), 956, 1058-68.

[39] Pedro M. Pruna and Armando García, *Darwinismo y Sociedad en Cuba. Siglo XIX*, (Madrid, CSIC, 1989). See also M. Rivero de la Calle, ed., *Actas de la Sociedad Antropológica de la Isla de Cuba*, (Havana, Comisión Cubana de la UNESCO, 1966), as well as M. A. Puig-Samper and F. Pelayo, "Darwin en Cuba. El Transformismo en la Revista de Cuba," *Revista de Indias*, 49 (1989), 423-435.

[40] J. Bosch Millares, *Don Gregorio Chil y Naranjo. Su vida y su obra* (Las Palmas, Cabildo Insular de Gran Canaria, 1971); F. Estévez, *Indigenismo, raza y evolución. El pensamiento antropológico canario, 1750-1900* (Santa Cruz de Tenerife, Aula de Cultura, 1987).

DARWINISM IN THE ANTHROPOLOGICAL SOCIETY OF SEVILLE

Antonio Machado y Núñez, physician from Cadiz, was the central figure in the introduction of Darwinism in Andalusia. Machado y Núñez was another of the "Men of 1868." His political activities during the six-year period revolutionary period included becoming rector of the University of Seville, having come there to occupy the Chair of Natural History, and mayor of the city. According to his biographers, he was one of earliest expositors of Darwin's theories in Spain, precisely through his university position in Seville. In 1869, he joined with the Krausist Federico de Castro in founding the *Academia Médico-Quirúrgica Matritense*, the principal organ of the diffusion of German idealist philosophy, Positivism, and evolutionism in Andalusia.[41] In one of the journal's first articles entitled "Apuntes sobre la teoría de Darwin" (1871), he recommended to his readers the immortal work on the origin of the species. After commenting positively on the concepts of natural selection, adaptation, and competition for survival, he concluded: "Darwin's doctrines have won such great acceptance among scholars, that according to one of his most impartial critics, his book on the origin of species has caused a revolution in Biology, as transcendent as that established in Astronomy by Newton's *Principia*."[42]

In the same year, Machado, once again with Federico de Castro, founded the Anthropological Society of Seville (1871-1873) with three sections: 1) Physical Anthropology directed by Machado himself; 2) "Psychic" Anthropology, where the Krausists concentrated under Castro's aegis; and 3) Social Anthropology with Antonio Machado Alvarez, son of Machado Núñez and father of the poets Antonio and Manuel, at the head.[43]

Machado y Núñez continued to promote Darwinism in his journal, with articles like "Teoría de Darwin. Combate por la existencia" (1872); "Teoría de Darwin. La selección natural" (1872); "Darwinismo. La edad de la Tierra" (1872), as well as commentaries on Haeckel's *Historia de la creación de los seres organizados según las leyes naturales* (1874) and Herbert Spencer's article "De la creación y de la evolución" (1874). Likewise Machado edited the translation of Haeckel's work, *El monismo como nexo entre la religión y la ciencia* (Madrid, 1893), making himself one of the German biologist's main defenders in Spain.

One of Machado y Núñez's most important colleagues in Seville was his own son, Antonio Machado y Alvarez, who introduced Folklore studies in Spain. He was a

---

[41] Antonio Machado y Núñez, *Páginas escogidas* (Seville, Ayuntamiento, 1989). Preliminary study by Encarnación Aguilar.

[42] A. Machado y Núñez, "Apuntes sobre la teoría de Darwin," *Revista Mensual de Filosofía, Literatura Ciencias de Sevilla*, 3 (1871), 461-470.

[43] M. Méndez Bejarano, "Sobre la Sociedad Antropológica de Sevilla," *Actas y Memorias de la Sociedad de Antropología, Etnografía y Prehistoria*, 7 (1928), 22-23; I. Moreno Navarro, "La antropología en Andalucía. Desarrollo histórico y estado actual de las investigaciones," *Ethnica*, 1 (1971), 109-144.

Krausist, a student of Federico de Castro with whom he actively collaborated in the journal *La Enciclopedia* and in the activities of the recently created Ateneo de Sevilla (1879).[44] In 1884, he wrote a prologue to a translation of a book by Hugo Magnus, professor of ophthalmology at the University of Breslau, entitled *Historia de la Evolución del sentido de los colores* (Madrid, Francisco Alvarez). The translation was published in the collection "Biblioteca Biológica" directed by Romualdo González Fragoso, who had already published Lanessan's work, *La lucha por la existencia y la asociación para la lucha*, with a prologue by Machado y Núñez. Machado y Alvarez considered himself a folklorist. As such, he felt somewhat distanced from the study of natural sciences: "I have unfortunately not been close to those fields cultivated by the men whom I esteem as the greatest in human history: Aristotle in ancient times and Darwin in the present." In his prologue to Magnus, Machado y Alvarez evaluated the search for historical proof in ancient civilizations for the possible evolution of the sense of color, an investigation limited in scope, as he admitted such a process, like all evolutionary ones, must have necessarily been slow. He concluded: "The sense of vision and of colors, like that of smell, for example, have required, in my judgment, many centuries to have passed from what reasonably must have been its pristine state that of perfection it now displays, as a result of natural selection." In similar fashion, Machado y Alvarez's translation of E. B. Tylor, *Antropología. Introducción al estudio del hombre y de la civilización* (Madrid, El Progreso, 1888) bears witness to the introduction of the evolutionist perspective in social anthropology.

The most prominent sociologist in the Andalusian evolutionist group was Manuel Sales y Ferré, who was born in Catalonia but moved to Seville to assume the Chair of Geography and History in 1874. That same year, he wrote the prologue to the translation of Quatrefages's book, *Historia natural del hombre*. His gradual evolution from an orthodox Krausist position towards Positivism led to a confrontation with Federico de Castro in the Ateneo de Sevilla. In 1879, he translated Hartmann's book, *La verdad y el error en el darwinismo*, and in the two following years, published in the Biblioteca Científico-Literaria, a collection which he created, his own books, *Prehistoria y origen de la civilización* (1880) and *El hombre primitivo y las tradiciones orientales. La Ciencia y la Religión* (1881). In both books, Sales y Ferré conterposed evolutionary explanation of human social development with that based on creationist principles.[45]

DARWINISM IN THE ANTHROPOLOGY SECTION OF THE MUSEO DE CIENCIAS NATURALES

After the death of Pedro González de Velasco in 1882, the locus of anthropological studies in Madrid passed to the Museo de Ciencias Naturales, where an anthropology

---

[44] E. Aguilar Criado, *Cultura popular y folklore en Andalucía. Los orígines de la Antropología*, (Seville, Diputación Provincial, 1990); R. Chabrán, "Antonio Machado Núñez and Antonio Machado Alvarez: A Family of Progressive Nineteenth Century Intellectuals," *Asclepio*, 36 (1984), 305-324.

[45] M. Núñez Encabo, *Manuel Sales y Ferré: los orígenes de la sociología en España* (Madrid, Edicusa, 1976); R. Pérez Jerez Mir, *La introducción de la sociología en España. Manuel Sales y Ferré: una experiencia frustrada* (Madrid, Ayuso, 1980).

section was created the following year. The section was directed by Manuel Antón Ferrándiz who had previously served as the museum's zoology assistant, thanks to the influence of his teacher Lucas de Tornos, a malacologist who had also participated in the founding of the Sociedad Antropológica. Antón Ferrándiz soon became the leading exponent of biological anthropology in the last third of the nineteenth century and of the early twentieth. He worked in the orbit of French anthropology, the result of his stay in Paris with Quatrefages and Vernau, who recommended that he study the Cro-Magnon race in Spain.

In 1885, Antón inaugurated an Anthropology chair in the museum. It was held as often by professional anthropologists like Federico Olóriz and Rafael Salillas as by Antón's own students, Telésforo de Aranzadi y Unamuno, Luis de Hoyos Sáinz, Domingo Sánchez, and Francisco de las Barras de Aragón, a group that would later form the leadership core of Spanish anthropology. The museum also had an anthropology laboratory, whose first years of activity were later summarized by Luis de Hoyos Sáinz:

> From the Museum's laboratory came the works of the chair-holder Aranzadi, such as *El pueblo Eskalduna* (1887), the first study of anthropometry published in Spain, along with my own *Notas para un estudio antropológico sobre el crecimiento* and *Técnica antropológica* (1893), the first general work written in Spanish; *Un avance a la Antropología de España* (1892, Hoyos y Aranzadi), in which we established the first division into regions according to cephalic indices, nasal and cranial modulation We were likewise the authors of *Lecciones de Antropología* (first edition, 1893), in which some early data on the distribution of height and eye color in Spain appear.[46]

After Antón's active participation in courses at the Madrid Ateneo, he secured the creation in 1892 of Spain's first official chair of anthropology, at the Universidad Central. We know the nature of the first classes thanks to the course notes of one of his disciples, a medical student.[47] The classes were entitled "Conferencias de Antropología" and in them Antón reveals his views on evolution. Antón was convinced that Darwin's theory had caused a greater revolution in the natural sciences than in any other although, following the French school, he preferred the term "transformism" and was partial to Lamarckian views. With regards to the concept of evolutionary *progress*, Antón Ferrándiz did not share the ideas of many of his contemporaries. While agreeing that in general there was an advance towards perfection in biological species as well as civilizations, he also held that natural selection and adaptation to the environment introduced an uncertainty factor that could give rise as readily to "progressions" as to "retrocessions." In similar fashion and in contrast to the paleontological arguments brandished by some anti-Darwinists—like his friend Vilanova—Antón commented in his *Conferencias* that the human remains found until then represented a sufficient

---

[46] L. de Hoyos Sáinz, "Notas para la historia de las ciencias antropológicas en España," *Asociación Española para el Progreso de las Ciencias. Congreso de Granada*, 1912, V: 76-77.

[47] The Museo Nacional de Antropología preserves a copy with the reference number R. 1892.

sample to confirm the phenomenon of species transformation, even if they did not yet fill out a complete evolutionary scale.

With regard to the origin of life in the form of primitive protoplasm, he inclined towards Haeckel's arguments about the existence of primitive Monera which arose from inert matter by physio-chemical processes. Antón leaned this way despite his familiarity with Pasteur's experiments disproving spontaneous generation, because he felt that the experimental techniques of the new discipline of chemical biology were still too rudimentary to permit a hard conclusion. He also favored monogenism as a logical corollary of Darwinian evolutionism, supported by the experiments of Darwin himself and of other naturalists like Quatrefages, who preferred physiological criteria for the definition of species.

As for man's possible simian origin, Antón asserted that Darwin had been excessively prudent. Haeckel had given overwhelming proof of this fact, especially by applying his biogenetic law to the human species. Embryonic development demonstrated the kinship between man and other primates, as did apparently useless organic survivals, like body hair, that were nothing but relicts from our evolutionary past.[48] With regard to mental evolution, Antón Ferrándiz admitted the intelligence of animals without reservation, basing himself on Darwin's work and that of George John Romanes, which he had translated into Spanish in 1886.[49] He proclaimed the quantitative evolution of animal intelligence up to the level of Haeckelian troglodytes incapable of articulating language. Religious prudence restrained him from specifying how the leap to *Homo Sapiens* had come about; he followed Huxley in claiming there was not sufficient evidence in this area. Consequently, Antón affirmed the compatibility of evolutionist ideas with religious beliefs inasmuch as "Transformism theory is the only one that explains both human and animal origins in a scientific and rational way."

The evolutionist ideas adumbrated in these first classes of Antón were shortly thereafter embodied in his *Programa razonado de Antropología* (1897) and were developed, with an impressive critical apparatus, in his *Antropología o Historia Natural del Hombre* (1912). This textbook would mark the fundamental lines of Spanish anthropology in the first third of the twentieth century, along with the *Lecciones de Antropología* (1899-1900) by his disciples Hoyos Sáinz and Aranzadi and *Notas para un curso de Antropología* by Barras de Aragón, his successor in the chair at the Central University.[50] Antón was a creator of institutions, establishing the *Museo de Antropología, Etnografía y Prehistoria* in 1910, followed in 1929 by the foundation of

---

[48] Manuel Antón would develop this theme in his *Los orígenes de la Hominación* (Madrid, Real Academia de la Historia, 1917).

[49] Gonzalo Jiménez de la Espada translated another work by G. J. Romanes, *La evolución mental del hombre*, in 1906 (Madrid, Daniel Jorro). The translator was the son of the famous naturalist who had particpated in the Comisión Científica del Pacífico.

[50] C. Ortiz, *Historia del pensamiento antropológico en España: Luis de Hoyos Sáinz*, unpub. doctoral diss., Madrid, Universidad Complutense, 1988. The only doubt that emerges regarding the line that Manuel Antón took with respect to Darwinian evolutionism appears in Aranzadi's notes to the translation of E. Frizzi, *Antropología*, 2nd ed. (Barcelona, Labor, 1928), in which he lightly qualifies the idea of natural selection following T. H. Morgan. Morgan´s work, *Evolución y Mendelismo*, had been recently translated by Antonio de Zulueta (Madrid, Calpe, 1921). Regarding the Morgan and the crisis of Darwinism, see Peter J. Bowler, *El eclipse del darwinismo* (Barcelona, Labor, 1985).

the Spanish Society of Anthropology, Ethnography and Prehistory. He was joined in the latter project by Rafael Salillas and his own students Hoyos Sáinz, Barras de Aragón, and Domingo Sánchez. Another of his students, Telesforo de Aranzadi, who held the anthropology chair in Barcelona, participated in the creation of the Associació Catalana d'Antropologia, Etnografia i Prehistória.[51]

Centro de Estudios Históricos, Madrid

[51] C. Ortiz, *Luis de Hoyos Sáinz y la antropología española* (Madrid, CSIC, 1987); M. A. Puig-Samper and A. Galera, *La antropología española del siglo XIX* (Madrid, CSIC, 1983); E. Ronzón, *Antropología y antropologías* (Oviedo, Pentalfa, 1991); L. A. Sánchez Gómez, "Sociedad Española de Antropología, Etnografía y Prehistoria (1921-1951)," *Revista de Dialectología y Tradiciones Populares*, 45 (1990), 61-87; L. Calvo, *La Antropología en Cataluña* (Barcelona, Universidad de Barcelona, 1989); L. Pericot, "Un episodio en la historia de la etnología en España. L'Associació Catalana d'Antropologia, Etnologia i Prehistória," *Revista de la Universidad Complutense de Madrid*, 98 (1975), 15-21; A. Goicoetxea, *Telesforo de Aranzadi. Vida y obra* (San Sebastián, Sociedad de Ciencias Aranzadi, 1985).

LAURA SUÁREZ Y LOPEZ-GUAZO

THE MEXICAN EUGENICS SOCIETY

Racial Selection and Improvement

INTRODUCTION

Based on family histories, i.e. heredity, and the use of multiple statistical methods, in 1865 Francis Galton defined eugenics as "the science dealing with the influences that improve inborn qualities or the raw matter of a race, and the influences that can be developed to attain the highest superiority."[1]

> Before correctly appraising the direction required to perfect each of the races, our minds must be freed from many prejudices. [...] The moral and intellectual wealth of a nation consists, to a great extent, in the multiple variety of talents of the men who comprise it. The assimilation of all its members into a common type would therefore constitute a retrogression in the process of perfection.[2]

Galton's conception on the heredity of talent was taken into account for the institution of public health policies in several European countries and America. It was also closely related to racism and the supposed racial degeneration of the lower classes, both fully established ideologies in the last decades of the nineteenth century and beginning of the twentieth.

The ideology of racial improvement was legitimized and strengthened upon scientific grounds in genetics and evolutionism. It was considered that the direction offered by both fields would lead to the progress or decadence of nations, while promoting the "scientific" interpretation of the "natural" causes behind social stratification.

[1] Lecture given by Francis Galton in May 1904 to the Sociological Society, School of Economical and Political Science, London University. In Francis Galton, *Herencia y eugenesia*, Spanish translation, introduction and annotations by Raquel Alvarez Peláez (Madrid, Alianza Editorial, 1988), p. 165.

[2] F. Galton, "Inquiries into Human Faculty and its Development" (1883) in Galton, *Herencia y eugenesia*, pp. 86-87.

Thomas F. Glick, Rosaura Ruiz, and Miguel Ángel Puig-Samper (eds.), The Reception of Darwinism in the Iberian World, 143-151.

HEALTH PROBLEMS IN POST-REVOLUTIONARY MEXICO

In the first decade of this century, the Mexican positivist oligarchy, known by the nickname of the "científicos" (scientists), was the driving force for the continuity of Porfirio Díaz's dictatorship as a guarantee of their privileged social position. This is why they declared that:

> ...whites, or prominent whitened individuals like *mestizo* Porfirio Díaz, were fitter... Since Indians would perish in the struggle for life, European settlers afforded the best means for Mexico's progress.[3]

Emilio Rabasa used evolutionist theses to scientifically justify and legitimize the concentration of wealth and power, and the existence of the thousands of unemployed, uneducated Indians who had been dispossessed of their lands.[4] Rabasa held that it was not possible for all Mexicans to have access to education, since the population was divided between those who were "able" and got the best positions in society, i.e. those whom the State should protect, instruct and to whom it should direct its interests, and those "incapable" of learning, i.e. the Indians. He considered that resources should not be indiscriminately allotted by the State for the education of all. In 1921 he wrote:

> Scientific notions taught in schools are useless to Indians, who are still isolated in their environment. Firstly because they do not understand them, and secondly because they do not apply them in their work nor are they useful in their daily relations.[5]

For Justo Sierra, Minister of Public Instruction at the time, the biological traits of each individual determined his position in society. According to him, "laws must ensure the natural evolutionary trends of society."[6]

In the face of health problems, particularly the high rate of mortality among children under three years of age, an aspect that became more serious after the Mexican Revolution (1910), the State was permanently concerned with legislating marital relations to avoid perpetuating physical illnesses such as syphilis, tuberculosis and other diseases which were considered infectious or hereditary. In 1917, the president, Venustiano Carranza, thus proposed to the Constituent Congress the approval of the Law of Family Relations, in light of the scant coverage of this area in the Civil Code. Several Darwinian concepts were enunciated in the presentation of this Law, wherein a series of social factors expressed the urgency and convenience of reforming family legislation:

> ...based on the more rational and righteous grounds that may elevate spouses in the supreme mission they have been endowed with by society and nature for the propagation of the species and the foundation of the family...
> ...it thus becomes indispensable to regulate marriage in such a way that the interests of the species and of the spouses themselves will be guaranteed...

---

[3] For the ideology of the "*científicos*," see L. Zea, *El positivismo en México: Nacimiento, apogeo y decadencia* (Mexico, F.C.E., 1968) quoted by R. Moreno, *Ensayos de historia de la ciencia y tecnología en México* (Mexico City, UNAM, 1986), p. 148.

[4] R. Ruiz, *El positivismo Mexicano: Introducción del Darwinismo en México* (Mexico City, UNAM, 1987), pp. 147-148.

[5] E. Rabasa, *La Evolución Histórica de México* (Mexico City, Frente Cultural, 1921), p. 326.

[6] J. Sierra, *Obras Completas: Programa de la libertad* (Mexico City, 1948), IV: 238.

> ...It is also likewise necessary, in the interest of the species, to increase the required marital age so spouses may be sufficiently capable of fulfilling the physiological and moral function entrusted to them. For the same reason it is also convenient to legally incapacitate those disabled by nature for marital functions, i.e. those suffering from incurable physical impotence, syphilis, tuberculosis or any other chronic and incurable conditions, which are also infectious or hereditary, as well as alcoholics, since anyone in the conditions cited can transmit pathological traits to their descendants, making them weaker and less able to work efficiently, both physically and intellectually, and in turn transmitting their weakness to future generations. All this is detrimental to our Nation, whose vigor depends on the strength of its children, and is likewise detrimental to the species itself, whose perfection requires a sane and judicious artificial selection added to natural selection so its rigors may be deflected and mitigated.[7]

NATIONALISM AND PROMOTION OF EUGENICS

The anticlerical revolutionary State made post-revolutionary Mexico increasingly receptive to new developments in both scientific and social domains. Some of the first Mexican publications supporting the importance of "racial improvement" are found in the *Gaceta Médica de México*, publication of the National Academy of Medicine in the last quarter of the 19th century. An article published in 1896 by Porfirio Parra cites the disadvantages of blood relations in the Mexican race, so common in our country, especially among the middle and higher classes due to historic and geographical reasons, particularly in rural areas.[8]

After 1900, the medical community began to express concern about public health problems related to hereditary factors, specifically the transmission of diseases such as syphilis, tuberculosis, epilepsy, as well as pre-conception conditions of women, such as chronic alcoholism.[9]

In 1910 Francisco Hernández published a brochure titled *Higiene de la especie: Breves consideraciones sobre la estirpicultura humana* (Hygiene of the Species: Brief considerations on human stirpiculture). The next year, the first article on the use of eugenics for racial improvement was published, based on the "feminist" statements of eugenicist Caleb Saleeby, where eugenics is described as a means of protection for women against venereal diseases and other conditions related to reproductive health. In December 1911, Blanche Z. de Baralt published a broad review of Saleeby's book, *Feminismo eugénico* (Feminist Eugenics).[10] According to Alfredo Saavedra, permanent president of the Mexican Society of Eugenics, this was the first time that some

---

[7] *Ley sobre relaciones familiares* (April 14, 1917) (Mexico City, Andrade, n.d.), pp. 1-4.

[8] Porfirio Parra, "La unión carnal entre consanguíneos puede por sí misma producir seres degenerados y predispuestos a muchas y diversas enfermedades," *Gaceta Médica de México*, 34 (1896), 544.

[9] In 1900, the *Gaceta Médica de México* published the article "Heredabilidad y Casualidad" by Dr. F. Smeleder. In 1907, another article was published on the transmission of diseases such as tuberculosis and syphilis. In the November 1909 session of the National Academy of Medicine, Dr. Joaquín Cosío presented a lecture on "The Heredity of Family Diseases," published in the *Gaceta Médica de México*, Jan. 10, 1910.

[10] Blanche Z. de Baralt, "El Feminismo eugénico," *El Diario* (Mexico City), Dec. 24, 1911.

statements were openly disseminated in the Mexican press.[11] Some of the most important principles mentioned in this article indicated:

> Women must consider themselves the main agents for the continuity and evolution of the race towards a higher physical, intellectual and spiritual level;... Education of girls and young women must prepare them for this great mission. Upon reaching marital age they must have such an elevated and clear notion about it that they should refuse to wed men with inferior physical, intellectual and moral conditions.
>
> ...Natural selection would not be wholly incompatible with love if these tremendous issues were studied and better understood throughout the whole world.
>
> The improvement of race dreamed by philosophers and preached by biologists would not be a monstrous violation of affection, if we become accustomed to edifying our love on solid moral and religious foundations.[12]

EUGENICS IN MEXICO

Unlike Argentina, where eugenics was constituted around social problems derived from immigration, Mexican society was mainly constituted by Indians and *mestizos*. Long-standing debates about the lack of real integration of Indians to national life and the problem of guaranteeing health for the poor led to ideas of "racial improvement," converging with eugenic doctrine.

In 1921, the First Mexican Congress of Childhood promoted eugenics, heredity and reproductive guidance for racial improvement. Félix Palavicini, director at the time of the daily *El Universal*, Constituent Deputy, ex-Minister of Public Instruction, vice-president of the Mexican Society of Eugenics in 1932, and president of the Athenaeum of Sciences and Arts of Mexico in 1934, organized and presided over the Childhood Congress where the eugenic sterilization of criminals was proposed. Although the measure was approved by a narrow margin of 7 votes, it had no practical nor legal consequences.

In 1929 the Mexican Society of Puericulture was founded in Mexico City with a special section for eugenics devoted to heredity, reproduction-related diseases, infantile sexuality, sex education and birth control. The future founders of the Mexican Society of Eugenics emerged from the former organization. The Mexican Society of Eugenics (MSE) was founded on September 21, 1931 with 130 scientific, political and medical members, closely associated with the group in power and public health authorities.[13] Some of its members, renowned for their studies on human genetics, included Fernando Ocaranza, pediatrician and physiologist, head of the Medicine School of the National University from 1925 to 1933 and José Rulfo, veterinarian, member of the National Academy of Medicine from 1931, and professor of physiology and pathology at the

[11] Alfredo Saavedra, permanent president of the Mexican Society of Eugenics, considered that this was the first article published in Mexico with important social impact: "Lo eugénico enunciado por primera vez en México," *Acción Médica*, no. 199 (September 1956), 16-17.

[12] Baralt, "El Feminismo eugénico" (note 10, above).

[13] José Rulfo, "Ponencia de Eugenesia" (Lecture on Eugenics), presented to the First National Congress on Internal Medicine, Mexico City; *Eugenesia* (publication of the Mexican Society of Eugenics), III, No. 31 (May 1942), 12.

University of California at Berkeley.[14] They introduced experimental methodology and Mendelian genetics in higher learning in Mexico during the 1930s.

The predominant notion in Mexico was that the middle and upper classes controlled their reproduction with the available programs of the time, whereas the "less desirable" lower class did not. This, according to the orthodox eugenicists who had direct influence on public health programs, was the reason behind the "degeneration" of the Mexican race.[15]

In the 1930s, the official ideology was deeply anticlerical, divorce was legalized, and elementary school was laicised. The fanatic anticlerical governor of Veracruz, Adalberto Tejeda, one of the last *caudillos* of the Revolution, promoted both sex education and sterilization. In July 1932 he encouraged the State Congress to pass Law 121, as well as to create the Eugenics and Mental Health Division of the General Health Ministry of the State. Tejada suggested that progress in positivist science and the success of genetics and biology in agriculture and cattle breeding should be applied to the "human race." His concept of responsibility as governor included promoting physical and mental improvement among the citizens of Veracruz by financing the scientific research required to reach this goal. This proved to be the only eugenic law for sterilization enacted in Mexico. A few months after it was passed, the struggle between radicals and conservatives for land ownership found its way into the political arena, and Veracruz's sterilization law was thus forgotten.[16]

The limitation of eugenics and opposition to sterilization were discussed in the Mexican Society of Eugenics as of 1933.[17] The following year, the eugenic objectives of the Nazi sterilization legislation were discussed and strongly criticized in the "Second Eugenics Week" held in Mexico City. Sterilization consequently became anachronistic in Mexico.

In the 1940s, as the Mexican Society of Eugenics grew in strength, many physicians and educators focused their attention on sexual education, combining modernist and conservative attitudes, such as the anti-alcohol, anti-feminism and anti-pornography campaigns. The first birth control and marital health programs were advanced, largely based on the 1932 "Plan for Sex Education and Venereal Disease Prevention" which was compulsory in government-financed schools for all children under 16 years of age.[18]

Eugenic groups focused their attention on the topic of racial consolidation and the adaptability of the Mexican nation. The members of the Mexican Society of Eugenics discussed nationality in terms of race and heterogeneity (Indians, Europeans, *mestizos*). They acknowledged the poverty and marginality endured by Indian groups and also

---

[14] Félix Palavicini, *México: historia de su evolución constructiva*, (Mexico City, Editorial Libro de S.R.L., 1945), IV: 25-26. The influence of French medicine on Mexican physicians is evidenced in this book. Several of the most renowned Mexican physicians of the first quarter of the twentieth century studied in Paris, and a lesser number in Vienna and Berlin.

[15] Agustín Mejia Hernández, "Ley de Eugenesia e higiene mental de Veracruz," *Eugenesia*, 23 (Mar. 10, 1933),.3-6.

[16] Romana Falcón and Soledad García, *La Semilla en el Surco: Adalberto Tejeda y el radicalismo en Veracruz, 1883-1960* (El Colegio de México and Gobierno del Estado de Veracruz, 1986), pp. 269-270.

[17] Fernando Ocaranza, "Límites de la Eugenesia," *Eugenesia*, 2 (Dec. 1933), 27-29.

[18] Alfredo Saavedra, *México en la educación sexual (de 1860 a 1959)* (México, Costa Amic, 1967), p. 34.

shared the revolutionary idea of biological virtues conferred by racial mixture. It is interesting to note that the concept of "hybrid vigor" had not yet been formally advanced in human biology, while its implementation in agriculture and cattle breeding was already a long-standing tradition.

Eliseo Ramírez, member of both the National Academy of Medicine and the MSE, stated that class and racial separation advanced in other countries was in opposition to the Mexican eugenic ideal because, although some combinations could decrease the best qualities from our ancestors, hybridization could also lead to excellent results, when there was affinity between the combined races.[19]

In relation to education and its influence on intellectual development, Adrián Correa,[20] member of the Mexican Society of Puericulture and Mexican Society of Eugenics, held:

> If psychic heredity is good, and the environment where the child lives is intellectually bad, the psychic development of the child is arrested. In contrast, a child with mediocre psychic heredity may enjoy good development if he lives in an intellectual environment.[21]

This position is, of course, far-removed from that held by Galton, who disdained the influence of education and only valued heredity.

In general terms, Mexican eugenicists shared José Vasconcelos' idea of the benefits acquired by the Indians because of their miscegenation with Europeans. Interesting exceptions to this controversy are found in anthropologist Manuel Gamio, for whom Europeans were those who benefited from their mixture with Indians.[22]

Gamio distinguished three groups within the Mexican population: aborigines, groups of European ancestry, and *mestizos*. He stressed the differences in immunity among the Maya and other aborigines from the Mexican lowlands, in contrast with Europeans:

> The physical development of autochthonous groups living in equal or similar social conditions to those characterizing the existence of the population of European ancestry would be incomparably better than these. ... Health programs accordingly developed and applied would have less requirements and be more efficient for the other population groups, who enjoy the inborn advantages derived from adaptation and natural selection.[23]

With respect to the group of European ancestry, Gamio considered its mixture with the aborigines necessary so it would adapt and benefit from the advantages conferred by the latter. In this regard, health policies in Mexico would have to be directed to this particular group "not only for its socio-political convenience, but mainly for the beneficial biological results involved."[24]

---

[19] Eliseo Ramírez, "Discurso," *Eugenesia*, 2 (Nov. 1933), 19-22.

[20] Adrian Correa, "Cómo debe impartirse la educación sexual en nuestro medio," *Revista Mexicana de Puericultura*, 2, no. 17 (Mar. 1932), 237-246.

[21] *Ibid.*, p. 238.

[22] Manuel Gamio, "Algunas consideraciones sobre la salubridad y demografía en México," *Eugenesia*, 3, no. 28 (Feb. 1942), 3-8. This work mentions the great advantages Europeans in Mexico enjoyed as a result of their marked mixture with Indians who, throughout the centuries, have adapted themselves to the climate and geography owing to natural selection (*ibid.*, p. 6).

[23] *Ibid.*

[24] *Ibid.*, p. 8

Rafael Carrillo, head of the Eugenics Section of the Mexican Society of Puericulture, who noted the immunological advantages of Indians and the tallest height of *mestizos* with respect to the Spanish colonizing race."[25] Carrillo considered it essential to study the ethnology of this variety of groups, particularly the physical constants (in anthropometrical terms), and the findings about their pathological features, immunity, evolution, etc. He also criticized the inability of modern ethnographers to establish the number of races in our species, in view of the difficulty in finding "pure" examples to make this determination. He nevertheless asserts, in general terms, that the multiple peoples in our country were comprised by the yellow race, represented by Indians; the white race, constituted by Europeans; and the black race.[26] The existence in Mexico of such a varied racial spectrum, and the eugenic need to learn how it could be directed, led Carrillo to recommend the study of the anthropometrical features that differentiated the races, their biological function and mentality, and their different abilities and moral values, within a broad ethnological framework.

Carrillo wrote that he would not be held back by the hypotheses of so-called heredity theories, such as those of Darwin, Weismann or by speculation over chromosomes, and only consider Mendel's and Galton's laws. In this respect he referred to several examples of breeding among purebred animals and mixtures of black and albino skin to deduce the quality of the two following generations (F_1 and F_2) as evidence of clear general knowledge about the principles of Mendelian heredity.

To apply the principles of Mendelian heredity to eugenics, Carrillo proposed a series of measures:

> The English Royal Commission has demonstrated that the percentage of births among those affected with mental feebleness is twice that of normal individuals. Hence the need for preventive means. There are two kinds of eugenicists: the positive ones and the negative ones. Negative eugenicists exclude the non-adapted classes whose reproduction needs to be prevented, towards their gradual eradication.[27]

According to Rafael Carrillo, three factors have determined the ethnographic composition of the Mexican Republic: immigration, race and heredity. Regarding immigration, he held that the first settlers from the Old Continent failed to fulfil eugenic requirements, because a number of them were undesirable individuals, given their inferior physiological and mental traits with respect to the autochthonous population.

> ...we do not wish to distance ourselves from our [eugenic] ideal..., ...We naturally do not think of selecting superior eugenic individuals; we do not seek a Marañón, a Shaw, a Mussolini, a Hindenburg nor an Edison, yet neither do we accept epileptics, alcoholics, feeble minded or syphilitics. We only wish to inject into Mexican *mestizos* of eugenically selected individuals who, according to the Galton's scale of values, do not ostensibly deviate from the mean.[28]

---

[25] Rafael Carrillo, "Tres problemas mexicanos de eugenesia, etnografía y etnología, herencia e inmigración," *Revista Mexicana de Puericultura*, 3 (Nov. 1932), 1-15.
[26] *Ibid.*, p. 2-3.
[27] *Ibid.*, p. 6.
[28] *Ibid.*, p. 12.

Carrillo thus held that governments should establish offices in ports and borders staffed with personnel familiar with eugenic issues.

> Health authorities should consider the eugenic perspective when foreign tourists cross to the Mexican side if only for a few hours—more than sufficient time for them to sow gonococci or Schauden, or spawn a feeble-minded child.[29]

In spite of the superiority conferred by Carrillo on the white race, he expressed a nationalist position when referring to the Indian population and its combination with different autochthonous groups.

> Should we prevent sexual unions with the individuals of that race, which in its origins attested to a rather advanced civilization, like the ancient Mexicans, Mixtecs, Zapotecs, and Maya, among others?
>
> ...judging the appreciation of particular cases, we have had Indian types that could have ranked well in Galton's scale, with some of them being thus considered superior.[30]

In 1940, Alfredo Saavedra, one of the most important orthodox eugenicists in Mexico, editor of the journal *Eugenesia*, wrote:

> Until the State attempts to deal with the primary cause behind the problem of selected births, combat hereditary factors, venereal diseases, and drug addiction, and isolate the feeble-minded, we will not be able to have a strong and healthy race in the precise sense of the word ... Integral protection of childhood will not be scientifically solved ...each child constitutes a useful investment for society... Mental health is to a great extent a product racial health.[31]

In his work, *La selección de los generadores humanos* ("The Selection of Human Progenitors"),[32] Saavedra suggested that the State should, through health authorities and by relying on scientific criteria, monitor genetic health.

> It seems otherwise intriguing that the criterion for sanitary policies has not as yet attempted to become involved in solving a question closely and basically related to racial health.
>
> ...Two actions must be considered: restrictions to prevent the reproduction of the unfortunate undesirables and to foster the reproduction of the better endowed.
>
> ...It would be so much for the better if the reproduction of positively endowed families were facilitated so that they could replace all the feeble minded that swarm everywhere.[33]

BY WAY OF REFLECTION

Most Mexican eugenicists had medical backgrounds and repeatedly appealed to scientific figures such as Mendel, Galton, De Vries, and Weismann, among other renowned biologists, to form a foundation for their opinions, yet they rarely discussed their ideas. Rather, they liked to state that the factors for the control and modulation of hereditary expression are "so complex" that it was more appropriate not to deal with them. Their notorious lack of knowledge about genetics was still evident in the 1940s; their eugenic propositions were as a whole suffused with environmentalism, a

---

[29] *Ibid.*, p. 14.

[30] *Ibid.*, pp. 14-15.

[31] "Editorial," *Eugenesia*, 1, no. 6 (Apr, 1940), 1-2.

[32] Alfredo Saavedra, "La selección de los generadores humanos," *Eugenesia*, 2, no. 7 (May 1941), 11-14.

[33] *Ibid.*, pp. 12-13.

perspective opposed to Galtonian precepts. Mexican eugenicists fostered all aspects of puericulture, much like eugenics in France, reflecting the tradition among Mexican intellectuals of extolling French culture. Education of future parents in relation to their descendants was especially valued. Marital and birth control problems and the importance of survival reflected a concern about the high rate of infant mortality in Mexico in the 1930s and 40s.

The influence of the Mexican Society of Eugenics on law-making was evident on several fronts, including the 1935 Prenuptial Certificate Act, which mandate pre-marital testing (which had been optional); the April 1940 regulations for the campaign against venereal diseases; the abolition of prostitution regulations; sex education programs; campaigns to disseminate and promote responsibility towards progeny, formally implemented in basic education by the Public Education Ministry; and the Public Health Department's campaigns for the prevention of venereal diseases and the transmission of psychopathological traits.

Mexican Galtonian and orthodox eugenicists reached consensus at the end of the 1950s, with Rabasa's thesis that the protection and benefits granted by the State to the weakest (and thus less fit) individuals was in opposition to natural selection.

An analysis of Mexican eugenics reflects, as in many other spheres of knowledge, the use of evolutionism to scientifically legitimize the dominant ideology. Inheritance of biological traits, such as disease resistance, larger chest capacity, etc., and of psychological traits, such as aggressivity or supposedly inborn criminality and delinquency, have abounded in the human sciences. This confirms that scientific arguments derived from genetics, anthropometry and evolutionary biology have acted as timely excuses for coming to grips with social inequality.

The use of concepts like adaptation, fitness, and struggle for survival, coupled with proposals by influential orthodox eugenicists in the public health domain to implement the careful "artificial selection of human progenitors" based on family histories, reflects an attempt to limit reproduction and perpetuate "socially desirable traits."

In spite of multiple declarations of intent, particularly during President Cárdenas' term when health and education programs for Indians were prioritized, the participation of the Mexican Society of Eugenics in these programs was virtually nonexistent. Mexican eugenics once again reveals that biology has aided medicine in the ideological arena to rationalize and sanction the exercise of power, concealing its social significance.

It is finally necessary to point that eugenics and Galtonian biometry expanded their influence to the fields of anthropology, criminology and sociology, with the application of concepts from the fields of biotypology, psychology and psychiatry The full circle was reached when Mendelian genetics was applied with a similar animus, with several studies on chromosomes undertaken in Mexico in the 70s in prisons and insane asylums. At the same time, different ethnic and social groups (such as peasants, workers, soldiers, and so forth) were studied in order to demonstrate the inborn character of delinquency, criminality, and a great variety of mental alterations, and to promote eugenic ideas based on Lamarckian and Darwinian evolutionist arguments.

National Autonomous University of Mexico

ARMANDO GARCÍA GONZÁLEZ

DARWINISM, EUGENICS AND MENDELISM IN CUBAN BIOLOGICAL EDUCATION: 1900-1959

The development of eugenics—defined towards the end of the nineteenth century by its creator Sir Francis Galton as "the science of lineage improvement" had its scientific foundations in the biological and medical advances of the day, especially with respect to inherited characteristics. Eugenics sought to follow these intertwining findings in medicine and biology to achieve the social and biological advancement of the human species. All of this was connected to the pretensions and interests of different countries in relation to issues of colonization, emigration and immigration, and the problems generated by criminality, prostitution, alcoholism, infant mortality, and the transmission of epidemics and diseases, among others. This meant that any legal system which sought to regulate these questions would have to intervene in the creation of a better population. Toward this end, doctors, biologists, sociologists and lawyers staked out significant roles for themselves in the area of social control.

Eugenics attained much greater prominence in the first half of the twentieth century than it had had in the latter part of the nineteenth, because all of the problems listed above had become more urgent, especially after World War I. In this way, it became quite an important movement, not only in the European nations, but also in the Americas in countries like the United States, Mexico, Brazil, Peru, Argentina, and Cuba. This importance translated into the elaboration of books, articles and scientific works on this matter as well as eugenic congresses, conferences, and workshops. Some prominent examples included the International Eugenics Congresses held in London (1912) and New York (1921 and 1932), the *Congreso Latino de Eugénica* in Paris (1937), the *Conferencias Panamericanas de Eugenesia y Homicultura* in Havana (1927) and Buenos Aires (1934), as well as various national eugenics workshops held

Thomas F. Glick, Rosaura Ruiz, and Miguel Ángel Puig-Samper (eds.), The Reception of Darwinism in the Iberian World, 153-169.

in Brazil in 1929, Spain in 1933, and Peru in 1939 and 1943. A number of sanitary-hygienic campaigns took place as well, producing significant results at times.[1]

Without entering into details already examined in other research projects,[2] it suffices to say that from the scientific point of view eugenics based itself on various Darwinian or evolutionary concepts. These included natural and artificial selection, the struggle for existence, Weismann's theory of germ plasm, and Mendelian laws. Dependence on these concepts helps explain the often-exaggerated view of heredity espoused by eugenicists. For them, the transmission of hereditary characteristics helped explain points like genius and moral conditions as well as discrete defects like mental retardation, epilepsy, criminality, prostitution, and alcoholism. Today, we see these problems as depending much more on socio-economic and environmental conditions than on "harsh and fatal" heredity. Eugenicists did concede some importance to the environment, relying on it to underscore certain aspects of education (general, health) where it was abundantly clear that the population's "degeneration" resulted from social and economic conditions. In some countries heredity was accorded predominance over environment, or even vice versa. However, this was not always the case in all countries or among all eugenicists. It is important to emphasize that, over the course of the twentieth century, many eugenicists came to recognize to an increasingly greater degree the necessity of improving man's environmental conditions, especially the hygienic or sanitary ones, if they wished to obtain a better population or "race."

In the present essay, I look at the teaching of biology to examine the use that Cuban eugenicists made of the theories of Darwin, Mendel, and Weismann. The fields of medicine and sociology produced similar applications as well. In the second half of the nineteenth century, the concern that Cuban doctors and biologists showed for heredity helped to disseminate the evolutionary ideas of Lamarck, Darwin, Wallace, Spencer, Haeckel and, somewhat later, Weismann, whose germ plasm theory enjoyed considerable acceptance by eugenicists in the early decades of the twentieth century.

In 1868, Francisco de Frías y Jacott, Count of Pozos Dulces, first introduced Darwinian ideas in Cuba in a speech made to the Real Academia de Ciencias Médicas, Físicas, y Naturales de la Habana.[3] The actual teaching of Darwinism in the Real Universidad de la Habana began with the university reform of 1880. The naturalists

---

[1] Some of the positive results arising from eugenic measures or the social and scientific atmosphere of the day included the creation of institutions dedicated to the care of pregnant or lactating women. In addition, sanitary campaigns helped increase public awareness through articles, the press, radio, television, and other venues.

[2] See in this respect: Raquel Alvarez, *Sir Francis Galton, padre de la eugenesia* (Madrid, CSIC, 1984; Cuadernos de Galileo, 4); *idem*, *Francis Galton: Herencia y eugenesia* (Madrid, Alianza, 1988); and Raquel Alvarez and Armando García González, *Eugenesia y Homicultura en Cuba* (in press). On the degenerationist theories of Morel and Magnan with which the eugenicists explained some of their proposals, see Rafael Huertas, *Locura y degeneración*, (Madrid, CSIC, 1987; Cuadernos de Galileo, 5) and D. Pick, *Faces of Degeneration: A European Disorder, c. 1848- c. 1918* (Cambridge, Cambridge University Press, 1989). On eugenics and immigration in Cuba, see Consuelo Naranjo and Armando García González, *Medicina y racismo en Cuba* (Tenerife, Taller de Historia, 1996).

[3] Regarding the introduction of Darwinism in Cuba, see Pedro M. Pruna and Armando García González, *Darwinismo y sociedad en Cuba. Siglo XIX* (Madrid, CSIC, 1989).

Felipe Poey and Carlos de la Torre who taught natural history in the reformed university were instrumental in these innovations. Evolutionism was addressed not only in these two institutions but also in the Sociedad Antropológica de la Isla de Cuba and in various periodicals of the period. Two works dealing with these themes from the second half of the nineteenth century were particularly significant. One was published in 1893 by the physician A. de Funes Morejón, who based his observations on Koehler and other researchers of spermatogenesis, ovogenesis, and fertilization in *Ascaris megalocephala*, highlighting the role of chromosomes throughout the process.[4] This appears to have been the first mention of cellular structures in Cuba.

Until that moment, Cuban biologists discussed only the most general aspects of Darwin's ideas, restricting themselves to natural selection and the struggle for existence. Pangenesis and other more specialized aspects of Darwinian heredity had received less attention. Admittedly, this might have been less the case in the teaching of biology at the Universidad de la Habana.

Funes views heredity idealistically, as the hand of an operator who draws a radius corresponding to each indiviual, surrounded by a sphere which constitutes the zoological or botanical species. This heredity includes the influence of environmental agents on the organism. Adaptation is not absolute but relative, and is a limited aptitude that requires favorable conditions. This aptitude is real although it does not manifest itself until conditions force it to, and it is also inherited. Therefore, for Funes, heredity involves both conservation and variation, the latter being coextensive with or "contained within" heredity *(en la herencia está envuelta la variación)*.

Funes views the relationship of these processes to the transformation of species in a way reminiscent of Lamarck:

The circle of variation is as extensive as that of heredity. In order for this proposition to be favorable to the transformist theory, it would be necessary to observe heredity passing beyond the limits of species existing today. Because it could then be reasonably sustained that both are lacking in observable limits. But inasmuch as practical observations demonstrate the opposite, and as no one denies that in the present order of things species manifest fixity; if variation is reduced to the same circle, then it is then clear that it cannot overstep the limits of species.[5]

He also interprets heredity as a force and believes that it may or may not come into play from one generation to the next and may as a result diminish—in the negative case—until extinguishing itself. Atavism represents latent heredity, a special aptitude for the development of a determined quality in a particular circumstance. The function of the sex cells is not for him "a mere effect of adaptation, but rather an example of the admirable plan that Nature reveals," so that Funes returns to the concepts of Geoffroy Saint Hilaire, very much in vogue in that century. In spite of his apparent fondness for

[4] A. de Funes Morejón, "Nuevos datos sobre la embriogenia," *Revista de Ciencias Médicas* (Havana), 8 (1893), 158-161.
[5] *Ibid.*, p. 160.

Lamarckian concepts, Funes in fact refers more to the most prominent figures of his time period like Darwin, Spencer, and Weismann.

One year later in 1894, Gastón Alonso Cuadrado, a Spanish doctor and chemist later nationalized as a Cuban, summarized the ideas of these three scientists in relation to the process of the transformation of the species. [6] Cuadrado too shows himself to be a supporter of the Lamarckism (inheritance of acquired characteristics), as developed by Spencer. He does not provide his own explanation of Darwin's pangenesis theory and Weismann´s germ plasm, but rather appends to his work translations of George J. Romanes' explanations of these concepts as well as a translation of Spencer on "the insufficiency of natural selection," from the *Revista Contemporánea*.

The first years of the twentieth century saw the continued discussion in Cuba of various works centered on these evolutionist ideas with the addition of the names of Hugo de Vries, Gregor Mendel, and others. The clash between neo-Lamarckism and neo-Darwinism began to gather momentum after the rediscovery of Mendel's laws in 1900, giving an impulse to, and in great measure changing, conceptions about heredity current until then.

López Sánchez asserts that until 1908 Cuban naturalists had been occupied with the polemic regarding transformism and not the problems of heredity, referring to the latter as only a means of transmission.[7] The work of José Nicolás Ferrer, professor of natural history at the Instituto de Segunda Enseñanza in Santiago de Cuba, who introduced the ideas of Hugo de Vries in 1908, provides one example of this.[8] As López Sánchez asserts, the introduction of Mendel in Cuba came through medicine and not botany or agriculture, as in other parts of the world. Anton Lutz, a Swiss doctor and occulist residing in Havana, had the honor of introducing Mendel to the Academia de Ciencias Médicas, Físicas y Naturales de la Habana in 1912,[9] although Mendel's laws had already been mentioned in Havana some years previously.[10]

---

[6] Gastón Alonso Cuadrado, "La ley de la selección natural en la lucha por la existencia," *Revista Cubana* (Havana), 9 (1894), 37-48, 102-120, 200-243. Three years earlier, Alonso had published in the same journal a work entitled, "La ley de la selección natural contra las creencias," *ibid.*, September-October 1891.

[7] José López Sánchez, "Significación histórica-científica de Mendel y el Mendelismo," in *Gregorio Mendel, sesquincentenario de su nacimiento* (La Habana, Academia de Ciencias de Cuba, Museo Histórico de las Ciencias Carlos J. Finlay, 1974), pp. 25-26, 49.

[8] José Nicolás Ferrer, "Nuevas orientaciones sobre el transformismo," *Revista de la Facultad de Letras y Ciencias* (Havana), 7 (1908), 314-352.

[9] Anton Lutz, "Sobre algunos árboles genealógicos y la aplicación de las Reglas de Mendel en Oftalmología," *Anales de la Academia de Ciencias Médicas, Fisicas y Naturales de la Habana* [hereinafter cited as *Anales*], 48 (1911-1912), 486-498. See also, Lutz, "Herencia unilateral: sobre la pérdida de la correlación hereditaria entre las dos mitades del cuerpo, de los aparatos pares y especialmente de los ojos y descripción de dos casos de la así llamada anisocoria fisiológica," *Actas del Cuarto Congreso Médico Nacional de Cuba* (Havana, 1918), II, 161; reproduced in *Archivos de Opthalmogía*, 47 (1918) and in *Revista Cubana de Oftalmología.*

[10] See the anonymous letter entitled, "Patología social III. La República enferma," *Vida Nueva* (Havana), 1 (1909), 165-167, where the author discusses the application of Mendelian laws to social questions. The paragraph in question states: "en este conjunto social que forma nuestro pueblo se puede comprobar la ley de Mendel, que explica cómo los carácteres hereditarios se transmiten y cómo hay oportunidades para que reaparezcan, en los descendientes, rasgos propios y determinados de los antecesores, así como se reproduzcan

Lutz presented his work accompanied by some "photoelectric projections" of some genealogical trees composed by other authors and by himself. These genealogies referred to various diseases like congenital night blindness, atrophy of the optic nerves, and others, highlighting differences in modes of transmission. As for Mendel, Lutz alludes to his rediscovery by De Vries, Correns, and Tschermak and adds that this permitted the recognition that certain illnesses were produced by particular laws rather than pure chance. Although these laws could not be applied to all cases, a broader understanding of them would be worthwhile. Lutz makes clear as well, with some reason, that these laws refer to the germ plasm—remember Weismann—and are only valid for truly hereditary diseases like Daltonism and hemophilia, but not those like tuberculosis and syphilis which are transmitted by placental or spermatogenetic infection. Despite this caution, eugenicists over the following decades continued to insist on the hereditary disposition of these diseases, including them in lists of disorders that had to be eliminated in order to obtain the betterment of the population.

Lutz explains Mendelian genetics (homozygotic and heterozygotic characteristics, phenotype, genotype, dominant and recessive genes, first and second generations, mosaic heredity, latency, genetic correlation, the significance for hereditary diseases of crossing and consanguineous marriages, and a bit about sex-linked heredity) using examples in plants, animals and even human beings. He mentions prominent geneticists like W. Bateson and C. B. Davenport. Among his conclusions, Lutz concurred with eugenicists that genetic laws "are of consummate importance to the hygienist and, in the final analysis, the legislator, owing to their relevance to the family."[11]

In this presentation, Lutz did not mention eugenics, but the following year, 1913, saw the publication of what appears to be the first work in Cuba tying Mendel to this theme. We refer to an article by the English author A. F. Tredgold, "El estudio de la Eugénica,"[12] in which he attributed the deterioration of the British people to a morbid heredity that had created degenerate lines of mental defectives, epileptics, alcoholics, criminals, tuberculosis sufferers, chronic poor, parasites, incompetents, and other "social disasters." In his view, biological weakness, not the environment was to blame, because the State had improved living conditions through sanitary measures and laws. For Tredgold heredity, as demonstrated by Galton, Mendel, and Weismann, clearly played a principal, deleterious role. This erroneous generalization was typical of European eugenicists who only conceded to the environment a capacity to produce "morbid" forms. They believed these forms had weakened "the vitality of the plasm to

---

a través del tiempo y de los cambios sucesivos, los carácteres dominantes en los ancestrales" (pp. 165-166). This article, most likely written by a physician, shows that Mendelian laws were known in Cuba by as early as 1909.

[11] "Sesión pública ordinaria del 26 de enero de 1912," *Anales*, 48 (1911-12), 484-485. See the reply to Lutz by Juan Santos Fernández, a medical opthalmologist who noted the importance of heredity research for both the general health of the population and social amelioration.

[12] A. F. Tredgold, "El estudio de la Eugénica," translated by Ramón de Armas y Colón from *The Quarterly Review* (July 1912), *Cuba Contemporánea* (Havana), no. 1 (January 1913), 191-218. Tredgold is the author of *Mental Deficiency: Amentia* (London, 1908) and *Eugenics and Future Human Progress* (London, 1912), and a contributor to *The Eugenics Review*.

the point of causing pathological variations of a transmissible nature." Environment could not create those conditions, which could only be transmitted via heredity.

These and other aspects of environmental and hereditary influences in the impoverishment of peoples appeared more frequently after 1914 in various national and foreign works touching on eugenics and heredity. This phenomenon did not occur by chance given that in different European countries, the United States, and even Cuba these issues were widely discussed in meetings, conferences, and workshops, and were disseminated in the general and specialized press. This process was aided as well by the publication of books and articles on Mendel's genetic discoveries and the contributions of other geneticists like De Vries and Morgan.[13] Among these conclaves was the First International Eugenics Congress (London, 1912), which took up issues like eugenics, racial hygiene, and heredity. The prominence of this kind of gathering helps to explain the resonance these themes would find in Cuba.

HEREDITY AND EUGENICS IN UNIVERSITY TEACHING

As previously mentioned, the naturalists Felipe Poey and Carlos de la Torre were pioneers in the study of heredity and evolutionism at the University of Havana in the second half of the nineteenth century, especially after 1880. In the twentieth century, these subjects continued to be taught with the progressive addition of new genetic knowledge regarding Mendelian heredity as well as eugenics. This last theme was a typical concern of medical doctors with an interest in biology, one of the most significant of whom was Arístedes Mestre y Hevia, son of Antonio Mestre, also a prominent physician. Arístedes was professor of biology and anthropology at the University of Havana for many years and wrote several textbooks and other works dealing with heredity, evolutionism, and eugenics.

Mestre first demonstrated his interest in these themes in 1893 when he presented his membership address to the Academia entitled "On Relations between Various Pathological States Considered in the Individual and in the Ancestral and Hereditary Series."[14] Despite touching on some hereditary diseases, the concepts and authors cited by Mestre in this work indicate that he was not aware of the most relevant advances of the day. For example, Funes y Morejón´s work in the same year showed greater knowledge of chromosomes and their role in heredity. Even in 1910, Mestre continued to be out of step with the most important new works on heredity, inasmuch as in his university-level text *Curso de Biología* (which purported to be completely up-to-date) he mentions Darwin, his precursors and some of his successors like Weismann, but neither Mendel nor De Vries.

Four years later (in 1914), however, Mestre began to assimilate more recent ideas on heredity and eugenics, as in his reviews of two books on heredity by Bateson—

[13] With regard to these authors see the bibliographies compiled by Arístides Mestre in the works cited below.
[14] Arístedes Mestre, "De las relaciones entre los diversos estados patológicos, considerados en el individuo y en la serie ancestral y hereditaria," *Anales*, 29 (1892-1893), 438-459.

Mendel's Principles of Heredity (1909) and *Problems of Genetics* (1913)-- as well as two others on heredity and eugenics (*Heredity in Relation to Eugenics*, by Charles B. Davenport, and *Heredity and Eugenics* by W. E. Castle, J. M. Coulter, C. B. Davenport, E. M. East, and W. L. Tower), published the year before. The former, he writes "describes recent scientific results of reasearch into the betterment of the human condition, aiming at the resolution of social problems in the light of biological science,"[15] while the latter "[synthesizes] the most recent studies relative to variation, heredity, evolution of plants and animals, and human improvement."[16]

Mestre used these same books to compose a chapter on heredity and eugenics in the second edition (1917) of his textbook, *Curso de Biología*. Here his bibliography also lists works by the well-known geneticists T. H. Morgan and Hugo de Vries, published between 1914 and 1916. Similarly, he stresses Galton's laws as an example of the study of "the manifestations of heredity through the application of statistical methods to biological questions and especially to the phenomena of variation." Galton had observed these facts "in numerous families, with reference to diverse physical and psychological characters, these laws being most applied to reproduction with the same variety or race."[17] In similar fashion, Mestre underscores the application of genetic laws to support man's physical, moral, and intellectual advancement through eugenics, all of which fit well with his anthropological conceptions regarding the degeneration of criminals and mental defectives, issues which received special attention in this period.

In 1918, Mestre demonstrated how thoroughly he had applied himself to the problems of heredity and evolution in his article "The Laws of Heredity and Applied Biology," where he discusses the ideas of Galton, Darwin, De Vries, Weismann, and Mendel.[18] Here, he emphasizes that heredity constitutes biology's central problem and, following Conklin, he recognizes that the "origin" of the individual represents the key issue of the twentieth century in the same way that the origin of the species had in the nineteenth. Regarding Galton, Mestre indicates his contribution in the area of statistical methods, calling him the founder of Biometry, with worthy successors like Bateson, Pearson, and even Davenport, who had carried out interesting research at the Station for Experimental Evolution in New York.

---

[15] Arístedes Mestre, *Curso de Biología* (Havana, 1915), pp. 245-246.

[16] Arístedes Mestre, "Bibliografia," *Revista de la Facultad de Letras y Ciencias*, 18 (1914), 354-357. One year later, in his *Curso de Biología*, Mestre added to his bibliography the titles: W. Bateson, *Mendel's Principles of Heredity* (Cambridge, 1913), and *Problems of Genetics* (Chicago, 1913); C. B. Davenport, *Heredity in Relation to Eugenics* (New York, 1913); T. H. Morgan, *Heredity and Sex* (New York, 1914); and others. In 1912 a Cuban journal published a review of the problems of applied biology examined in the Cuarta Conferencia de Genética which took place under the auspices of the French National Society of Horticulturists; López Sánchez, "Significación de Mendel," p. 51. The review in question summarized research introduced at this meeting on heredity in plants and animals (e.g., heredity of sex) and certain human characteristics (e.g., brachicephaly), of which Mestre may have been aware.

[17] *Ibid.*, p. 251. In 1928, Mestre published a third edition of this work which included nothing new.

[18] Mestre, "Las leyes de la herencia y la biología aplicada," *Revista de la Facultad de Letras y Ciencias*, 28 (1918), 163-193.

Mestre also sums up Galton's two laws: the first, that "there exists for each generation a mean and constant level with respect to the variations of a character or faculty," and the second, that "ancestral heredity, which establishes the relations of degree to which each generation contributes to the constitution of a determinate individual." Here, Mestre correctly insists that these laws certainly cannot be considered absolutes or overly generalizable since "they are applicable rather to reproduction within a variety or race, as has been established."[19] Mestre considered eugenics to constitute a new direction in biology, a new field that had achieved—he says—"an extraordinary level of development given the extent of its domain and the beneficial action of its principles." He goes on to mention various institutions devoted to eugenics,[20] refers to the First International Eugenics Congress, and lists the sections planned for the Second (suspended in 1915 due to World War I). [21] Eugenic methods are complemented by "Eutecnics." a system of individual improvement via immersion in a favorable environment. He understands as well that "in spite of great obstacles, positive eugenic measures rest on logical applications" and insists once again on the importance and utility of knowing the laws of heredity, so as to free the human species from the anomalies, defects, and morbid processes harmful to physique, intellect, and morality. Mendel's laws are to biology what Dalton's theory is to chemistry.

Mestre's discussion of agricultural genetics and biology provides some interesting information on their state of instruction. He begins with the importance of studying both Mendel and De Vries in order to master animal husbandry and plant breeding. He then describes his adamant effort over twelve academic years to arrange for students in the Agronomy School to learn biology: they had not been taught anything about individual variation, adaptation, artificial selection, nor other biological processes. Finally biology was incorporated into the discipline of agronomy by the Law of 1916. According to López Sánchez, however, the key date for the incorporation of genetics in agronomy and biology was 1918, when the Congreso Agrícola de Santiago de las Vegas was held. There, Mario Calvino presented a paper summarizing a book later published as *Multiplacación de las plantas*, in which he refers both to mutations and Mendel´s laws.[22]

At the University of Havana itself, not just biology students but many others in education and law received instruction in heredity and eugenics. The latter two groups learned about these subjects if they took anthropology, which at the time included general and juridical anthropology. This chair was held from its establishment in 1899

[19] *Ibid.*, p. 174.

[20] Eugenics Laboratory in University College (K. Pearson, director), Eugenics Record Office in Cold Spring Harbor, Long Island, New York, in connection with the Eugenics Section of the American Genetics Association. Mestre received, according to his references, this institution's *Journal of Heredity* which was published monthly and which by this date had reached its ninth volume.

[21] The sections were: 1) Heredity (physical, experimental, and descriptive bases); 2) Factors of decline; 3) Selection, fertility; 4) Eugenics in relation to national prosperity, race, and immigration; 5) Genealogy and history; 6) Analytical methods.

[22] López Sánchez, "Significación histórica de Mendel," p. 53.

by Luis Montané. When twenty years later Montané petitioned for permission to retire, the dean of the Faculty of Letters and Sciences gave the chair to Arístides Mestre who took possession on November 8, 1920, even though he had occupied it on a temporary basis since the previous year. Between 1900 and 1904 Mestre had given open classes on criminal anthropology and other subjects, such as mental and nervous diseases which he taught at clinics which he organized at Aldecoa and Numero Uno hospitals, where he served as a psychiatrist. After 1904, Mestre gave occasional classes in the university as a substitute for Montané whenever the latter was absent. He also oversaw final examinations and took students to prisons or mental hospitals to study various patients. In addition, he gave several lectures on legal medicine and psychiatry to law students during the 1918-1919 school year. Mestre related these themes to degeneration, eugenics and, in a general sense, human betterment, as is clear in his works from the decade of the twenties.

In 1920, Mestre published *Antropología Jurídica*, a summary of his lectures on that subject. In some of the chapters, Mestre mentions laws of heredity established by Darwin, Mendel, and de Vries and even refers to his own work "Las leyes de la herencia y la biología aplicada" of 1918. This article deals with pathological heredity, the central core around which eugenically-mined doctors, biologists, lawyers, and sociologists would rally:

> Mendel's laws have been confirmed for the human species, especially in the area of inherited diseases, recording interesting facts about anomalies, monstrosities, as well as organic and functional diseases. In this respect, Professor Davenport has contributed important studies on epilepsy, madness, organic and dynamic conditions of the nervous system, etc., interpreting such phenomena in the context of Mendelism.[23]

Mestre emphasizes how eugenics was created out of genetics, whose discovery had served to unite various subjects:

> The sum total of the laws of heredity and everything that affects the physiology of reproduction constitute the science of Genetics. General appreciation of how predispositions are manifested, how the most diverse morbid states (especially those presenting themselves in the brain and its functions) are transformed and consolidated, and what determines the different forms of degeneration, has. . . led to the creation of Eugenic science, whose objective is the physical, intellectual and moral betterment of man. This has been a very favorable development with respect to mental illness, insanity, the perversion of the instincts, and criminal tendencies. Eugenics is an eminently prophylactic science, founded on the solid base of Genetics, which seeks to obliterate the germs of psychopathy and delinquency.[24]

Mestre also included in another of his textbooks, *Curso de Antropología General* (1924),[25] various aspects of evolutionary theory, transmission of hereditary

[23] Mestre, *Antropología Jurídica* (Havana, La Propagandista, 1921), pp. 199-200.

[24] *Ibid.*, p. 200

[25] Mestre, *Curso de Antropología General (Notas de Clase)*, (Havana, Rambla, Bouza, 1924), pp. 2-4 (Chapter: "Antrpología pedagógica").

characteristics, and anthropological applications. In the first of these sections he refers to figures like Lamarck, Saint-Hilaire, Darwin, Wallace, Wagner, and Naegeli all of whom examined evolutionary factors influencing the human race. In the second, he lists Mendel's laws, referring as well, in the third, to these same laws in relation to the "hybrid" (mixed) features that are observed in the anthropological traits of the face, skin, and hair.

In the following years, other figures besides Mestre who taught anthropology classes at the University dealt with evolution and genetics. The most prominent among these were the physicians Julio Morales Coello (forensic anthropology), Victor J. Rodríguez, Luis Howell Rivero, Carlos García Robiu, René Herrera Fritot, Teresa Gurri (physical anthropology), and especially the biologist Julio Fernández de la Arena. Arena taught embryology, human and general Genetics, and Eugenics at the University's Escuela de Biología from 1943 through the sixties. In 1954 he gave a course entitled "Human Genetics and Eugenics" in the Department of Morphology and Genetics," lamenting that the benefits of the subject matter would not reach medical students on this occasion.[26] The importance he gave to these topics emerges clearly in a series of eight lectures he delivered at the Universidad de la Habana between May 15 and June 8, 1956. The last of these was titled "Selection and Counter-Selection in the Human Species," in which he also touched on aspects of hygiene, genetics, eugenics, and human-directed evolution.[27] Moreover, Fernández de la Arena was a great popularizer of these disciplines, a task which he promoted in the newspaper *El Mundo*, as well as on radio and television during the period 1956-1957. Finally in 1960 and 1962, he taught an introductory course on medical genetics.

Since 1936 Fernández de la Arena had stressed the distincttion between hereditary and congenital diseases. His concern stemmed from confusion in medical and biological studies on pathological heredity which held that hereditary diseases were incurable. In similar fashion, he described the make-up of the chromosomes and genes, following T. H. Morgan.[28] In his judgment, many of the errors surrounding these hereditary issues could be corrected if general physicians had sufficient preparation in basic biology, the most important subjects being cellular biology, chemical embryology, cytogenetics, and general and human genetics. Fernández de la Arena also stressed the value of genetics for preventive medicine (particularly with regard to

---

[26] Julio Fernández de la Arena, "Genética y medicina," *Memorias de la Sociedad Cubana de Historia Natural Felipe Poey,* 22 (1955), 1-14.

[27] The other lectures were: a) Genetics, the science of biological heredity: its practical applications. False conceptions; b) The bases of human nature. Genes and characters. Destiny's dice. Heredity in action; c) Gregor Mendel, father of modern genetics. His works. His laws; d) We are not equals. Variability and individuality in the human species; e) The heredity of abnormal characteristics, physical and mental defects; f) Susceptibility and resistance to diseases. Genetic and non-genetic factors; g) Heredity and environment. Biology of twins. See Julio Fernández de la Arena, Expediente administrativo, no. 11180, Archivo Histórico de la Universidad de la Habana.

[28] Fernández de la Arena, "El concepto de la herencia y las enfermedades pseudohereditarias," *Medicina de Hoy* (Havana), 1 (1936), 131-133; and in *Memorias de la Sociedad Cubana de Historia Natural Felipe Poey,* 10 (1936), 25-28.

diseases like tuberculosis and certain kinds of anemia), the legal determination of paternity (as had already been demonstrated through the typing of blood groups), and the relationship between microorganisms and antibiotics.

Fernández de la Arena viewed eugenics as that branch of genetic studies which applied the principles of genetics to human improvement. Like some eugenicists, he felt that it was possible to apply the same genetic understanding of plants and animals to man without violating any ethical or moral codes. Pathological mutations among primitive peoples are presumably eliminated by natural selection. However, the therapeutic gains achieved in modern times have increased the likelihood that individual carriers of hereditary defects might survive to transmit their "dangerous and undesirable genes" to their descendants. Consequently, in Fernández de la Arena's judgment, there existed a tendency for these undesirable genes to accumulate and reach high levels of concentration. While diseases of infectious origin have responded to medical advances, those of genetic origin have increased in significance. Even if it proved impractical to achieve spectacular results for the human race, he still believed that "at least we are obliged to prevent the deterioration of races through the action of eugenic factors. To be inclined to the gradual elimination of undesirable genes and intend to preserve and reproduce favorable ones, is the aim which should unite us promote the obligation we have to future generations."[29]

In consequence, Fernández de la Arena proposed a program to "limit voluntarily" the reproduction of those who carried defective genes. Geneticists would play key roles in this process by informing prospective parents if a particular anomaly were hereditary and by calculating the probability of its appearance among any children. In Fernández de la Arena's view these were primarily biological, rather than medical, problems and so physicians and biologists would have to work in tandem. To this end, he proposed the creation of a Department or Institute of Human Genetics which would work in connection with well-equipped hospital centers to attend to patients with problems of a genetic nature.

As in the earlier efforts of Galton and his followers, Fernández de la Arena proposed the detailed study of families and the elaboration of corresponding genealogies with the most complete data possible. This would require the close collaboration of families and medical specialists for the particular affliction in question, with university geneticists providing consultative and instructional support relative to the diagnosis and prevention of hereditary diseases. Academic geneticists would also teach students, inasmuch as only some medical schools provided this education, and even these tended to focus more on infectious diseases. However, the greatest causes of mortality at that time, in the Cuban biologist's view, stemmed primarily from hereditary factors. Another function of the Department would be scientific research and dissemination of results. No large staff would be necessary. The Department could consist of one geneticist, one physician (or physician-geneticist), a secretary, and a laboratory technician. As for the physical plant, a waiting room and a few offices would

[29] Fernández de la Arena, "Genética y medicina," p. 11.

suffice. This proposal seems quite modest in comparison to the great aspirations and grandiose projects put forward by other eugenicists (both in Cuba and abroad) in the first decades of the twentieth century. However the creation of a biogenetic institute would have to wait until the Cuban Revolution.

HEREDITY, EUGENICS AND EVOLUTION IN SECONDARY EDUCATION

The study of genetics and eugenics formed part of the secondary school curriculum as well. We do no know precisely when this kind of instruction was introduced. Still, the ideas of Lamarck, Darwin, and other evolutionists appear in condensed form in many of the textbooks written in the first decades of the twentieth century. Eugenics, oddly enough, began to be covered in datil in the 1940s and 50s, when the subject was in full decline elsewhere.

The twentieth century also witnessed a profusion of differing positions with respect to evolution. Among these positions were those which defended religious dogmas against the transformation of the species and the scientific explanation for the origin of man, expounded in biology and geology textbooks written by Jesuits. Specific titles include *Geología Moderna* by Miguel Gutiérrez (Barcelona, 1927), ***Evolución de los seres vivientes*** (1914-1924) by Fordham University professor Gustavo A. Caballero, *Biología* by Pelegrín Franganillo (Havana, 1944), and *Fundamentos de Biología* by Faustino García (Havana, 1957). Caballero´s book includes a chapter summarizing the theories of Lamarck, Darwin, and Weismann as well as the ideas of Hugo de Vries and Gregor Mendel. He concludes, however, that Darwin's theory had been discarded and that none of the other figures mentioned had resolved the problem of the origin of the species.[30]

The book by Franganillo—the man who issued Caballero's *Nihil obstat*, incidentally—was the fifth-year ***bachilerato*** textbook in Cuban secondary school education and covered some aspects of genetics and eugenics, especially basic concepts like genotype, phenotype, and hybridism, types of heredity (mixed, mosaic, and alternative), and the transmission of acquired characteristics. He refrains from including under the last-mentioned rubric accidental modifications, mutilations, and infectious diseases like syphilis, but does describe certain external agents as capable of modifying genes and therefore heredity. In addition, he summarizes Mendel's laws and those of Galton on ancestral heredity (parents determine half of the child's characteristics, the grandparents provide one-fourth each, *etc.*), heredity in man, and something of sex-linked heredity. Here, Franganillo commits the error, common at the time, of describing mulattos as hybrids. Similarly, he states that Mendelian laws do not apply to skin coloration.[31]

---

[30] Gustavo A. Caballero, *Evolución de los seres vivientes* (Havana, Burgay, 1927), pp. 45-64. The work does not cover eugenics.

[31] Pelegrín Franganillo, *Biología* (Havana, Cultural, 1944), p. 124.

With respect to eugenics, Franganillo expresses the idea that no one could deny that the human races, even the most civilized, had entered into a period of complete degeneration. This could be seen in the United States, France, and Germany (Gratham had declared that 23 million people in Germany are hereditary cripples). These exaggerations, of course, make us think of the Nazi extermination campaigns, also based on racial and eugenic presuppositions. However, writing in 1944, Franganillo alludes to these extremes only in the most euphemistic way. He notes that with the better understanding of hereditary mechanisms came the idea of using this knowledge to obtain healthier and happier progeny, an idea received sympathetically by both Church and State. This positive idea had been adulterated, however, by irresponsible and superficial people whose materialistic conception of life constitutes the only cause of the degeneration of the human races:

> In this scientific conception are rooted all the factors of the degeneration of races: the low birthrate of the elite classes, the degradation of women, the great obstacles facing those people best fitted for starting families, sexual heredity, unheard of precocity in sexual relations, alcoholism, syphilis. The biologist knows that the annihilation of the race lies at the end of this road.[32]

In this sense, Franganillo criticizes the "materialists" whose eugenic literature proposed two ways to improve the human race: "Malthusian" selection (use of contraceptive methods) and the sterilization of undesirables. The latter seems to him an unjust abuse, and he asks, with reason, who would determine which individuals were undesirable. Franganillo observes that some who are far from normal due to specific defects still prove useful to society and may be highly talented. Consequently, Franganillo proposes the following measures: a) elimination of factors which bring about the appearance of defective, vice-ridden individuals: alcoholism, drugs, physical and moral burdens; b) "purification" of the environment in which children and youths are raised (elimination of pornography and application of a reasonable censorship of public entertainment and the press), c) sex education; d) well-thought-out spousal selection; and e) integrity in the fulfillment of conjugal obligations.

Franganillo's text remained in use in secondary school education until the 1950s, so that it provided the principal source of those students' ideas on genetics and eugenics. In 1957, it was replaced by Faustino García Peralta's more extensive and up-to-date work, ***Fundamentos de Biología***. This contained a prologue by Julio Fernández de la Arena who had been García's mentor. Both men agreed on the necessity of reconciling science with religion, assuming an agnostic attitude with regard to the metaphysics of these vital phenomena. However, García opposed scientific explanations of the evolution of species and the origin of man, remaining faithful to his religious ideology.

García addresses the same aspects of genetics that Franganillo does but in greater detail, including as well the heredity of blood types. He also spends a bit more time on eugenics. García states that Galton was one of the first who managed to reduce the mass of observations on heredity and variation to a system by establishing certain

---

[32] *Ibid.*, pp. 127-128.

principles through statistics, studying individual characters, and introducing quantitative measures.[33] García also summarizes Galton's investigation of selected traits like genius and intellectual capacity, stature, eye color, and others which permitted him to establish his principles or laws of ancestral heredity and of filial regression or the tendency towards the mean.

García criticizes these statistical methods, claiming that Galton failed to distinguish between hereditarily and environmentally-determined characteristics. In addition, García sees these studies as having little scientific value, since they were based on a sample of many families and lines not closely related genetically. He adds as well a quote by Jennings (1910) suggesting that Galton's laws are the product of a failure to distinguish between two divergent factors: hereditary fluctuations on one hand and genotypical differentiation on the other. For García, plants and animals permit large-scale trials to determine if the differential cause of a trait is hereditary, environmental or both, but human subjects do not. Consequently, he does not see Galton's laws as being particularly helpful for understanding the fundamental principles of heredity.

Precisely who would set goals and standards represented a central problem for eugenics, given the movement's objective of perfecting certain racial traits while reducing defects to a minimum. The materialist would treat eugenics and its applications to man as if referring to mere animals, following the same rules that would apply to animals for the conservation and selection of races. For his part, the spiritualist would run the risk of paying too little attention to the body. Faustino García finds both of these attitudes overly one-sided and therefore defective. Instead, following his Catholic ideology, he sees man as composed of body and soul in complementary and mutually reinforcing fashion. For García, then, the scientific basis of eugenics is to be found in Bionomics or Human Ecology which has "as its objective the discovery and study of the specific laws that govern the activities of each species," an objective which also applies to the human species. Man, due to his ability to reason and his intelligence, is spiritual. Since animals lack these faculties, man therefore occupies the peak of the zoological scale. Consequently, eugenics has to take into account this special feature by focusing not just on the body but also the soul.

The factors that should influence human eugenics are human relations based on the special status of man (relations with God, and social, individual, and moral relations); and biological heredity, which forms part of the eugenic complex which insists that progenitors be healthy, strong, and free from defect or vice. This takes into account the fact that the "spermatogenetic tissues" can suffer alterations and infections by germs and viruses that could infect the procreator. In the end, García proposes the same means to achieve the improvement of the human races as Franganillo, using many of the same words.

Many eugenicists were concerned about whether the State could or could not prohibit the marriage of individuals deemed defectives. On this question, García declares the State should limit itself to explaining why these marriages should not take

---

[33] Faustino García, ***Fundamentos de Biología***, (Havana, Minerva, 1957), pp. 447-450.

place, but avoid legal bars or penal sanctions. Above all, the State should refrain from sterilization or mutilation, actions he viewed as a violation of human rights and an unjustified violence against the individual. In fact, this was the general position adopted by Catholic medical eugenicists, some of whose articles were published during this period.[34] Catholic eugenicists not only opposed sterilization, but also decried other measures advanced by many eugenicists, such as divorce, abortion, and birth control. Catholic principles also inspired protests against the promotion and application of Malthusian and Darwinist criteria and Mendelian laws and even, on occasion, against the racist and discriminatory views which they generated or implied. Thus, during the short time that García´s work was used as a textbook, Cuba's secondary school students learned about genetics and eugenics informed with a certain religious sentiment. With the Cuban Revolution (1959), new biology texts, stripped of this connotation, were adopted. For example, in 1963, the eighth-grade textbook included a chapter on genetics and evolution but not eugenics, now discredited in its original version.

Not all of the biology textbooks used before 1959 had such obvious religious content, however. Some limited themselves to simply presenting the different evolutionary theories without taking a particular stand (again reflecting the pervasive influence of the Church). An example of this noncommittal approach is *Elementos de Biología* by Araceli Calderón de Rabina, professor of Natural History at the Instituto de Segunda Enseñanza del Vedado in Havana, whose second edition (probably written at the end of the 1950s) contains chapters on genetics, the origins of life and of species, evolution, and ecology. In the first of these chapters, one finds the concept of the gene (here given a hypothetical character), types of heredity, the chromosomal theory of heredity, Mendel's laws, sex-linked heredity, Galton's laws and the role of eugenics. This last chapter then discusses human heredity concretely. With respect to Galton, Calderón says that he "he gave a great impulse to Biometry or the science of the measurement of living beings founded in 1845 by Quetelet" and provides a summary of his laws. He explains that eugenics is the science dealing with heredity and the elimination of hereditary diseases or predispositions so as to "increase the number of healthy births and diminish that of unhealthy ones."[35]

Many of the books on the natural sciences published in Cuba in the first half of the twentieth century do not even allude to evolution, in order to evade the obstacles

---

[34] A particularly interesting article in this respect is that of José F. Ferrer y Rovira, president of the Asociación de Médicos Católicos de Cuba, who opposed the methods of sterilization and castration proposed by eugenicists and applied by the Nazis in Germany. In addition to his very appropriate criticism of Nazism, Ferrer right after summarizing the Mendelian laws attacked the eugenicists who tried to apply these laws to human beings in a strict fashion. He pointed out the difficulty of predetermining favorable hereditary characteristics from phenotype. In similar fashion, he highlighted the role that education (environment) exercises over phenotype, "aun cuando su acción sobre el genotipo sea nula para algunos e imperceptible para otros," and the importance that phenotypic modifications can have for the successful socialization of those who bear them. He alludes as well to the recessive nature of the heredity of mental illnesses along with the biological, psychological, and social risks incurred by sterilization. See José F. Ferrer, *La castración de los degenerados* (Havana, Compañía Editora de Libros y Folletos, 1941).

[35] Araceli Calderón de Rabina, *Elementos de Biología* (Havana, Ciencias, n.d.), pp. 332-358.

already mentioned. To a certain degree, this tendency was favored by the conventional divisions established in the biological sciences curriculum: botany, zoology and anatomy, physiology, and hygiene. The study of man from the evolutionary point of view only had to be touched on when referring to his placement within the group of mammals. However, even here discussion was limited to man's place in nature, the human races and some of the anthropological traits that distinguish them (often expressed in discriminatory terms).[36]

Something similar can be observed in geography textbooks that touch on, although in even briefer fashion, these anthropological issues. Often these geography manuals would take these sections from outdated biology books, resulting in the inclusion of topics like man in his own distinct kingdom (the Hominal), the correlation of intelligence with Camper's angle, and other discarded ideas. As in the case of the biology textbooks, the geography manuals written by Catholics displayed an openly anti-evolution bias. For example, at times they refer to a religious monogenism based on the first human couple. The more discerning authors, however, either assumed an agnostic stance or dodged the thorny subject of evolution entirely when discussing man's zoological position.[37]

At the University of Havana, Roberto Agramonte fully incorporated heredity, eugenics, and evolution into the teaching of sociology starting at the end of the 1930s, and especially in the 1940s. Other professors did so as well at the secondary school level when sociology was included in the curriculum in the 1940s. Agramonte´s book, *Sociología*, which includes a chapter on Mendelism and Darwinism was important in Cuba and was distributed in other Latin American countries as well. Agramonte explains social phenomena through natural, sexual, and artificial selection as well as eugenics, adopting at times a critical tone in discussing the latter.

Cuban eugenicists demonstrated considerable interest in heredity and its influence on the degeneration of the population. As a result, they proposed artificial and sexual selection as scientific measures to combat this process. Even so, many Cuban eugenicists—and it appears this characterizes those in other Latin American countries as well—were aware of the importance of improving environmental and sanitary conditions if they wanted to obtain a biologically superior population. In general terms, the campaigns in favor of ameliorating these conditions became more intense after the fall of President Gerardo Machado in 1933. Physicians like the obstetrician José Chelala-Aguilera and sociologists like Roberto Agramonte, both of whom were socialists, played an especially important role in this movement. These eugenicists used Darwinian concepts, Mendelian laws, and evolutionism to support their arguments but placed greater emphasis on social factors. This approach contrasted with that of the group led by the Domingo Ramos. Between 1910 and 1933, his group made greater use of these scientific currents for its campaigns in favor of sanitation, hygiene,

[36] Armando García González, "Los grilletes de la ciencia: Discriminación y antiracismo en la enseñanza de la biología, la geografía y la sociología en Cuba" (unpublished ms.)
[37] *Ibid.*

puericulture, and immigration control. These campaigns were not always well formulated and at times verged on extreme solutions with respect to gametogenic control of immigration (determining by gametes whether immigrants were desirable or not), sterilization of criminals and the mentally retarded, obligatory prenuptial medical certification, and other such measures.

With respect to Cuban biologists, it appears quite clear that evolution, Mendelism and Darwinism attracted their attention and efforts. Biologists involved in education were the most engaged with these issues, which were often closely tied to Galtonian and post-Galtonian eugenic currents. All of this fit with the enthusiasm of the day for "racial", social, political, and economic betterment. In this view, the correct application of eugenics would eradicate the conditions of impoverishment and "degeneration" observed in so many communities. Of course, we know now that these conditions derive from the great political and economic inequalities which exist between nations and between social classes, and not in any great degree to biological laws.[38]

Centro de Estudios de Historia de la Ciencia y la Tecnología, La Habana, Cuba

---

[38] This chapter was translated by John-Marshall Klein.

RICARDO CAMPOS MARÍN AND RAFAEL HUERTAS

THE THEORY OF DEGENERATION IN SPAIN (1886-1920)

Only two years before the appearance of *On the Origin of Species*, the French alienist B. A. Morel unveiled his theory of the degeneration of the human species. He conjoined anthropological and philosophical concepts of Rousseau and Buffon, Prosper Lucas's notion of "dissimilar" heredity, plus some others drawn from Lamarckian evolutionism, and put them all into a theocratic framework that goes all the way back to "original sin" as the prime cause of "*la déviation maladive du type normal de l'humanité*," that is, the degradation of a perfect primitive being created by God.[1] Years later, in 1895, V. Magnan and P. M. Legrain substantially modified the Morelian concept of degeneration by introducing in their own argument the Darwinian idea of the struggle for life, displacing the mystical-religious concepts present in Morels' work—principally the myth of the "fallen angel"—and producing a theory more in line with the orthodox scientific thinking of the times.[2] To Magnan and Legrain, the "ideal" and the "perfect" were not absolute concepts, but rather are in constant relation to both earlier and later evolutionary states. This leads to an understanding of degeneration not as a deviation from an hypothetical "ideal" man, but as "a progressive movement from a more perfect state to a less perfect state."[3] Degeneration is thus defined as

> a pathological state of the organism that in comparison with its most immediate ancestors is constitutionally inferior in its psychophysical resistance and does not satisfy completely the biological conditions for the hereditary struggle for life. This inferiority, transformed into permanent traits, is essentially progressive, except for intermittent regeneration, but apart from this exception it ends more or less rapidly in the annihilation of the species.[4]

---

[1] B. A. Morel, *Traité des dégénérescences physiques, intellectuelles et morales de l'éspecie humaine et des causes qui prosuissente ces variétés maladive*, (Paris, Baillière, 1857), p. 47.
[2] V. Magnan and P. M. Legrain, *Les Dégénérés. Etat mental et syndrome épisodiques* (Paris, Rueff, 1895); on this subject, see R. Huertas, "Madness and Degeneration, I. From "Fallen Angel" to "Mentally Ill," *History of Psychiatry*, 3 (1992), 391-411.
[3] *Ibid.*, p. 76.
[4] *Ibid.*, p. 79.

Thomas F. Glick, Rosaura Ruiz, and Miguel Ángel-Puig-Samper (eds.), The Reception of Darwinism in the Iberian World, 171-187.

Given such a hypotheses it is no wonder that the theory of degeneration played a role in the debates over Darwinism and that discussions of degeneration in the literature of psychiatry and public health should reflect the impact of evolutionary theory.[5] But, despite its undoubted relevance of degeneration to the evolution polemic (the role of biological inheritance, physical stigmas, its supposed incurability, etc.), its acceptance in Spain was late and uneven. The reasons for this delay is in part due to the predominance of opposing theories, such as the concept of monomania,[6] but also to the peculiarities of psychiatric care in Spain,[7] and to the low level of institutionalization of psychiatry as a discipline.[8]

It was not until the 1880's that the first signs of support for, and interpretations of, the concept of degeneration appear in Spanish medical literature. Degeneration began to be introduced in the courts, in forensic reports by alienists to demonstrate the fragile mental state of some defendants.[9] At the same time, psychiatrists and hygienists introduced the concept to advance both for clinical and socio-medical objectives.

1. DEGENERATION IN SPANISH ALIENISM

The acceptance by Spanish alienists of the theory of degeneration was not by any means unanimous. In fact, except in forensic medicine, the most outstanding representatives of nineteenth-century Spanish psychiatry considered Morel's theory of little use in psychiatric nosology, those persons showing the most interest being relatively minor figures. Many of these early references are so confused and contradictory that it is difficult to assess how well the original doctrine was understood. What can be said is that interest in degenerationism in the last two decades of the Nineteenth Century was eminently clinical. Attention was centered on inheritance as a vehicle for degeneration,

---

[5] See, e.g., J. Hochman (1992), "La théorie de la dégénérescence de B. A. Morel, ses origines et son évolution," in P. Tort, ed., *Darwinisme et Société* (Paris, PUF, 1992), pp. 401-412; C. Bénichou, "Dégénération, dégénérescence", in Tort, ed., *Dictionnaire du Darwinisme et de l'évolution* (Paris, PUF, 1996), I: 1151-1157.

[6] J. Martínez-Perez, "Problemas científicos y socioculturales en la difusión de una doctrina psiquiátrica: la introducción del concepto de monomania en España (1821-1864)," in E. Arquiola and L. Martínez-Perez, eds., *Ciencia en expansión. Estudios sobre la difusión de las ideas científicas y médicas en España (siglos XVIII-XX)* (Madrid, Complutense, 1995), pp. 489-520.

[7] J. Espinosa Iborra, *La asistencia psiquiátrica en la España del siglo XIX* (Valencia, Cátedra e Instituto de la Historia de la Medicina, 1966); J. M. Comelles, *La razón y la Sinrazon. Asistencia psiquiátrica y desarrollo del Estado en la España Contemporanea* (Barcelona, PPU, 1988); J. M. Bertolin, "Dispositivos de asistencia psiquiátrica en la España contemporánea del período de 'Entresiglo'," *Asclepio*, 45 (1993), 189-215.

[8] R. Huertas, "La psiquiatría española del siglo XIX. Primeros intentos de institucionalización," in *Un siglo de psiquiatría en España* (Madrid, Extra, 1995), pp. 21-47.

[9] This was one of the most important means of entry into Spain for the theory of degeneration, as well as for Lombrosian criminal anthropology. See A. Galera, *Ciencia y delincuencia. El determinismo antroplógico en la España del siglo XIX* (Seville, CSIC, 1991). Also, R. Huertas and J. Martínez, "Illness and Crime in Spanish Positivist Psychiatry," *History of Psychiatry*, 4 (1993), 459-481.

on the description and evaluation of physical and mental symptoms, and in problems surrounding the classification of madness, though always restricted to degenerate individuals and their families. Unlike the situation in France until well into the Twentieth Century, Spanish psychiatrists were disinterested in sociological readings of degeneration, let alone entering into apocalyptic discussions of the dangers that degeneration supposed held for the "race" and the State. They all believed in the Morelian idea of the self-regulatory capacity of nature, whereby hereditary degeneration never lasted more than four generations and resulted in the eventual extinction of the line.[10]

This clinical perspective—without excessive theoretical concerns—also explains why Spanish psychiatry failed to take a stance in the controversy over evolution in which there was a central disjuncture between modern thought—it was rational and secular—and the traditional values of believers and theocrats.[11] There were, of course, both idealist and materialist psychiatrists, but the discussion of degenerationism and indeed all mental illness always took place purely within the sphere of clinical *praxis*. This does not mean to say that, in other spheres of their scientific and social activity, there were no alienists that were not openly Darwinian. Such a psychiatrist was José Maria Escuder, who in the famous debate on evolutionism that took place in the Athenaeum of Valencia in 1878 stated that "There is an antagonism, a complete opposition between revelation and science."[12] However, when this same author turned to degenerationism in his book *Locos y anómalos* (1895), he makes no reference to the relationship between Science and Religion although he does mention biological evolutionism.[13]

Therefore, support for Morel was never based upon a possible congruence of religious beliefs with the French author, but rather on the arguments regarding inherited diseases that can be found in his works. In this sense, Prosper Lucas's concept of *hérédité dissimilaire* is especially pertinent, one of the mainstays of the Morelian theory of degeneration.[14] Lucas introduced "dissimilar heredity" to explain the hereditability not only of physical traits, but also of mental and moral ones, to account for the etiology of mental disorders, an essential argument upon which Morel's idea of the hereditary

[10] For examples of the incidence of this idea in the Spanish literature, see L. Dolsa y Ramón, *Concepto de la degeneración y responsabilidad legal de sus productos mentales* (Barcelona, Heinrich, 1895); V. Ots y Esquerdo, *Neurosis y Degeneración* (Madrid, Revista de Medicina y Cirugía Prácticas, 1897), as well as titles mentioned below.

[11] D. Núñez, *El Darwinismo en España* (Madrid, Castalia, 1997); T. F. Glick, *Darwin en España* (Barcelona, Peninsula, 1982).

[12] The debate is summarized in *El Mercantil Valenciano*, June 15, 1878; see Glick, *Darwin in España*, pp. 29 and following.

[13] J. M. Escuder, *Locos y Anómalos* (Madrid, Rivadeneyra, 1895).

[14] P. Lucas, *Traité philosophique et physiologique de l'heredité naturelle dans les états de santé et de maladie du système nerveuse avec l'application méthodique de lois de la procréation au traitement général des affections dont elle est le prinicipe* (Paris, Baillière, 1848-50).

transmission of mental disorders leading to degeneration was based.[15] For this he describes a series of "means" of inheritance (direct, indirect, retrospective, by impregnation, etc.) that psychiatrists found useful in interpreting their clinical observations.

In accordance with this conceptualization, V. Garrido indicated in 1888 that heredity is not always transmitted in similar fashion, but is "a series of strange and as yet unknown conditions that influence and oppose the mathematical expression of a biological law," ensuring that "similar reproductions or unhealthy transformations that renew or modify the hereditary principle. . . reveal a common extent of imbalance, the degeneration of the nervous centers."[16] Along the same lines M. Bonafonte, author of a 1900 doctoral thesis titled *Degeneración y locura* and which is probably the best Spanish text on the subject,[17] goes into greater depth on heredity, insisting that it need not always be

> similar or equivalent, and may consist of the transmission of an original vice, a state of abnormality consequent to various conditions or diseases of ancestors. These conditions are generally of a nervous nature: organic disease of the nervous system, neuroses, various mental illnesses, chronic intoxications, etc. . . .It is the role of neuropathological heredity in the degeneration of the species, stressing the laws of hereditary transmission, that Morel endeavored to study.[18]

In view of this, the so-called "laws of hereditary transmission" do not deserve that name. In the final analysis, they refer to "types" or "means" of heredity formulated in a speculative manner, which although initially useful in the explanation of hereditary disorders with no nosological relationship among them, could not be maintained (as Magnan understood) when faced with the development of Darwinian evolutionism or, later, Mendelian laws.[19] Like so many other medical men of his time, Bonafonte was unaware of the true "laws" of heredity, and his lack of knowledge could explain why he persisted, even in 1900, with outdated theories like that of dissimilar heredity. Still more alarming is that, as late as 1919, E. Vilches could still refer to atavism which he describes as "no more than one of the forms of heredity, that is to say the transmission of characters jumping one or more generations—a backward jump—and from here stems the term 'returning heredity' that it has been given by P. Lucas."[20]

---

[15] On Prosper Lucas and his influence on the Morelian theory of degeneration, see Huertas, *Locura y degeneración* (Madrid, CSIC, 1987), pp. 31 and following.

[16] V. Garrido, *La carcel o el manicomio* (Madrid, Administración Casa Editorial, 1888), pp. 231-232.

[17] M. Bonafonte, *Degeneración y locura* (Zaragoza, Manuel Ventura, 1900); on this subject, see Huertas (1995) "Sobre la recepción del degeneracionismo psiquiátrico en España: la obra de Mateo Bonafonte," in Arquiola and Martínez-Perez, *Ciencia en expansión,* pp. 521-534.

[18] Bonafonte, *Degeneración y locura*, p. 71. The allusion to Morel is direct. Based on Bonafonte's citation style, we can surmise that his comments on dissimilar heredity are taken from this author and not from the original text of Lucas, who is not quoted here.

[19] On the reception of Mendelism in Spain, see J. García Martínez, "Aportaciones a la historia de la genética española (1920-1936)," Madrid, unpub. lic. thesis, 1984.

[20] E. Vilches, *Degeneración y Atavismo* (Madrid, Tordesillas, 1919), p. 43.

This view, according to which mental illnesses develop in a complex and polymorphic manner thanks to faulty heredity was established empirically as the continued result of clinical cases and reports in which psychiatrists looked for family background as a determining factor in the "degeneration" of an individual, thereby giving mental illness a generational dimension in addition to an individual one. In this way, madness was biologically labeled not only by the conviction that it was the physical manifestation of a psychological conflict but also that by means of hereditary transmission it could be transformed and take on new forms, becoming a "stain" on an entire lineage—one that is destined to disappear. The conceptual identification of hereditary madness with degeneration was one of the aspects that most discussed among degenerationists. These discussions reached their peak during the sessions of the Societé Médico-Psychologique of Paris in 1886, when Magnan defended that identification by proposing the term *héréditaire dégénéré*. In Spain, although no proper scientific discussion was held, there were people who took a position on this question. In this respect, Bonafonte wrote that "all degeneration is not necessarily hereditary" (admitting the existence of acquired degeneration), but "neither are all hereditary beings degenerates."[21] In this way he drew attention to the confusion caused by the term "hereditary":

> Degenerates are known by this name and for many people the terms "hereditary" and "degenerate" are synonymous, but the confusion should merely be apparent as the expression "hereditary" is no more than an aetiological notion and the expression "degenerate" is a symptomatic one.[22]

He concludes with a criticism of the "scant scientific basis" of Magnan's characterization of a degenerate person: "Magnan gives degeneration a wide field . . . the absence of sufficient motive, the lack of a manifest etiology are (for Magnan) revealing signs of degeneration."[23] This is closer to Morel, who believed in acquired degeneration and said that many individuals can be born perfectly normal "without any hereditary blemish" and suffer, before their brain has completely developed, from "the destructive effect of perturbations that can affect their eventual and complete development, as a result of an acute or chronic infection, traumatisms, *etc.*"[24] In these cases there was always doubt that the hereditary blemish might not be manifested and that predisposition[25] and imbalances[26] are what are transmitted from parent to child, crucial elements in the definitive systematization of the theory of degeneration proposed by Magnan. Even during moments of profound change in psychiatric nosology when,

---

[21] Bonafonte, *Degeneración y locura*, p. 27.
[22] *Ibid., loc cit.*
[23] *Ibid.*, p. 31.
[24] *Ibid.*, pp. 27-28.
[25] Magnan and Legrain define predisposition as "the initial state of the degenerate in the absence of complication," *Les Dégénérés*, pp. 55 and following.
[26] An idea that might be taken as a "lack of harmony" between different organic functions (*ibid.*, pp. 116 and following).

during the early years of the Twentieth Century, the concept of premature dementia and Kraepelin's paradigm were being implanted, the persistent hold of a degenerationism that was already beginning to decline gave rise to comments such as that made by Tomás Maestre, professor at the Central University of Madrid, who wrote in 1906:

> I do not accept so-called premature dementia as a defined and distinct form of mental nosology, inasmuch as this process of dementia is no more than the end result of various different forms of madness that come about during adolescence or even in younger brains that are marked by hereditary law with a degenerative blemish.[27]

It is precisely the doctrine of degeneration that, according to this physician, permits the very rapid evolution of mental conditions contracted at an early age towards dementia when he asserts that:

> What is characteristic about madness in adolescence and youth is the rapid way in which these forms of insanity run their course and reach the final dementive process, a permanent and chronic state that resembles true imbecility. This rapid evolution is easily explained if we bear in mind the incomplete, weak and unbalanced state of the brains of degenerate children.[28]

Such an imbalance, in accordance with degenerationist principles, would give rise to a regression towards less advanced stages on the scale of phylogenetic and, in this case, ontogenetic evolution: "when an organ stops developing before it reaches the biological rank of its species a regressive process is immediately identified within it that tends to take it back towards its embryonic state."[29]

Not until the second and third decades of the Twentieth Century can one see an overturn in Spanish psychiatry: Kraepelinian nosology, the relationship between psychiatry and the neurosciences, reforms in treatment practices, mental hygiene, *etc.*, which were taken up by the generation of Lafora, Sacristán, Sanchis Banús, Mira y López, Safocarda, Escalas, *et al.*, who had few connections with nineteenth-century alienism. Their generation was strongly influenced by the neurology of Ramón y Cajal and Simarro—both evolutionists, the latter a staunch Darwinian—and by new trends in psychiatry diffused from abroad.[30]

And yet in spite of all, the degenerationist paradigm, although in decline, refused to disappear. Authors of standing such as Gonzalo Rodríguez Lafora or José Sanchis Banús continued to refer to it, although from a updated point of view.[31] Lafora, in 1917, refers to modern genetics when he states that "in the modern *eugenic* concept only that which has transformed a germ or seed in some way and this is then transmitted to its descendants in accordance with certain biological proportional laws (Mendel's Laws)

---

[27] T. Maestre, "Formas y patogenía de la demencia precoz," *Revista Frenopática Española,* 4 (1906), 183.
[28] *Ibid.*, p. 348.
[29] *Ibid.*, p. 349.
[30] There is an abundant bibliography on these subjects, e.g., A. Albarracín, "Las ciencias médicas en España," in R. Huertas, A. I. Romero, and R. Alvarez, eds., *Perspectivas psiquiátricas* (Madrid, CSIC, 1987), pp. 7-18.
[31] J. Sanchis Banús, *Estudio médico-social del niño golfo* (Madrid, Excelsior, 1916).

can be accepted as hereditary."[32] For this reason, given the ingrained tendency to include any modification transmitted through heredity under the rubric of degeneration, he cautiously states that:

> This is an abuse of the term degenerate, which has already met with much reaction, with some people seeing degeneration in all abnormalities. *Not all those with inherited problems are degenerates, only those who, as well as having intellectual perturbations and a badly developed brain, show various true degenerative signs.*[33]

Lafora distinguishes between hereditary blemishes and congenital defects—something that alienists of previous generations did with difficulty—and describes them as "causes that behave more congenitally:" alcoholism, syphilis, tuberculosis, and endocrine afflictions. It is a good example of the hygienic and social tone that Spanish psychiatrists began to adopt during the first three decades of the Twentieth Century, assimilating a socio-medical school of thought that for some time had related degeneration to various "white plagues" identified by the public health movement.

2. DEGENERATION AND SOCIAL HYGIENE

Although, as we have just seen, Spanish alienists offered individual interpretations of the theory of degeneration, one cannot forget that the work of Morel had important social and hygienic implications.[34] Although he recognized heredity as a cause of degeneration, Morel referred its ultimate cause to external factors. In 1860 in a letter to the Lower Prefect of the Seine, he declared the need to study and fight the causes of madness and degeneration, focusing his attention on society. In order to unravel the causes, he suggests an analysis of such diverse factors as morality, nutrition, hygiene, culture, the harmful effects of industry, criminality, alcoholism, *etc.*, within whatever sector of the population that is to be studied. In order to carry out this program of "prophylactic hygiene" to prevent "the indefinite spread of madness and of all human degeneration," it is, in his opinion, necessary to "look closely at the social life of a given locality" and to take note of its physical and moral hygiene, a task he considered could only be carried out under the patronage of government.[35]

Until well into the Twentieth Century Spanish alienists do not seem to have focused on this aspect of Morel's work. Elsewhere we have tried to demonstrate how the reality

---

[32] G. R. Lafora, *Los niños mentalmente anormales* (Madrid, La Lectura, 1917), p. 72. On Lafora's eugenic ideas, see R. Alvarez, "Herencia, sexo y eugenesia," in Huertas, *et al.*, *Perspectivas psiquiátricas*, pp. 203-218.

[33] Lafora, *Niños mentalmente anormales*, p. 72, emphasis in original.

[34] Hochmann, "Théorie de la dégénérescence;" R. Huertas, "Entre el nihilismo terapéutico y la higiene social: la asistencia psiquiátrica en el positivismo francés," in A. González de Pablo, ed., *Enfermedad, clínica y patología. Estudios sobre el origen y desarrollo de la medicina contemporánea* (Madrid, Complutense, 1993), pp. 301-314.

[35] B. A. Morel, Le non-restreint ou de l'ábolition des moyens coercitifs dans le traitement de la folie (Paris, Victor Masson, 1860), p. 103.

of care provided for the insane and the professional interests of alienists combined to dissuade psychiatrists from exploring the relation between social hygiene and madness. Only at the beginning of the century was there a change in attitude that would culminate in the 1920s with the creation of a Mental Hygiene League.[36] It was only then, with growing interest in social medicine, that psychiatrists turned to the social and collective aspects of degeneration.

The hygienists focused on the problems of the working class and proposed a series of measures to resolve them that were directed towards ameliorating the environment in which they lived as well as inducing them to alter their self-destructive customs and lifestyles.[37] With this knowledge, reinforced by the belief that their particular science was the most appropriate for resolving these problems and neutralizing social conflict, it is logical that the hygienists' reading of degeneration was centered on classes, not on individuals. The strictly clinical and individual aspects of degeneration were of minor importance to the social physicians and were superseded by collective aspects. Their principal preoccupation was the deleterious consequence of degeneration on the species, which reinforced their views interventionist proclivities.[38] In 1909 N. Fernandez Cuesta wrote that hygiene

> seeks intellectual perfection and through it, pure and practical morality, passing from individuals to collectivities in order to reach into public life, formulating carefully considered principles. . . which, when enacted through the reform of customs and races, will assure society of material regeneration which will lead to moral, social, and political progress.[39]

In just that period, J. Juderias wrote that hygiene "is not understood to mean simply the art of preventing illness and conserving health, but also the most important art of improving the race." In order to reach this objective he emphasized the State's right to impose the precepts of hygiene, "constricting individual freedoms in the interests of the higher, supreme aim of the defense of the race."[40]

The collective dimension of degeneration found in biological heredity one of its basic pillars. In opposition to the individual focus of the psychiatrists, hygienists gave

---

[36] R. Campos Marín, "Psiquiatría e Higiene Social en la España de la Restauración" in *Un siglo de Psiquiatría en España*, pp. 53-66.

[37] On social medicine in Spain, see E. Rodríguez Ocaña, *La constitución de la Medicina social como disciplina en España (1882-1923)* (Madrid, Ministerio de Sanidad, 1987); *idem*, "Paz trabajo, higiene. Los enunciados acerca de la higiene industrial en la España del siglo XIX," in R. Huertas and R. Campos Marín, eds., *Medicina Social y clase obrera en España. (Siglos XIX y XX)* (Madrid, FIM, 1992), II: 383-400; P. Trinidad Fernández, "Trabajo y pobreza en la primera industrialización," in *Historia de la Acción Social Pública en España. Beneficiencia y Previsión* (Madrid, Ministerio de Asuntos Sociales, 1990), pp. 101-133; Campos Marín, "La Sociedad enferma: Higiene y moral en España en la segunda mitad del siglo XIX y principios del XX," *Hispania*, 3 (1995), 1093-1112.

[38] In the case of France, see G. Vigarello, *Le Sain et le malsain. Santé et mieux-ètre depuis le moyen âge* (Paris, Seuil, 1993), pp. 219 and following.

[39] N. Fernández Cuesta y Porta, *La vida del obrero en España desde el punto de vista higiénico* (Madrid, 1909), p. 139.

[40] J. Juderias, *La Higiene y su influencia en la legislación* (Madrid, M. Minuesa, 1911), pp. 5, 10.

degeneration a social spin that put the whole of the Spanish "race" in permanent danger. There is no doubt that this reading of degeneration had a lot to do with the sense of disaster that befell Spanish cultural and political elites in the wake of Spain's defeat in the Spanish-American War of 1898. A good number of the non-psychiatric texts relating to degeneration appeared after "The Disaster," which many of these authors attributed to the degeneration of the Spanish people and their loss of vigor.[41]

Questions such as the reduction in size and the increase in the births of sickly and rachitic children, predisposition to illness, *etc.*, that had always worried the hygienists were, in the latter part of the Nineteenth Century, increasingly related to degeneration and heredity. Descriptions of weakened human beings, whose physical and psychological symptoms were related to social deviance, took on an anguished tone as physicians came to accept the idea of the biological decadence of the Spanish people. Illness, driven by heredity and crystallizing in degeneration acquires an unsettling character when it manifests itself beyond the individual. Madness, alcoholism, tuberculosis, syphilis, *etc.*, became illnesses that did not disappear with the death of the subject, but were transmitted to their progeny, degenerating the race. Philip Hauser, one of the most important hygienists in nineteenth-century Spain, identified mental illnesses, addiction to morphine, opium or tobacco, alcoholism, and syphilis as racially degenerative diseases, and B. Malo de Poveda considered alcoholism, tuberculosis, and syphilis as "the true degenerative threesome in our race."[42] Likewise, high mortality, especially infant mortality, reduction in stature, misery, poor working conditions, the spread of prostitution, the incidence of crime and vice, *etc.*, are at the same time both cause and symptom of degeneration. This mixed bag of specific illnesses, constitutional weaknesses, and pitiful social conditions comprised the unsettling signs of the degeneration of a race.[43]

---

[41] D. García Guerra and V. Alvarez Atuña, "Regeneracionismo y Salud Pública. El bienio de Angel Pulido al frente de la Dirección General de Sanidad," *Dynamis,* 14 (1994), pp. 23-41.

[42] P. Hauser, "El siglo XIX considerado bajo el punto de vista médico-social," *Revista de España,* no. 101 (1884), 202-217; B. Malo de Poveda, "Causas que contribuyen a la mortalidad de los niños. Medios de remediarlas. Estadísticas comparativas," in *Actas y Memorias del IX Congreso Internacional de Higiene y Demografía, Madrid, 10-17 abril 1898* (Madrid, R. Rojas, 1900), IX: 42.

[43] As examples, see J. Gómez Ocaña, *La vida en España. Discurso leido en la Real Academia de Medicina* (Granada, Escuelas Ave María, 1900); C. Bernardo de Quirós and J. M. Llanas Aguilaneida, *La mala vida en Madrid, Estudio psico-sociológico con dibujos y fotograbados del natural* (Madrid, B. Rodríguez Serra, 1901); A. Martínez Vargas, *Nuestras madres y el agradecimiento patrio. Discurso pronunciado por el Dr...el 12 de mayo de 1906, al presidir el certamen médico* (Barcelona, Jaime Vives, 1906); Ubeda Arce, *El trabajo de la mujer en la industria: condiciones en que se efectúa y sus consecuencias en el porvenir de la raza. Medidas de protección necesarias* (Madrid, Sáez, 1913); A. López Núñez, *La acción social de la mujer en la higiene y mejoramiento de la raza* (Madrid, M. Minuesa, 1915); A. Marioni and L. Huerta, *Cartilla Popular de Medicina Social* (Madrid, Instituto de Medicina Social, 1919); M. Molina Collada, *El llamado modernismo en la vida actual. Su influencia higiénica y moral en la educación de la juventud de ambos sexos* (Madrid, M. Minuesa, 1924).

The individual, therefore, is the carrier of the "resources" of a species and their actions do not exclusively concern them, because heredity transfers their debilities to future generations:

> It could be said metaphorically that when man comes into this world he receives for his own use the property of the species, from way back, linked to its primitive trunk. He receives aboriginal traits as if the entire series of his ancestors had no more than their use. In the course of his lifetime, he passes these traits on to his children, perhaps not the complete package, but to the extent that it has not been consumed, transformed, or improved by his own contributions.[44]

This inheritance could put the whole of the race in danger:

> From the political and social point of view, if it is important for a people to look out for the health and lives of its citizens, it is even more important that the race conserve the integrity of the hereditary elements which spawn new generations. The flow of life can be contaminated or corrupted in two different ways: either in its origins or along the course that it takes. Illnesses produced by external causes can finish off an individual, but if they do not die they may regenerate; on the other hand those causes that attack the germ plasm cannot regenerate. Nothing can be inherited that is not in the cellular germ.[45]

Thus the physician must extend his compass from the study of isolated cases to the "collective case" which is "of greatest importance at present" because, as stated in the first issue of *La Medicina Social Española*, it "touches many lives," and "bears on the vigor, health, and longevity of families, townships, or aggregates of them, reflecting on the generations to come."[46]

3. ALCOHOLISM AND TUBERCULOSIS AS CAUSES OF DEGENERATION

The relationship between alcoholism and degeneration is paradigmatic of socio-medical concerns and is one of the key sites in the degenerationism debate. Indeed the theory of degeneration was probably best expressed in Spain in anti-alcohol manifestos. Alcoholism was considered one of the most horrible vices suffered by society and was also linked to the working class. In the Nineteenth Century it served as a convenient indicator of the moral stigma afflicting that class and was also used as an instrument for arguing in favor of intervention in moral hygiene.[47]

From the 1890s on, anti-alcohol propagandists centered their attention on the relationship between alcoholism and degeneration, stressing the role of hereditary alcoholism as a source of practically all individual and social ills. The publication in France during the latter half of the previous decade of numerous works and articles relating to degeneration and alcoholism, in which hereditary alcoholism was studied in

---

[44] A. de Redondo y Carranceja (1919), De la degeneración y de la regeneración de nuestra raza. Discurso leido en la solemne inauguración del curso académico del 1918-1919 (Madrid, Colonias, 1919), p. 78.
[45] M. Martín Salazar, *Problemas sanitarios sociales* (Madrid, Nieto, 1923), p. 72.
[46] "Saludo," *La Medicina Social Española*, 1 (1916), 7-8.
[47] R. Campos Marín, *Alchololismo, Medicina y Sociedad en España (1876-1923)* (Madrid, CSIC, 1997).

detail from a degenerationist point of view, were reflected in Spain a few years later, and in a short space of time became one of the principal arguments in the fight against alcoholism.[48] Thus in 1896, Rafael Cervera Barat referred to the "laws of heredity in alcoholism" in order to demonstrate that "the ravages of alcoholism accumulate from one generation to the next, until complete families are wiped out."[49] Following Legrain's data in *Dégénérescence sociale et alcoolisme,* Cervera established and classified in three groups (weak intelligence, impulsive and moral madness) a series of mental defects inherited from alcoholism over three generations. "These heirs," he explained, "suffer an infinity of disorders and defects of the mind, the multiple effects of which can be summarized scientifically in one word: mental degeneration."[50]

Belief in polymorphous heredity, by means of which mental disorders could be transformed and aggravated from generation to generation until they reached the point of exhausting and wiping out the stock, was universal among physicians studying alcoholism. The disease was transmitted hereditarily, increasing the degeneration of descendants who displayed "all sorts of irregularities of the nervous system."[51] The result was that the descendants of an alcoholic ended up forming "a true museum of disorders in which all sorts of moral, physical and intellectual monstrosities can be studied, above all when both the father and the mother have abused spirits."[52] The species, the race, and indeed the nation, all paid a high price for individual excesses which, through heredity, had given rise to all sorts of social problems. Likewise, alcoholic degeneration was tightly correlated with madness and crime, two of the most dangerous manifestations of alcoholism.[53]

Social hygienists also established a connection between tuberculosis and degeneration. Hauser stated at the Ninth Congress on Hygiene and Demography (Madrid, 1898) that tuberculosis was a disease that degenerated the human species. As late as 1934, when the theory of degeneration had supposedly fallen into disrepute, C. Díez Fernández presented a paper at the First National Health Congress entitled "Tuberculosis a Modifying Factor of The Race."[54] However, as tuberculosis is an infectious disease, the relationship with degeneration was not as evident as was the case with alcoholism. Koch's discovery in 1882 of the bacillus that caused the disease

---

[48] See R. Huertas, "Madness and Degeneration, II. Alcoholism and Degeneration," *History of Psychiatry,* 4 (1993), 1-21.

[49] R. Cervera Barat, *Alcoholismo y Civilización* (Valencia, A. Cortés, 1896), p. 22.

[50] *Ibid.*, p. 23.

[51] A. Piga and A. Marioni (1904), *Las bebidas alcohólicas. El Alcoholismo* (Barcelona, Soler, 1904), pp. 142-143.

[52] A. Muños Ruiz de Pasanis, *Alcoholismo. Su influencia en la degeneración de la raza latina* (Madrid, Ginés Carrión, 1906), pp. 84-85.

[53] R. Campos Marín, "La recepción de la teoría de la degeneración francesa en el discurso antialcohólico español (1890-1915)," in Arquiola and Martínez, *Ciencia en expansión*, pp. 449-467.

[54] P. Hauser, "La defense sociale contre la tuberculose," in *Actas, IX Congreso Internacional,* III: 109-125; C. Díez Fernández, "La Tuberculosis como factor modificativo de la raza," *Actas del I Congreso Nacional de Sanidad. Madrid, 6-12 de mayo de 1934* (Madrid, 1935), II: 335-337.

shattered the traditional hereditary thesis.[55] Nevertheless, despite the impact of bacteriology, a large group of phthisiologists with socio-medical interests continued to subscribe to the existence of a hereditary predisposition towards the illness. Some speakers at the Third Spanish Tuberculosis Congress (San Sebastián, 1912) invoked this concept, insisting on the degenerative capacity of tuberculosis and on the need to take measures to fight the degeneration of the race. Dr. Borobio, when analyzing the hereditary scope of tuberculosis, rejected the existence of congenital tuberculosis, while also stressing a hereditary predisposition to tuberculosis among those who were "more susceptible to consumption" due to having "inherited an organic site well prepared for the germination of the seed."[56]

At the same meeting, B. Malo de Poveda expressed himself in a similar manner, although he indicated that the children of consumptives might also be subject to other diseases:

> Now, if the direct transmission of tuberculosis from parent to child during the intrauterine stage can be discounted due to its exceptional nature, there can be no argument, due to its notoriety and the thorough manner in which it has been studied, about the logical and natural fact of the transmission of the parent's organic weakness, exhibited in the child as positive vulnerability not only to tuberculosis, but to other illnesses.[57]

Throughout this Congress there was mention of the danger that tuberculosis held for the vigor of the race. In the inaugural session, the Minister of State, M. García Prieto, noted that the mission of intelligent politicians was to "try to reduce futile expenditure in order to devote more to the improvement of the race, as this will lead to the growth of the Nation."[58] In order to reach this goal, to safeguard the "firm rights of the species," many were of the opinion that it was necessary to prevent marriages involving people with tuberculosis or with a predisposition to it. R. Royo Villanova, for example, the importance of cultivating the "new science of good breeding and its promotion among healthy people." If it were not cultivated

> there will be brought into the world unhealthy people who, instead of being honorable will be criminals, instead of healthy, sick, and who instead of fulfilling the aspirations of Humanity will be a disgrace to it, fodder for cemeteries, prisons and hospitals.[59]

Malo de Poveda insisted in his paper on Tuberculosis and Matrimony that to prevent the disease legislation was required so that the conception of new human beings might be governed by rational and scientific criteria, inspired by the rights and duties of

---

[55] On this point, see J. Molero Mesa, "Historia Social de la Tuberculosis. (1889-1936)," unpub. doctoral diss., University of Granada, 1989, pp. 31-42.

[56] Borobio, "Alcance hereditario de la tuberculosis", *III Congreso Español de la Tuberculosis*. Segundo con carácter internacional celebrado en San Sebastián 9-16 de septiembre de 1912 (San Sebastián, Sociedad Española de Papelería, 1913), I: 554.

[57] B. Malo de Poveda, "Tuberculosis y Matrimonio," *ibid.*, I: 453.

[58] "Sesión Inaugural," *ibid.*, I: 23.

[59] R. Royo Villanova, "Política antituberculosa," *ibid.*, I: 156.

society, which should be composed of strong and vigorous individuals, capable of competing successfully in the unavoidable struggle for existence.

The contrary would be a crime against humanity, a contribution to the "degeneration of the species itself."[60] Accordingly, he proposed that measures should be taken to prevent marriage between individuals who might transmit an unhealthy constitution. The Congress adopted, after the appropriate discussions, a resolution favoring the passage of a law that to "establish limitations on the contracting of marriage amongst individuals with tuberculosis."[61]

This was not a new question. Malo de Poveda himself had referred to it previously at the Ninth International Hygiene Congress. In 1903, B. González Alvarez proposed in his work "Prophylactic Hygiene of Children with Respect to Heredity," the creation of a legally binding health certificate prior to contracting marriage to be issued by a public health tribunal, as a means of alleviating the degeneration and extinction of the race.[62] Later, in 1915, this same doctor, in his capacity as Senator representing the Royal Academy of Medicine of Madrid, proposed a law making medical certification before marriage obligatory. In the event that either party suffered from "an illness or organic state that is transmissible, either hereditarily or congenitally to their descendants," the marriage would be prohibited.[63] This proposal gave rise to a great deal of debate in the pages of *La Medicina Social Española.* During the 1920s and 30s the number of such proposals increased dramatically, reflecting the rising tide of Spanish eugenics.[64]

Hygienists and their allies found in degeneration a basic argument for social action. The interests of the species, the bloodline, in fact of the nation were above those of the individual and justified intervention with the aim of extirpating the roots of degeneration. "Purifying the blood of parents is an official duty of the State," a respected lawyer claimed in 1912.[65] In any case, although degeneration might be inheritable, heredity itself did not necessarily have to be unhealthy; it could also promote racial vigor. Once again. Malo de Poveda expressed it clearly:

> Organic heredity, recent or remote, maternal or paternal, convergent or neutralizing, by means of which robustness or weakness of constitution, illnesses or an aptitude for them may be transmitted, represent, according to each case, an important cause of the vigor or degeneration of a race, and is therefore to a great degree responsible for infant mortality.[66]

---

[60] Malo de Poveda, *loc cit.*

[61] "Sesión de conclusiones," *ibid.*, I: 339.

[62] B. González Alvarez, "Higiene profiláctica del niño respecto de la herencia" *El Siglo Médico,* 1903, pp. 183-185 and 582-585.

[63] The law is included in B. Malo de Poveda, "De Legislación Sanitaria," *La Medicina Social Española,* 1 (1916), 60.

[64] R. Alvarez Peláez, "Introducción al estudio de la eugenesia española (1900-1936)," *Quipu,* 2 (1985), 95-122.

[65] G. Doval, "Intervención del derecho en los matrimonios de los tuberculosos," *III Congreso Español de la Tuberculosis,* II: 451.

[66] Malo de Poveda, "Causas que contribuyen a la mortalidad de los niños," p. 38.

Heredity need not always presage catastrophe. If the necessary hygienic and moral measures were undertaken, the degenerative tendency could be corrected. The somber fate of the Spanish race, proven by the country's decadence, was more of a rhetorical conceit, a warning of what could lie ahead in the absence of intervention. Degeneration was not limited to the problem of biological heredity, but was impacted by social and moral causes as well. For many physicians the social aspect had the same, or more, importance than did heredity:

> We believe that degeneration of the human race has no fixed outcome and cannot be represented as a continuous, descending progression as some mentalists—among them, Morel—have thought. The counterpoising regenerative elements that man possesses can be favored by the environment in order to mitigate any inherited disease.[67]

The causes of degeneration could be also be moral. Pornography, the cinema, coarse public spectacles, vice, onanism, could produce moral, physical and intellectual degeneration.[68] So too would the prolonged influence of poor hygienic conditions at home or at work, laying "the right ground for the germination" of tuberculosis and "those illnesses known as degenerative to the human race."[69] In fact, social misery was singled out by some doctors and organizations as the cause of the propagation of degeneration. Referring in 1918 to the consequences of the misery of the proletariat, F. Murillo wrote:

> It is clear that the damage does not end with the burial of the victims, and that if there are many innocent orphans whom society has failed to protect, there are many more who survive the shipwreck but who will live forever with the stigma of degeneration, propagating it by the continuous crossing of offspring.[70]

From this point of view degeneration could best be arrested by public health measures: Education, social improvements, the heightening of morality, and cleaning up the environment were important tasks to be undertaken in order to mitigate or prevent the disaster of degeneration.

---

[67] T. Sanz, "Problemas médico-sociales," *El Siglo Médico*, no. 2016 (1892), 519.

[68] J. Molero Mesa, "La tuberculosis como enfermedad social en los estudios epidemiológicos españoles anteriores a la guerra civil," *Dynamis*, 9 (1989), 185-223; F. Vázquez and A. Moreno, *Sexo y razón. Una genealogía de la moral sexual en España (Siglos XVI-XX)* (Madrid, Akal, 1977).

[69] Hauser, "El siglo XIX considerado bajo el punto de vista médico-social," pp. 219-220.

[70] F. Murillo, *La defensa social de la Salud Pública. Discursos leídos en la Real Academia de Medicina para la recepción pública del académico electo (...)* (Madrid, Nicolás Moya, 1918), p. 45 "Our mortality is terrifying. And despite everything that is not the worst part. The worst part is the pauperism, exhaustion, and physical and mental degeneration which bring as a consequence our unimaginable living conditions and place us in a position of obvious inferiority to undertake the work of civilization, and condemn us to walk behind--a century behind--the more advanced nations;" "Instituto de Medicina Social," *La Medicina Social Española*, 4 (1919), 503.

4. SOCIAL DEGENERATION

Some observations on the sociological uses of degeneration conclude our survey. Turn-of- the-century literature introduced into Spanish culture a feeling of decadence which, reinforced by the events of 1898, ended up confirming "degeneration" as a diagnosis of the ills afflicting Spanish society.[71] Thus, socialists and anarchists alike accused the bourgeoisie of being "degenerate" because of their inability to complete their historical mission and for contraverting the "laws of nature" by creating an immoral system based upon exploitation.[72]

In medicine, as well as in sociology and other disciplines, it was common to compare society with an organism that was susceptible to cure by the right therapy. This reading of social ills influenced both the first hygienist movement and the discipline of sociology, "because those who cultivate the terms 'social' illness or disease referred to problems in the economic, political, or moral order of society—crime, begging, prostitution, strikes, revolution."[73] At the turn of the century, this concept of social illness found degeneration theory fertile soil in which to develop.

At the beginning of the century, the Argentinean psychologist C. O. Bunge published a famous book on education (*La educación*), the second volume of which contains a long chapter on the education of degenerates, in which he outlines two types of degeneration: "the medical type, individual and absolute, and the sociological type, general and relative." In his opinion, only the first had received the attention of doctors, whereas the second had escaped their analysis because just to perceive it requires "a knowledge of history—a metaphysical concept of history, I would almost say. And there is nothing more opposed to metaphysics than the positivist spirit of medicine."[74] In Bunge's opinion, continuous clashes between "healthy germs and germs weakened by heredity" have produced the degeneration of society "in all its spheres."[75]

Later on, A. de Redondo y Carranceja, in his lecture inaugurating the 1918-19 academic year, noted a correlation between the individual degeneration of Spaniards and social degeneration. Convinced that the misgovernment of Spain was a degenerative

---

[71] On this question, consult L. Maristany, "El artista y sus congéneres. Diagnósticos sobre el fin de siglo en España," unpub. doctoral diss., Barcelona, 1985; also, L. Litvak. (1990), "La idea de la decadencia en la crítica antimodernista en España (1888-1910)" and "Temática de la decadencia en la literatura española de fines del siglo XIX: 1880-1913," in her *España 1900. Modernismo, anarquismo, y fin de Siglo* (Barcelona, Anthropos, 1990), pp. 111-127 and 245-258.

[72] On socialists, see R. Campos Marín, "Herencia biológica y medio social en el discurso antialcohólico del socialismo español (1886-1923)," in Huertas and Campos Marín, *Medicina Social y clase obrera de España*, I: 67-91; A. Girón Sierra, "Evolucionismo y Anarquismo: La incorporación del vocabulario y los conceptos del evolucionismo biológico en el anarquismo español (1882-1914)," unpub. doctoral diss, Universidad Complutense de Madrid, pp. 456 and following.

[73] E. Rodríguez Ocaña, "El concepto social de la enfermedad", in A. Albarracín, ed., *Historia de la enfermedad* (Madrid, Saned, 1987), p. 345.

[74] C. O. Bunge, *La Educación. Educación de los degenerados. Teoría de la educación* (Madrid, Daniel Jorro, 1903), p. 81.

[75] *Ibid.*, pp. 75-76.

factor of the first order, Redondo compared the human body with society in order to analyze the ills that had befallen the country:

> Society, a collection of human beings, displays characteristics so similar to those of the members that comprise it that the use of metaphors likening one to the other is not strange. Nor should it be strange to hear of static and dynamic societies, comparing them with the individual, or words like birth, infancy, and decrepitude, nor of vegetative functions and their relationships, psychology, sensibilities, movements, organs, development, decadence, *etc*. In fact, that society should be compared to a human being, to man, demonstrates their similarity.[76]

From this proposition he goes on to compare the different parts of the body with the organization of society. The government is the brain, the proletariat are the cells, and the intellectuals are the nervous system. The incapacity shown by each of them in carrying out their roles in society was, in his opinion, a clear symptom of social degeneration spreading throughout Spain.

On more than one occasion, this conservative, moralizing view of degeneration was used to discredit the revolutionary activities of the working class movement. In 1921, Angel Pulido, in full clamor of social conflict, published a book called *La Degeneración del Socio-Sindicalismo. Necesidad de su regeneración higiénica y moral* [Degeneration of Socio-Syndicalism. The Need for its Hygienic and Moral Regeneration] in which he took this view of degeneration to its limit. The work is a 435-page collection of insults and reprimands directed at the working class and its organizations, using Social Medicine as a shield. When referring to social conflict in Barcelona, Pulido attributes it to "a degenerative and malignant infection of the worker" that has converted the city into fodder for "every senseless, criminal, deadly radicalism . . . .to the point where there is no city. . .in this most politically turbulent and neurotic of nations that can be compared to it."[77] Even though he believed many of the workers' claims to be justified, he was also convinced that they then suffered from

> degenerations, hypertrophies and monstrosities of revolutionary and disruptive origin that lead to catastrophic barbarities, rather than to a truly healthy development of civilization which would be beneficial to both social democracy and the proletarian classes, which have an indisputable right to betterment.[78]

The heart of the matter for Pulido was that societies are ruled by biological laws, the transgression of which leads organisms and societies to "a state of degeneration, atavism, sterility, and finally to death." The principles upon which social life is based are patriotism, love of work, and respect for authority and religion. Non-compliance by the working classes was a sign of "the moral decadence into which they have fallen," in short, of their degeneration. To get things back on track, he prescribes social legislation

---

[76] Redondo Carranceja, *Degeneración y regeneración*, pp. 25 and following.

[77] A. Pulido, *La Degeneración del Socio-Sindicalismo. Necesidad de su regeneración higiénica y moral* (Madrid, M. Nuñez Samper, 1921), p. 186.

[78] *Ibid*., p. 224.

under a dictatorship, which he likens to the strong medicines required to cure tackle serious illnesses.[79]

This type of reasoning, expressed in such a radical manner by Pulido, was shared by other doctors. In 1916 M. Martínez Vargas referred to the need to apply "Social surgery, because in order to cure the ills of our country it is necessary to get rid of the ineffective remedies currently in use and seek more vigorous measures, surgical removal without hesitation"[80] And Martín Salazar in 1923, the year of General Primo de Rivera's coup d'état, insisted that non-compliance with the laws of nature was a reason for the degeneration of society, and he accused the workers' movement of contributing to it:

> A society built upon a base of radical trade unionism, with its principles of equality, is at odds with the laws of nature, with its opposition to all individual initiative. . . with no sense of personal responsibility, which is the true origin of all moral perfection in Man. Such a society is doomed to degenerate and disappear over a lesser or greater number of generations for not obeying the law of natural selection and due to the absence of those important psychological resorts that over the life history of the species have promoted the organic perfection of the individual.

The "small experiment of radical trade unionism" would relax moral principles and disturb the social economic order, producing an organic impoverishment that the race would have to suffer "running the risk, if it followed that path for a long time, of degeneration and extinction."[81]

Biology represents morality in this line of thought. Those individuals, groups, and societies which contravene its laws are destined for degradation and decadence.[82]

Centro de Estudios Históricos, Madrid

---

[79] *Ibid.*, p. 212.
[80] Martínez Vargas, "Alcance de la Medicina Social Española," *La Medicina Social Española,* 1 (1916), 6
[81] Martín Salazar, *Problemas sanitarios sociales*, p. 12.
[82] This study was carried out within the framework of DGICYT Project Pb94-0060.

ALVARO GIRÓN

THE MORAL ECONOMY OF NATURE

Darwinism and the Struggle for Life in Spanish Anarchism (1882-1914)

The acceptance of Darwinism in Spain, as Thomas F. Glick has observed, is clearly associated with the 1868 Revolution and the scientific institutions created after it.[1] Moreover, during the six-year period of Democracy an event of great importance occurred: in 1869-70 part of the Spanish working-class movement broke tactically with Federal Republicanism, and Bakuninist Anarchism appeared on the scene.[2] However, it is not really possible to speak of significant acceptance of Darwinism within Spanish anarchism at this time. We need to wait, rather until the 1880s, from which moment it would be fair to say that anarchists were also Darwinists.[3]

[1] Thomas F. Glick, *Darwin en España* (Barcelona, 1982). p. 25.

[2] The ideological dependence of Spanish anarchism with respect to Federal republicanism has been emphasized by Antonio Elorza, "Utopía y revolución en el movimiento anarquista español," in B. Hofman, P. Joan i Tous, and M. Tietz, eds., *El anarquismo español y sus tradiciones culturales* (Frankfurt and Madrid, 1955), pp. 79-108, on p. 84. Other authors, while recognizing this dependence, see clear differences between republicans and anarchists. These discrepencies centered on the libertarian aim of abolishing the State, in the anarchist ideal of collectivizing the means of production, and in the evident divergence of tactics derived from the anarchist rejection of politics. See José Alvarez Junco, "El anarquismo en la España contemporánea," in *El movimiento obrero en la historia de Cádiz* (Cádiz, 1988), pp. 41-51, on p. 44.

[3] There are very few references to the debate over Darwin in the libertarian press until the year of his death (1882). There are some isolated references to Darwin in *La Revista Social*, e.g., "Un proletario de la ciencia," *La Revista Social*, 75 (1874), 299. Even more indirect--referring in a generic manner to the possible animal origins of man--are those that appear in *El Trabajo*, e.g., "La ciencia en la Revolución," *El Trabajo*, 1 (1872) 1. Proposals advocating change in the education system based on the introduction of an evolutionist point of view are more interesting; see A. Lorenzo, *El proletariado militante* (Madrid, 1974), p. 261; "Las ciencias al alcance del pueblo," *La Revista Social*, 114 (1875), 453; C. E. Lida "Educación anarquista en la España del ochocientos," *Revista de Occidente*, 97 (1971), 33-47; *idem*, *Anarquismo y revolución en la España del XIX* (Madrid, 1972), p. 152. Some of the Bakuninists did at first maintain a close relationship with the supporters of Darwinism. One such case is that of the physician Gaspar Sentiñón, who both knew and translated Ludwig Büchner. He even edited a freethinking magazine called *La Humanidad*, in which he published Büchner's reflections on Darwin's theory (nos. 31, 33 and 41, 1871). However Sentiñón retired from public life early on, and there is nothing to indicate that there was any serious discussion of Darwin's theory itself in the 1870s. See Alvarez Junco's annotation of Lorenzo, *El proletariado militante*, p. 446 n. 21; J. Termes, *Anarquismo y*

Thomas F. Glick., Rosaura Ruiz, and Miguel Ángel Puig-Samper (eds.), The Reception of Darwinism in the Iberian World, 189-203.

Although "Darwinism" is an ambiguous term, the Spanish anarchists were as much Darwinian as the majority of their contemporaries. That is, they were evolutionists: they believed that the origins of life, of man, and of the diversity of species could be explained exclusively by the action of natural law. They also subscribed to an idea of progressive unilineal evolution that inevitably culminated in civilized man. Of course, this last proposition was at odds with fundamental aspects of the approach developed by Darwin, but it is no less true that what was understood by "Darwinism" in the latter decades of the nineteenth century was mainly a loose acceptance of the general idea of evolution that radically avoided the anti-teleological line of *The Origin of the Species.*[4]

It could be said that Spanish anarchists were Social Darwinists[5] in that they tended to establish a minimum distance between Nature and Culture.[6] Yet, Spanish libertarians did not fully accept the conventional definition of Social Darwinism inasmuch as, from the mid-1880s on, they rejected what they called the "struggle for life theory."[7] Their reasoning was clear: middle class scientists were converting Darwinism into a justification of inequality which brought socialism into disrepute. The starting point, especially in articles written by the patriarch of Spanish anarchism, Anselmo Lorenzo, was typically a

---

sindicalismo en España. La primera internacional (1864-1881) (Barcelona, 1972), p. 126 n. 10; J. J. Morato, *Líderes del movimiento obrero español* (Madrid, 1972), p. 24 n. 6.

[4] The idea that Darwinism was a loose consensus has been argued by various authors, including D. Hull, "Darwinism as a Historical Entity: A Historiographical Proposal," in D. Kohn, ed., *The Darwinian Heritage* (Princeton, 1985), pp. 773-812; J. Moore, "Deconstructing Darwinism: The Politics of Evolution in the 1860s," *Journal of the History of Biology*, 24 (1991), 353-408; M. Di Gregorio (1996), "Darwinisme anglo-saxon," in P. Tort, ed., *Dictionnaire du darwinisme et de l'evolution* (Paris, 1996), pp. 866-889. Other authors have gone further, claiming that the so-called Darwinism of the end of the nineteenth century can be defined as a finalist evolutionism basically disconnected from the logic of Darwin's theory: D. Becquemont, *Darwin, darwinisme, evolutionisme* (París, 1992); P. J. Bowler, *The Non-Darwinian Revolution* (London and Baltimore, 1992).

[5] There is not, at present, any widely accepted definition of so-called Social Darwinism. See, on this score, A. La Vergata, "Images of Darwin: a Historiographic Overview," in Kohn, *Darwinian Heritage*, pp. 901-972, on p. 948. La Vergata also claims that Social Darwinism, taken as a historiographical category, can only lead to confusion: "Biologia, scienze umane e 'darwinismo sociale': riflessioni contro una categoria storiografica dannosa," *Intersezioni*, 2 (1982), 77-97. It seems preferible to continue with the broad, conventional definition of Social Darwinism, according to which social laws are laws of nature and the struggle between individuals or social groups is a source of social and biological progress.

[6] This is not strange if we bear in mind that one of the principle sources of Spanish anarchism insofar as these subjects are concerned, the Russian anarchist Peter Kopotkin, forged a strongly naturalist vision of social processes. According to Diego Núñez, Kopotkin made the same gnoseological error as the liberal social Darwinists, only changing the social lesson derived from Darwinist theory; "El impacto del naturalismo y del evolucionismo en el pensamiento liberal y socialista," *Anthropos*, 16-17 (1982), 66-72, on p. 71.

[7] The rejection of this supposed theory was not applied to Darwin himself. A line was usually drawn between the English naturalist and the middle class that used his theory for its own benefit. There are examples to be found during the 1880s and 1890s: "Arte y ciencias. Carlos Darwin," *La Revista Social*, 52 (1882), 3-4, on p. 4; "Darwinismo social," *La Alarma* 19 (1890), 1. It was even said that "an aristocratic tendency could not be attributed to Darwinism, as middle class commentators wished, but a socialist one:" "Darwinismo social," *La Alarma*, 21 (1892), 1-2. There were critics, but not many: J. Montseny, *La ley de la vida* (Reus, 1893), p. 37; C. Jacquinet, "Ciencia falsa," *El productor*, 59 (1904), 1. In fact, the introduction of Kropotkinian solidarism during the early twentieth century reinforced this distinction between Darwin and his middle class manipulators: see A. Lorenzo, *El derecho a la evolución. Conferencia sociológica* (Buenos Aires, 1928), p. 24; "Carlos Darwin. Número extraordinario," *Humanidad Nueva*, 1 (1909), 10.

partial echo of the famous debate which took place in 1877 between Rudolf Virchow and Ernst Haeckel.[8] Haeckel defended himself against the accusation that Darwinism was related to socialism by confirming the aristocratic nature of the former: social hierarchies are no more than the projection of natural hierarchies.[9] Therefore, attempts by socialists to establish an artificial equilibrium between the strong and the weak leads directly to overreproduction of the latter and, therefore, the degeneration of the species.[10] Spanish anarchists saw these arguments as a new dogma or theology, legitimizing inequality by invoking science. Federico Urales, for example, said that writers exerted themselves "to perpetuate prevailing injustice, some using Jesus and others Darwin for support."[11]

The anarchist answer to the Haeckelian and similar arguments is well known and has been analyzed by Alvarez Junco: according to Spanish anarchists there is no such isomorphism between Nature and Society, since the middle classes have not really shown themselves to be the most fit in the struggle for life, but rather have competed from a privileged position.[12] Such statements were usually accompanied by a rejection of the supposedly innate superiority of the middle classes and the aristocracy and emphasized instead the supposed degenerative features displayed by members of those classes. An example of this type of literature is found in this curious comparison of the respective capabilities of the proletariate and the ruling classes in *La Idea Libre* in 1894:

> Leave some to their own pursuit of the scents with which to cover the stench of scrofula and syphilis. . . Fertilize others with the sweat of the fields. . . There is no hope for the morally depraved and the physically rotten. At the same time, the healthy and the physically strong, are. . . like the raw material from which the socially emancipated free man has been

---

[8] On the Virchow-Haeckel debate, see B. Rupp- Eisenreich, "Le darwinisme social en Allemagne," in Tort, ed., *Darwinisme et société* (Paris, 1992), pp. 169-236, on pp. 192-193, 206-207; M. Di Gregorio, "Entre Mephistophélès et Luther: Ernst Haeckel," in *ibid.*, 239-283, on p. 276.

[9] The assertion of the aristocratic character of Darwinism takes shape in Haeckel's *Freie Wissenschaft und freie Lehre* (1878). Here he responds to the accusation of socialism that Virchow had heaped on Darwinism; see Di Gregorio, "Entre Mephistophélès et Luther," pp. 275-276. It seems that Haeckel supported his aristocratic vision of Darwinism in formulas taken from Spencer, especially his interpretation of the expression "survival of the fittest;" see Rupp-Einsenreich, "Darwinisme social," p. 206. Haeckel's "aristocracy," however, was middle-class, an aristocracy of talent as opposed to that of lineage; see B. Muniesa, "El impacto del darwinismo en el pensamiento social," *Anthropos*, 16-17 (1982), 81-84, on p. 83. Some authors have interpreted this insistence on the aristocratic character of Darwinism as a reflection of a conservative trend in German liberalism; R. Weickart, "The Origins of Social Darwinism in Germany," *Journal of the History of Ideas*, 54 (1993), 469-488, on p. 483.

[10] Anselmo Lorenzo was the first to quote Haeckel, when he tried to refute his arguments in 1886. For the general lines of his argument, see his article "Refutación de un sofisma," *Acracia*, 7 (1886), 57-64. He also published a series of articles responding to Haeckel in nos. 32-45 of *La Asociación*, reproduced eighteen years later as "Pasado, presente y porvenir," *La Revista Blanca*, 149 (1904), 129-135.

[11] F. Urales, "Socialismo y cristianismo," *La Revista Blanca*, 14 (1899), 387-390, on p. 387. In the same line, see also "Los productos de la tierra," *Bandera Social*, 6 (1885), 1-2; A. Lorenzo, "El dogma burgués," *La Idea Libre*, 23 (1894), 1; *idem*, "Opinión autorizada," *Tierra y Libertad*, 53 (1908), 1.

[12] J. Alvarez Junco, *La ideología política del anarquismo español* (Madrid, 1991), p. 144. Anselmo Lorenzo was the most insistent on this line: see "Refutación de un sofisma," p. 59; "Pasado, presente y porvenir," p. 134; and "El dogma burgués," p. 1.

> made. Since, if, according to Darwinist theories, the strongest and best prepared must prevail over the weak, those freaks of hysteria can now prepare themselves, since the others, strong of nerves and pure bred, promote ideas ready to use force tomorrow, and the fight between the premature runt and the equally robust and illustrious laborer is easy to predict. [13]

However, the real threat posed by the struggle for life was that a literal interpretation of the metaphor could be used to attack the image of a harmonic and provident Nature, the truely fundamental crux on which the anarchists built their perception of a utopian counterworld.[14] The essence here was not how to explain evolutionary change, but rather the relative presence of good and bad in society and nature, a question of theodicy, that is. Peter Kropotkin, the Russian naturalist and one of the most important anarchist leaders, emphasized this in 1905: "When Darwin proposed the idea of the 'struggle for life,' representing this struggle as the great impulse of progressive evolution, he stirred up the age-old question of the moral and immoral aspects of Nature."[15] Certainly this was an old question. And there is no doubt that many of the social interpretations of Darwinism that flourished at the end of the nineteenth century and the beginning of the twentieth took it up again. As Antonello La Vergata has indicated, many social Darwinists of the end of the nineteenth century did nothing if not revive a key point of natural theology from the eighteenth: the question of equilibrium and the order of Nature normally associated with the general interpretation of the benificence of Creation. In this manner of thinking, negative aspects of life (lesser evils like suffering or the apparent useless waste of life) had to be rationalized and justified within the framework of a harmonic and just order.[16]

From this perspective, the real enemy was more the evolutionist philosopher Herbert Spencer than the biologist Ernst Haeckel.[17] Spencer had also tried to justify the negative aspects of social and natural reality, asserting that evil inherent in the process of adaption led to a biologically superior form of humanity. The suffering and misery visible in society

---

[13] "Veraneando," *La Idea Libre*, 10 (1894), 2. The accusation of a degenerate and biologically inept middle class was extremely frequent. An example in *Bandera Social* in 1885 is quoted in D. Núñez, *El darwinismo en España* (Madrid, 1977), pp. 338-339, and there are many more examples through the years: e.g., "En l'Ateneu dels senyors," *La Tramontana*, 305 (1887), 2-3; A. Lorenzo, "La revolución es la paz," in *Segundo Certamen Socialista* (Barcelona, 1890), 147-156, on p. 156; J. Llunas, "Qüestions socials. Educació y capacitat de les classes obreres," *La Tramontana*, 502 (1891), 2; F. Urales, "Leyes de la Evolución," *La Revista Blanca*, 41 (1900), 465-469.

[14] According to Diego Núñez (*Darwinismo en España*, p. 56) there was a tension between a Darwinian theory that did not sit well with the "harmonizing" conception of Spanish anarchism and the secularized potential of this same theory. On the concept of provident nature, as an operative criticique of the existing social order, see Alvarez Junco, *Ideología política*, pp. 47-52; and A. Girón, *Evolucionismo y anarquismo en España 1882-1914* (Madrid, 1996), pp. 17, 34.

[15] P. Kropotkin, "La necesidad ética del presente," *La Revista Blanca* (1905), 422-425, on p. 423.

[16] A. La Vergata, *L'equilibrio e la guerra della Natura* (Naples, 1991), pp. 515-614.

[17] The confusion between Spencerian social evolution and the perspective developed by Darwin himself has been identified on various occasions: R. C. Bannister, *Social Darwinism, Science and Myth in Anglo-American Social Thought* (Philadelphia, 1979); J. C. Greene, *Science, Ideology and World View. Essays in the History of Evolutionary Ideas* (Berkeley, 1981); R. Hofstadter, *Social Darwinism in American Thought* (Boston, 1959). A number of authors have seen the true source of social Darwinism in Spencer's work, and not in that of Darwin: e.g., D. Freeman, "The Evolutionary Theories of Charles Darwin and Herbert Spencer," *Current Anthropology*, 15 (1974), 211-234; Y. Conry, *Darwin en perspective* (Paris, 1987), p. 87; P. Tort, in *Darwinisme et societé* (Paris, 1992), 2-7, especially p. 2.

at the end of the nineteenth century, were not, according to the English philosopher, gratuitous. They were predicted by Malthusian laws of population and had a positive role to play: they forced all individuals to compete, the majority aquiring increasing skills measurable in terms of greater initiative and self control.[18] These skills, when transmitted by heredity, accounted for the progress of the species.[19] But the struggle for life, as Spencer understood it, was not eternal, but would tend to weaken until it finally disappeared in the highest stages of human evolution.[20] Meanwhile, as he asserted in *Man versus the State* (as quoted by Anselmo Lorenzo in *Acracia*), "it is important to follow the process and accept the suffering."[21] The misery of the weak and infirm is a necessary cost of the progressive evolution of the species.

The Spencerian critique of socialism—and generally of all state intervention in favor of the weak—was based on the supposition that it altered the rate and nature of the process of adaption. This line was especially virulent in *Man versus the State*, published in 1884. To try to improve the conditions of life by decree could not be more counterproductive and was contrary to the very laws of Nature. The misery of the poor was nothing more than the product of their own bad behavior.[22] Any creature that did not have enough energy or character to suffer should disappear. The propagation of the inept becomes a burden on both themselves and on humanity and therefore does nothing for the happiness of the majority.[23] The general outline of Spencer's argument was familiar to Spanish anarchists by 1886,[24] and produced a kind of generic critique of socialism which directly attacked the basis of Spencer's argument: the organization of a future society in which all human necessities were covered, would imply a return to barbarism and biological decadence inasmuch the stimulus—suffering and misery—that pushes individuals to compete was

---

[18] Conry, *Darwin en perspective*, p. 90; D. Becquemont, "Aspects du darwinisme social anglo-saxon," in Tort, *Darwinisme et societé*, pp. 142-143.

[19] This possibility requires that the Lamarckian inheritence of acquired characters be a reality. It has been frequently argued that Spencer's sociological system rested, in reality, on Social *Lamarckism*: Conry, *Darwin en perspective*, p. 84; P. J. Bowler, *The Eclipse of Darwinism. Anti-Darwinian Evolution Theories in the Decades Around 1900* (Baltimore, 1983), pp. 69-75; *idem*, *Charles Darwin. El hombre y su influencia* (Madrid, 1995), pp. 38-39.

[20] Becquemont, *Darwin, darwinisme, evolutionisme*, p. 144; A. La Vergata, "Guerre, biologie et évolution," in P. Viallaneix and J. Ehrard, eds., *La bataille, l'armée et la gloire* (Clermont-Ferrand, 1985), pp. 433-438.

[21] Lorenzo, "Refutación de una sofisma," pp. 37-38.

[22] In fact, as J. R. Moore has demonstrated, Spencer, in agreement with the middle class, saw Nature as an ordered system of legal rewards and punishments that inevitably promoted "material and moral progress." Poverty resulted from "improvidence," and wealth was a sign of "individual value;" "Herbert Spencer's Henchmen: The Evolution of Protestant Liberals in Late Nineteenth Century America," in J. Durant, ed., *Darwinism and Divinity* (New York, 1985), pp. 76-100, on p. 80.

[23] Spencer's book is bathed in pessimism, a negative response to growing State intervention in the form of labor, education, public health, and other legislation (Becquemont, *Darwin, darwinisme, evolutionisme*, p. 149).

[24] See comments by Lorenzo on the theme of the individual against the state: "El individuo contra el Estado. Spencer y *La Revue Socialiste*," *Acracia*, 4 (1886), 28-30; "El individuo contra el Estado. Spencer y *La Revue Socialiste*," *Acracia*, 5 (1886), 34-36.

supposedly the guarantor of human progress. As proposed in anarcho-communist publication *La Justicia Humana* (1886):

> The middle classes. . . have seized this theory of "the struggle for life" demonstrating—at least so they believe—that it has caused all human progress by compelling individuals to always be alert. . . . According to them, this state of affairs must last forever, since if the individuals saw themselves in a social state in which they were safe and in which all their needs were satisfied and in which everyone were equal, there would be no emulation, and, therefore, neither would there be initiative; such a society—according to the working classes—would not take long to fall into rapid decline.[25]

The Spanish anarchists did not take long to respond to Spencer, first, by denying the reality of Malthusian population laws in the human world: misery is not the necessary partner of the free and natural action of these laws, but rather the perverse result of the monopoly of the means of subsistence by the privileged.[26] Then, second, they resorted—at least during the 1880s—to one of the more habitual forms (according to Antonello La Vergata) of toning down the apparently brutal aspects of the metaphorical *struggle for life*, and which, paradoxically, had Herbert Spencer as one of its main advocates. The Spanish libertarians therefore affirmed that gory struggles could have had a role in the early phases of human evolution, but as we delve deeper into its advanced stages, the struggle became more moderate or is deflected to other areas.[27] Civilized man no longer fights like primitive man or animals. Society is not, like Nature, a battlefield where biological progress is a kind of by-product of the ferocious fight between individuals. The fight becomes, as the conscience of *Homo sapiens* grows, a collective fight by the whole species against hardship. Thus as Ricardo Mella, one of the foremost theorists of Spanish anarchism,[28] asserted at the *Segundo Certamen Socialista*: "the general idea of association for combat" comes about when man understands that the force used in destruction is reciprocal, and it would be better "to direct it towards nature, joining all opposing

---

[25] "La sociedad al día siguiente de la Revolución (1)," *La Justicia Humana*, 5 (1886), 2. The fragment quoted is, without doubt, a translation of a text by the French anarchist Jean Grave; cf. Grave, *La sociedad futura* (Valencia, n.d.), I, 31-32. Anselmo Lorenzo makes similar points: Lorenzo, "Pasado, porvenir y presente," p. 135.

[26] There are innumerable anti-Malthusian pronouncements. See, among many, "Los Productos de la tierra, *Bandera Social*, 6 (1885), 1; "La ley de Malthus," *La Nueva Idea*, 3 (1895), 1-2; J. Prat *¿Competencia o Solidaridad?* (Barcelona, 1903), p. 26; A. Lorenzo, *El banquete de la vida. Concordancia entre el hombre, la naturaleza y la sociedad* (Barcelona, 1905), pp. 7, 9; J. Alarcón, *La esclavitud moderna* (Madrid, 1906), p. 6.

[27] According to La Vergata, the "time-honored works, in this era. . . which argue that the brutal aspects of evolution, after having played a beneficial role in the distant past, have gradually given way to less bloody methods, are innumerable," "Les bases biologiques de la solidarité," in Tort, ed., *Darwinisme et societé*, pp. 55-87, on p. 83. Among anarchists, Bakunin supported this position: D. Velasco Criado, *Etica y poder político en M. Bakunin* (Bilbao, 1993), pp. 56-58. The general lines of Bakunin's approach were known by Spanish libertarians: see "Consideraciones sobre la situación obrera," *La Solidaridad*, 52 (1889), 1. The French anarchist Emile Gautier, inventor of the term Social Darwinism, also said similar things: *Le darwinisme social* (Paris, 1880), pp. 29-31.

[28] On Mella, see W. Muñoz, *Antología ácrata española* (Barcelona, 1974); A. Segarra, *Federico Urales y Ricardo Mella, teóricos del anarquismo español* (Barcelona, 1977); A. Fernández Alvarez, *Ricardo Mella o el anarquismo humanista* (Barcelona, 1990); J. A. Lobo, "El anarquismo humanista de Ricardo Mella," *Estudios filosóficos*, 38 (1979).

forces."[29] Concretely the Spanish anarchists thought that if war were to be found at the beginning of the path, sociability, and especially, solidarity (that is, the highest forms of morality) are found at the end.[30]

As we can see, the discrepancy with Spencer is not based on the general scheme of human evolution that was being used: the English philosopher had also predicted a tendency towards a final state in which war and violent forms of struggle would give way to higher forms of morality and cooperation. Rather, it was derived from a different interpretation at the point of determining the phase of human progress deemed to be found in European society at the end of the nineteenth century. Spencer thought that it was still necessary to accept a certain level of inter-individual competition—and therefore inequality and suffering—because it assured social and biological progress. The Spanish anarchists, on the other hand, looked upon inter-individual combat in modern societies as the *artificial* perpetuation of a primitive form of the struggle for life, typical of lower organisms.[31] What they called the *social struggle for life*, was not, however, the manifestation of a natural law that dominated the world of the living. It was, on the contrary, the pathological, unnatural, and contingent manifestation of a form of vicious and unjust social organization.[32]

In the specific case of those who remained attached to anarcho-collectivist orthodoxy (for example, Josep Llunas and Ricardo Mella in the 1880s), the coincidences with Spencer were even greater. This partly explains how they could use the expression *struggle for life* without any critical reserve in a moment in which other libertarians - Anselmo Lorenzo for example - were already beginning to reject it.[33] In 1889, in the *Segundo Certamen Socialista*, Josep Llunas asserted:

---

[29] R. Mella, "El colectivismo. Sus fundamentos científicos," in *Segundo Certamen Socialista* (Barcelona, 1890), pp. 309-336. It is possible that Mella borrowed the expression "unite to fight," directly or indirectly from Jean-Louis Lanessan. Lanessan was one of the first French biologists to lend his voice to the "socialist creed," a phrase which he coined at a meeting in 1880; J. Lanessan, *Etudes sur la doctrine de Darwin: la lutte pour l'existence et l'asociation pour la lutte* (Paris, 1881). See also L. Clark, *Social Darwinism in France* (University, Alabama, 1984); La Vergata, "Les bases biologiques de la solidarité," pp. 78-81.

[30] The idea of the shift from individual struggle to collective struggle against Nature was extensive, persisting into the early 1890s: "La sociedad al día siguiente de la Revolución," *La Justicia Humana*, p. 2; "En L'Ateneu dels senyors," *La Tramontana*, 293, 2; J. Llunas, "Sobre anarquismo. Consideraciones a vuela pluma," *La Tramontana*, 566 (1890), 2-3; R. Mella, "De la solidaridad," *Ciencia Social*, 6 (1896), 167-170.

[31] *El Productor* even states that "the principle cause of the revolution is the struggle for life taken to its most brutal level" and strongly advises modifying "the manner of producing this fight;" "Revolución," *El Productor* (1891), p. 1.

[32] In reality, everything seems to indicate that, in this, the Spanish anarchists followed the line that Ludwig Büchner had established years earlier: see Büchner, *L'homme selon la science* (Paris, 1870), 329-332; *idem*, "La sociedad," *La Anarquía*, 24 (1891), 1-2. It is very possible that the Spanish libertarians had indirect access to Büchner's arguments via translations of articles of the Frenchman Jean Grave: cf. "La sociedad al día siguiente de la Revolución," *La Justicia Humana*, p. 2; Büchner, *L'homme selon la science*, p. 332; Grave, *La sociedad futura*, I, 32-33, 58-59. The text by Grave cited was repeatedly cited in the Spanish press (*La Anarquía, El Productor, La Justicia Social* and *La Alarma*).

[33] Even referencing neo-Lamarckian authors in the orbit of French socialism: cf. Lorenzo, "La revolución es la paz," pp. 155-156; L. Dramard, *Transformisme et socialisme* (Paris, 1892), pp. 13, 16, 22.

> We have said that the struggle for life. . . is a source of life and civilization. This is true. In the zoological scale of the whole animal kingdom, we see that the only species that have survived the struggle for life have perfected their inferior structure, exchange it for others succesively superior. . . And this is exactly what has happened with human beings.[34]

In fact, the necessity of a struggle for life is one of the arguments used to support their conception of a future society based on anarchist collectivism. It is that which distinguished them from a libertarian communism that they considered utopian.[35] The anarcho-communists, on the other hand, understood that perfectly. The collectivist defence of the fight for life became, for them, a source of additional criticism of anarcho-collectivism. A widely published pamphlet of 1890 states:

> Collectivist socialism aims to found an organization in which men, having fought amongst themselves, receive a prize for winning. Nothing for the loser . . . this socialism does not see men, only different forces, gladiators, or competitors fighting amongst themselves with greater or lesser ferocity, the prize for the winner being the greatest objective. All so that—and this is the bottom line for all schools of thought based on survival of the fittest—progress might not stop.[36]

To study the rationale of the debate in depth one must go to the nucleus of the discussion between libertarian communists and anarcho-collectivists of the 1880s: the question of the criteria that should determine the distribution of wealth to individuals in a future society.[37] The collectivists imagined a postrevolutionary horizon where the ownership of the means of production should be collective, but not so the products. These would be distributed in individual lots equivalent to the work done by the individual. The libertarian communists, anxious to eliminate all trace of competition in future society, wanted both the instruments of production and the product of collective work to be common property. In a hypothetical new world, where technological progress promotes abundance, work is agreeable, and the working day short, it does not seem difficult to establish the necessary link between work and consumption: the principle of to each according to his work, typical of the anarcho-collectivists, is replaced, in anarcho-communism, by that of to each according to his needs.[38]

The anarcho-collectivists tried to discredit the libertarian communists' approach by denouncing their great anthropological naïveté. If the link between work and consumption were eliminated, what motivation would the individual in a future society have to work?[39] Without a stimulus, a reward that satisfied the individual interest, human activity would

---

34 Llunas, "Sobre anarquismo," p. 335. Similarly, R. Mella, "La reacción en la revolución. III," *Acracia*, 23 (1887), 391-395, on p. 393.

35 "A 'El Obrero,'" *La Tramontana*, 163 (1884), 1-3, on p. 2.; Llunas, "Sobre anarquismo," p. 3.

36 E. Hugas. and V. Serrano, *Diálogos del calabozo. El socialismo colectivista y el comunsimo anárquico* (Barcelona, 1890), pp. 29-30. This pamphlet is, according to Alvarez Junco, the best Spanish publication on the controversy between the anarcho-communists and the anarcho-collectivists, "albeit from the communist perspective;" *Ideología política del anarquismo*, p. 361.

37 For more details on this controversy, see M. Nettlau, *La Première Internationale en Espagne (1868-1888)* (Dordrecht, 1969); J. Piqué i Padró, *Anarco-col.lectivisme i anarco-comunisme. La oposició de dues postures en el moviment anarquista catalá (1881-1891)* (Barcelona, 1989).

38 Alvarez Junco, *Ideología política del anarquismo*, pp. 354-358.

39 Piqué i Padró, *Anarco-col.lectivisme i anarco-comunisme*, p. 71; Alvarez Junco, *Ideología política del anarquismo*, p. 362.

cease and progress would stop definitively. The Catalan Josep Llunas thought that the libertarian communists, believing they could "do away with material stimuli," put too much faith in the "kindly impulses of the heart."[40] The argument is reinforced with biological arguments that lead to a conception of the struggle for life not so different from that of Spencer. We have seen how, in Spencerian sociology, the contemplation of evil associated with failure—misery—becomes the spur that pushes individuals to compete, and, therefore, to become ever fitter. Anarcho-collectivists like Josep Llunas change the type of stimulus but not the nature of the process.[41] It is evident that an anarchist cannot conceive a future society in which misery—derived in the final analysis from Malthusian laws of population—is responsible for assuring social and biological progress. In the new world the stimulus that will push human beings to aspire to become fitter is the hope or prospect of reward, that is, the individual's expectation that he will own a proportion of produce strictly proportional to the amount of work done. In other words, only the establishment of a necessary relation between work done and individual perception (the more the work, the greater the participation in the product) can assure the framework of psychological motivation in a future society so that people *struggle for life,* and so develop the necessary strengths that determine the social and biological progress of the species.[42] This is precisely what libertarian communism does not guarantee.

This struggle, in any case, will be a *transformed* struggle, reduced, according to Llunas, "to the limits to which it must be reduced by a rational person." This would have to be assured in the future society through the "abolition of individual ownership of land and the great instruments of work."[43] This would impede the parallel phenomenon of the aglomeration of wealth and misery, which is the last mechanism whereby modern society preserves a primitive form of the struggle for life. It is in the society of the future where *normalization* will occur, that is to say, conditions will be such that the combat between individuals, typical of lower animals, becomes a collective struggle against nature, typical of advanced stages of evolution.

Now, the scheme of evolution shared by the anarcho-collectivists and the anarcho-communists in the 1880s—progression from the fierce forms of the struggle for life to higher forms of cooperation and even solidarity—poses disturbing questions from the point of view of a doctrinal framework where Nature occupies a consecrated position. If what came before, that is to say, the natural, can be described as a framework of brutal processes where conflict occupies a privileged position, how does one preserve the image

---

[40] J. Llunas, "Qüestions socials. VI. Comunisme y colectisme," *La Tramontana*, 478 (1890), 2.

[41] Llunas' position is prototypical of Catalan collectivism. Other non-Catalan collectivists, Ricardo Mella for example, linked their defense of the struggle for life to the need for individual differentiation and to the struggle between individuality and collectivity (Mella, "El colectivismo," p. 324), reflecting as much the influence of Proudhon as of English individualistic anarchism; see Alvarez Junco, *Ideología política del anarquismo*, p. 324.

[42] *La Tramontana* asserts that it is "indispensable garantir l'estimul individual qu'engendra'l desitj de un mes enlla que determina la lluyta per la existencia." The end of this stimulus is to give "a cada un lo fruyt de sons esfors;" "A 'El Obrero,'" p. 2.

[43] Llunas, "Bases científicas," pp. 335-336.

of a just, harmonic, and provident Nature that opposes point by point the visible chaos of present society? The question of the good and evil of Nature did not only worry Spanish libertarians. An important group of Russian naturalists (led by Karl Fedorovich Kessler) directly opposed the idea of a moral or amoral Nature.[44] On the contrary, they thought that morality, far from being a late fruit of the latter stages of the evolutionary process, was itself a determining factor of progressive evolution from the lowest rungs of the ladder.[45] One of these Russian naturalists was Peter Kropotkin.[46] And it was, without a doubt, Kropotkin who most effectively helped the Spanish anarchists assimilate and lay the "scientific" bases of the idea that solidarity—and not competition between individuals—was the dominant factor in the Economy of Nature.[47]

Although Kropotkin had previously dealt with this question in both newspaper articles and papers, it is in his book *Mutual Aid: A Factor of Evolution*, where he most fully explained his ideas with respect to the social and ethical implications of evolution. It is important to look at Kropotkin's work, in fact, as part of a specific debate on this subject. *Mutual Aid* was based on T. H. Huxley's article (published in 1888) called "The Struggle For Existence: A Programme." In it Huxley began to formulate a position on the relationship between evolution and ethics which he would examine more thoroughly in later years. According to him, Nature, far from being a harmonious whole, could be described as an aggregate of violent and brutal processes. Nature is neither moral nor immoral, but amoral: there is nothing in it, that is, on which to base our moral and ethical conceptions. Furthermore, natural and social processes are opposed. Social—or ethical—behavior is not only characterized by putting limits on the amorality of natural processes, but by opposing them.[48]

Kropotkin, unlike Huxley, did not want to renounce a natural foundation for ethics.[49] However, in articles published between 1890 and 1896 in the magazine *The Nineteenth*

---

44 Kropotkin, for example, acknowledges the decisive influence of Kessler in his paper "La moral anarquista. VI," *El Socialismo*, 71 (1890), 1-2. It seems that he learned of Kessler's theory in 1883 in Clairvaux prison: G. Woodcock and I. Avakumovich, *El príncipe anarquista* (Madrid, 1975), p. 179; Kropotkin, *El apoyo mutuo. Un factor de la evolución* (Valencia, 1906), I, p. viii.

45 On Darwinism in Russia: D. Todes, *Darwin Without Malthus. The Struggle for Existence in Russian Revolutionary Thought* (Oxford, 1989); A. Vucinich, *Darwin in Russian Thought*, (Berkeley, 1988); J. Harvey, "Essay Review: Russian Darwinism," *Journal of the History of Biology*, 23 (1990), 523-527.

46 On the political dimension of Kropotkin: Woodcock and Avakumovich, *El príncipe anarquista*; C. Cahm, *Peter Kropotkin and the Rise of Revolutionary Anarchism (1872-1886)* (Cambridge, 1991); G. Crowder, *Classical Anarchism: the Political Thought of Godwin, Proudhon, Bakunin and Kropotkin* (Oxford, 1991).

47 In the recent bibliography of Kropotkin's work as a naturalist, there is a fundamental discrepancy between those that see his ideas in the context of Russian biological thought (Todes, "Darwin's Malthusian Metaphor and Russian Evolutionary Thought 1859-1917," *Isis*, 78 [1987], 537-551; Vucinich, *Darwin in Russian Thought*), and those who stress the determining role of the international and British context during the 1890s: R. Kinna, "Kropotkin's Theory of Mutual Aid in Historical Context," *International Review of Social History*, 40 (1995), 259-283.

48 On this aspect of Huxley's work, see La Vergata, "Bases biologiques de la solidarité," pp. 57-61; G. Lanero, "Il giardino della civiltá: Thomas Henry Huxley e 'l'etica dell'evoluzione.'" *Rivista di Storia della Filosofia*, 47 (1992), 125-166, on p. 151. On Huxley in general: M. Di Gregorio, *T. H. Huxley's Place in Natural Science* (New Haven, 1984); A. Desmond, *The Devil's Disciple* (London, 1994).

49 According to him, there was no other alternative than a naturalistic ethic or a return to old theological ideas: Kropotkin, "La necesidad ética del presente," p. 423.

Century (assembled in the definitive volume published in London in 1902), Kropotkin did not limit himself to copying Huxley.[50] He criticizes in a general way those whom he calls disciples of Darwin, who have focused on the most brutal aspects of the struggle for life and understand it exclusively as a fight to the death. Kropotkin recognizes that this type of struggle exists, but warns that Darwin also spoke of a metaphorical struggle for life, seeing it as a collective fight that each species maintains against the adverse conditions that get in its way.[51] In fact, this last type of struggle is much more important in the Economy of Nature than direct combat between individuals.[52] Now, in the indirect or metaphorical combat, that is, in the collective fight between different species and the hostile environment, the fittest are those that develop to the highest degree those social instincts that are oriented towards the practice of mutual aid, that is, the practice of solidarity.[53]

Moreover, sociability expedites the development of the highest faculties (intelligence and moral conscience).[54] The conclusions of such an approach are clear: the predominant factor in the world in which we live is not the fight between equals, but solidarity, which is the favored weapon in the fight between the different species and the environment. Mutual aid—solidarity—is the progressive factor in evolution, since a high level of sociability promotes an ever-increasing development of the higher faculties.

How did Spanish anarchists assimilate this conceptual framework? First, it is important to note that the progressive introduction of Kropotkin's ideas as the predominant factor in the Economy of Nature is noticeable during the 1880s and the early 1890s: his articles appear in *Acracia* (1887), *El Socialismo* (1890), and a significant portion of the articles on which *Mutual Aid* was based appeared in translation in *El Productor* in 1892.[55] However, there is little evidence that his ideas on this question were explicitly incorporated into the political discourse of the Spanish anarchists right away. It is only in the early 1900s that a deeper appreciation of Kropotkin's position can be detected.[56] Spanish libertarians adopted two fundamental aspects of Kropotkin's framework: a) solidarity has a bigger place in the

---

[50] The articles appeared between September 1890 and June 1896 and gave rise to the first English edition in 1902. The first Spanish translation was by Josep Prat (1902).

[51] Kropotkin, *El apoyo mutuo*, I, 20-22; La Vergata, "Bases biologiques de la solidarité," p. 71.

[52] Kropotkin does not deny the existence of individual competition, but it has, according to him, less importance than the Darwinists or even Darwin liked to think it has. This is so because true *Malthusian situations* rarely arise in Nature: Kropotkin, *El apoyo mutuo*, I, 83, 92.

[53] *Ibid.*, II, 171-172.

[54] *Ibid.*, I, 80-82.

[55] The 1887 articles in *Acracia* (especially those in nos. 23 and 24) are translations of an essay which appeared in *The Nineteenth Century* in 1887. Those that appear in *El Socialismo* (Cádiz) seem to correspond to those published in *La Révolte* between March and August 1890, and reappeared in a pamphlet in 1891. In 1892 a series of articles published in *The Nineteenth Century* which later became *Mutual Aid* began to appear in Spanish (*El Productor*, nos. 304-305, 311-312, 314-317, 341-346).

[56] In 1894 there are already indications of acceptance: see "Raciocinos," *La Idea Libre*, 33 (1894), 1. Anselmo Lorenzo developed the main lines of the Kropotkinian argument in "Falsedad de la lucha por la existencia," *La Revista Blanca*, 43 (1900), 529-533. This was mainly due to the translation of fragments of an article by Elisée Réclus, published in *Humanité Nouvelle* in February 1898.

Economy of Nature than the fight to the death between individuals;[57] b) mutual aid—solidarity—is the truely progressive factor in evolution.[58] Moreover, the reception of *Mutual Aid* stimulated a positive review of Darwin's work. On the one hand, Darwinists were criticized for not having stressed the metaphorical sense that Darwin had given the struggle for life. On the other, the connotations of the metaphor were embodied in an almost normative usage: the fight for life should be understood as the fight of the whole species against obstacles posed by the environment.[59] This permitted the elucidation of a new axis of discontinuity between the form of social and natural organization: a contrast between a vicious social organization that causes individuals to fight to the death, and a just natural order where different species of living beings collectively fight—through mutual action—those difficulties caused by both environment and other species. Federico Urales confirms the existence of "a repugnant social struggle" that is nothing more than "the caricature of Darwin's struggle for life." In animals "the survival of the fittest" is the result of both "the struggle maintained by the whole species against nature and with other species," while in human society the struggle has been portrayed as one of "man against man," the winner being the one who "has the guile to kill and injure."[60]

What next happened was that the symbolic polarization between the struggle for life and mutual aid dominated discussions of behavior.[61] Struggle and mutual aid became, from opposite poles, guides to conduct that represented middle class selfishness and the selflessness of the proletariate, respectively.[62] As stated in *El Porvenir del Obrero*, "political power, authority, private property, industrial and commercial competition, war, and everything that obstructs the well-being and prosperity of the human species are born of the idea of the struggle for life," while from "the principle of mutual aid come the revolutionary ideals of liberty and equality." And it was mutual aid that would definitively bring "peace and prosperity to the earth."[63]

---

[57] Lorenzo, *Las Olimpiadas de la Paz y el trabajo de mujeres y niños* (Madrid, 1900), p. 19; C. Jóvenes, "Comentarios," *Natura*, 6 (1903), 85-86; "El apoyo mutuo. Un factor de la evolución," *El porvenir del obrero*, 236 (1906), 3.

[58] F. Tárrida del Mármol, "Crónica científica," *La Revista Blanca*, 114 (1903), 554-557; Lorenzo, *El banquete de la vida*, p. 69.

[59] Lorenzo, "Darwinismo burgués," *El Productor*, 58 (1904), 1; *idem*, *El pueblo* (Valencia, 1909), p. 26; Prat *¿Competencia o Solidaridad?*, p. 26; anon., "Autonomía y solidaridad," *Natura*, no. 36 (1905), 178-181; anon., "Autonomía y solidaridad," *Natura*, no. 37 (1905), 195-198, on p. 195.

[60] F. Urales, "Anarquismo. Crítica de la sociedad presente. El exterminio por la vida," *La Revista Blanca*, 118 (1903), 678-681, on p. 678. Similar comparisons are used by Prat, *¿Competencia o Solidaridad?* p. 28, and Ricardo Mella in the pamphlet *La ley del número* (1900): see also B. N. Cano, ed., *El pensamiento de Ricardo Mella* (México, 1979), p. 206.

[61] Mutual aid would *neutralize* or overcome the effect of the struggle for life: Lorenzo, *El banquete de la vida*, p.59; *idem*, *El pueblo*, p. 23. Mutual aid is even given an identifying tone: the middle class is defined by its support for the *struggle for life*, while socialism defends the idea of "unite and fight:" J. Prat, *La burguesía y el proletariado* (Barcelona, 1937), pp. 112-113. Ricardo Mella asserts that we know the struggle as *appearance*, but solidarity is *reality*: "Por la anarquía," *Natura*, no. 46 (1905), 343-346, on p. 345.

[62] Mutual aid as a guide for conduct is not only applied to the moral domain, but also to proletarian strategy: mutual aid and workers' associations are made equivalent. This is the best weapon the proletariate has against its class enemies: A. Lorenzo, "Cartas de propaganda," *El porvenir del obrero*, 353 (1906), 2.

[63] Juan Cualquiera, "El apoyo mutuo," *El porvenir del obrero*, 336 (1913), 1. In similar terms: A. Lorenzo, *El hombre y la sociedad* (Barcelona, 1901); A. Galfe, "Una conferencia de A. Lorenzo," *El porvenir del obrero*, 353 (1903), 2.

This kind of assertion, while seemingly ingenuous, was but a symptom of a much more significant process. Little by little mutual aid became for the Spanish anarchists what Kropotkin called "the foundation of our ethical conceptions."[64] The mutual dependence ethic, thus construed, was diffused on a popular level when the major anarchist sector[65] and a small group of libertarians attracted by Nietzsche entered the fray.[66] This conflict started during the years 1903-1905, in a moment when the scientific spirit of previous decades was subjected to a profound critical revision. In fact Kropotkin identified an overreaction to this crisis of faith in Science in the spread of ideas so apparently antithetical to as neo-mysticism and radical Nietzschian individualism. It was Nietzsche who, according to Kropotkin, had posed the burning question, for he had rejected the Spencerian model of evolution with its progressively weakening struggle for life. From this point of view, the cessation of the struggle for life could only lead to biological decadence. What is more, the Nietzschians asserted that the struggle for life, apart from being the source of all progress, is in itself good. If this were so, all those moral values that attempt to limit or lessen the struggle for life are called into question.[67]

This new connection between Darwinism—again, in its loosest sense—and Nietzschian ultra individualism[68] did not go unnoticed among Spanish anarchists.[69] The spread of this radical individualism was perceived as a middle class infiltration within the libertarian fold with the object of attacking the definitional basis of authentic socialism: the

[64] Kropotkin, *El apoyo mutuo,* II, 27. The deepest reflections on the possible applications of mutual aid to the moral domain are developed by Josep Prat. This is especially true of his controversy with the republican Adolfo Marsillach in the pages of *El Progreso*. That controversy is reproduced in its entirety in J. Prat and A. Marsillach, *Una Pólemica* (Barcelona, 1919).

[65] A majority sector of "mutually binding," possibly "Puritan" tendency, and of "clear Christian inspiration," in the opinion of Alvarez Junco, *La ideología política del anarquismo español*, p. 124.

[66] The best summary of the development of the conflict is Alvarez Junco, *La ideología política del anarquismo español,* pp. 146-163. On the interaction between Stirner and Nietzsche in anarchistic individualism: P. Joan i Tous, "Sade y Stirner o la tradición imposible del anarquismo español," in B. Hofman, P. Joan i Tous, and M. Tietz, eds., *El anarquismo español y sus tradiciones culturales* (Frankfurt and Madrid, 1995), pp. 163-175. On more general aspects: G. Sobejano, *Nietzsche en España* (Madrid, 1967); U. Rusker, *Nietzsche in der Hispania* (Berne, 1962).

[67] Kropotkin's approach on the question in: Kropotkin, "La necesidad ética del presente," *La Revista Blanca*, 156 (1904), 353-355, 422-425 and 157 (1905), 386-388. On the crisis of faith in science: G. Lanaro, "La controversia sulla 'Bancarrota della scienza' in Francia nel 1895," *Rivista di Storia della Filosofia*, 1 (1993), 47-81. This change of intellectual trend is also visible in both the Spanish libertarian and non-libertarian press: A. Posada, "El año sociológico," *La España moderna*, 120 (1898), pp. 46-69; P. Corominas, "Bibliografía crítica," *Ciencia social*, 7 (1896), 218-222, on p. 219.

[68] A more general perspective on the question: G. Scarpelli, "L'aquila e il serpente: Darwinismo, nietzschianism e socialismo," *Bolletino Filosofico*, 9 (1991), 135-150.

[69] In an introductory comment to the translation of *Mutual Aid* in the pages of *El porvenir del obrero*, a link between Nietzschian individualism and a Hobbesian and Malthusian version of Darwinism is established: Kropotkin, "El apoyo mutuo. Principal factor de la evolución," *El porvenir del obrero*, 254 (1906), 1-2. Federico Urales was of the same opinion: "La evolución de la filosofía en España," *La Revista Blanca*, 109 (1903), 385-389.

principle of solidarity.[70] The establishment of the *true* sense of Darwin's work played a fundamental part in both the attack on this principle of solidarity and the defense against it. The clearest example can be found in the debate between members of staff of the magazine *Natura* and the Nietzschian José Comas Costa in 1905.[71] The basic strategy of the latter was clearly provocation: the struggle for life was a value in itself and mutual aid an ideology that shackles the free development of powerful instincts.[72] As we scend the scale of nature, moreover, the struggle for life becomes more intense.[73] *Natura*'s response was based on Kropotkin's version of Darwinism: solidarity can not obstruct any supposedly powerful instinct since it is a condition—also instinctive—of the survival of the individual and the species. The struggle for life must only be understood in the metaphorical sense: it is the solidarity of individuals, not the struggle between them, that is the true progressive factor of organic and human evolution.[74]

In conclusion, Spanish anarchists, like most of their contemporaries, directed part of their efforts to exorcising the negative implications, both ethical and political, of the struggle for life as a war or fight between gladiatorial combattants. Curiously enough, this led to the promotion of two opposing views of the evolution of *Homo sapiens*. In the 1880s, human evolution was presented as a continual progression from the fiercest forms of the struggle for life—true to our animal past—to the higher forms of cooperation, solidarity and altruism. The advantage of describing the history of the species in this way was clear: it promoted the view that the struggle for life in modern societies was not just a lesser evil, the guarantee of ultimate progress, but rather the artificial and pathological preservation of a form of conduct true only of the more primitive bio-social forms of humanity. The disadvantage was equally clear: how could the image of a just, harmonic, and provident Nature be preserved if it was asserted, at the same time, that the primitive or the *natural* were pictured as a concatenation of brutal processes? In the 1890s an opposing point of view colaesces, under the influence of the Russian anarchist Peter Kropotkin. Solidarity, not the fight between like species, is the dominant fact in the Economy of Nature. Social life and ethical norms are not the result of a process that distances us from our *natural* past, but have their source in social instincts inherited from our animal ancestors. In short, what mattered here was not the question of the explanation of evolutionary change, the reason that species and *Homo sapiens* change, but the old

---

70 Kropotkin, "El apoyo mutuo. Principal factor de la evolución," p. 1; prologue by Josep Prat to C. Jacquinet, *Ibsen y su obra* (Valencia, 1907), p. ix.

71 F. Núñez Florencio considers him a "radical Nietzschian individualist:" *El terrorismo anarquista* (Madrid, 1983), p. 109 n. 12.

72 In this he follows Nietzsche: Y. Quiniou, "La morale comme fait d'evolution," in Tort, ed., *Darwinisme et societé*, pp. 47-54, on p. 53.

73 I extract the arguments contained in J. Comas Costa, "La agonía de los dioses," *Natura*, no. 12 (1905), 190-191; *idem*, "El individuo como único valor real," *Natura*, no. 34 (1905), 150-153 and no. 42, 276-283, 192-196.

74 Based on the following editorials in *Natura*, all from 1905, with the title "Autonomía y solidaridad:" no. 36, 178-181; no. 37, 195-198; no. 38, 209-213; no. 39, 225-232, and "Nota de la redacción," *Natura*, no. 41 (1905), 260-261. The following also figure in the debate against Comas Costa: C. Jacquinet, "Lección de cosas," *Natura*, no. 44 (1905), 316-318 and R. Mella, "Por la anarquía," *Natura*, no. 48 (1905), 369-372.

question of theodicy: the relative presence of good and evil in society and nature. What was at stake, therefore, was an authentic *Moral* Economy of Nature.

Centro de Estudios Históricos, Madrid

MARTA IRUROZQUI

"DESVÍO AL PARAÍSO": CITIZENSHIP AND SOCIAL DARWINISM IN BOLIVIA, 1880-1920

The present work focuses on the progressive stigmatization experienced by the Bolivian mestizo, particularly with respect to citizenship.[1] The change in category from mestizo to the denigrating one of *cholo* permits a reflection on the social and political circumstances which inhibited or encouraged the reception of social Darwinism in Bolivia between 1880 and 1920. One can define this current as a conglomeration of ideas centering on the recognition of inequality between men, races, and classes as a fact, even a right, and which depicted social evolution as a constant struggle between victors and vanquished.[2] Judging by the research of Ramiro Condarco Morales,[3] Guillermo Francovich,[4] Juan Albarracín Millán,[5] and Marie Danielle Demelas,[6] Bolivia's libraries and bookstores offered access to texts representative of the multiple tendencies which gave rise to Darwinism.[7] In the last third of the nineteenth century, Bolivian elites demonstrated familiarity with European scientific, political, and philosophic debates, and were sensitive to the ambiguity of their relevance to Bolivia. In effect, in the decade of the 1870s, a need to concretize Bolivia's national self-image led to an effort characterized by studying the region's colonial past, exploring Bolivian

[1] This work is part of Research Project PB94-0060 (DGICYT). John-Marshall Klein provided the translation.
[2] Dossier on social Darwinism presented in *L'Avenç: Revista d'Història* (Barcelona), 1980-85; Consuelo Naranjo Orovio and Armando García, *Racismo e inmigración en Cuba, siglo XIX* (Madrid, Doce Calles, 1997).
[3] Ramiro Condarco Morales, *Historia del saber y la ciencia en Bolivia* (La Paz, Academia Nacional de Ciencias en Bolivia, 1981).
[4] Guillermo Francovich, *La filosofia en Bolivia* (La Paz, Juventud, 1987), p. 206.
[5] Juan Albarracín Millán, *El gran debate, Positivismo e irracionalismo en el estudio de la sociedad boliviana* (La Paz, Universo, 1978).
[6] Marie Danielle Demelas, "Darwinismo a la criolla: El dawinismo social en Bolivia, 1880-1910," *Historia Boliviana*, 1 (1981), 55-82; "Notas sobre el Darwinismo a la criolla," *Historia Boliviana*, 2 (1982), 212-214; and "El sentido de la historia a contrapelo: el darwinismo de Gabriel René Moreno (1836-1908)," *Historia Boliviana*, 4 (1984), 65-80.
[7] In the Sucre public library is a Spanish translation of the *Origin of the Species* (Madrid, 1887), as well as an 1887 French edition.

Thomas F. Glick, Rosaura Ruiz, and Miguel Ángel Puig-Samper (eds.), The Reception of Darwinism in the Iberian World, 205-227.

territory, and writing the first geography texts. These initiatives were carried out under the principles of scientific worth, determinism, and the progress of the human spirit as envisioned in the Positivist method.

The need to understand the laws of nature in order to improve society justified the appearance in Sucre and La Paz of groups of intellectuals who supported Positivism. The Sucre nucleus included such prominent intellectuals as Samuel Oropeza, Valentín Abecia, and Ignacio Terán, but it was in La Paz where the cultural associations had the biggest impact. There the Círculo Literario was founded in 1876, the Círculo de Amigos de las Letras in 1880, and the Sociedad Geográfica de la Paz in 1889. These efforts produced centers where, under the double auspices of English and Chilean influences, positivist and liberal ideas were discussed by figures such as Julio Mendez,[8] Felix Reyes Ortiz, Agustín Aspiazu, José Rosendo Gutierrez, Francisco Nicolás Armentia, Carlos Bravo, Daniel S. Bustamante, Eduardo Idiáquez, and Manuel Vicente Ballivián. In general, although Comtean doctrines spread via not entirely homogeneous groups of intellectuals located in Sucre and La Paz, they managed to erode the influence of Enrique Ahrens' juridical ideas which had been publicly supported in 1879 by José R. Mas.

The effort to remain up-to-date with European scholarship manifested itself as well in the decade of the 1900s with the presence of scientific expeditions in Bolivia. A French expedition, under the direction of Créqui Montford and Senéchal de la Grange, arrived in 1903 to undertake anthropometric studies of Aymaran prisoners tried at the court in Mohoza (1901-1904).[9] The task of establishing differences—both socio-cultural and physical—between Aymaras, Quechuas, and mestizos was completed in 1911 by a Belgian expedition led by Georges Rouma.[10] Likewise, the influence of Spencer was reflected in the work of Sucre-based Luís Arce Lacaze (1872-1929) and Daniel Sánchez Bustamante (1871-1933) in La Paz. Sánchez Bustamante—influenced first by Spencer, then by Guyau—advanced the notion that in addition to the struggle for self-interest there existed a fight for morality and justice in which only those whose ideals were highest would emerge victorious.[11] That is, the natural law of the "survival of the fittest" would be replaced by that of the survival of the most moral, thanks to whose upright behavior progress would be possible.

The confusion of Darwinism with Comteanism and Spencerism was typical of Bolivian intellectuals interested in the study of the laws that govern societies. Concepts from the natural sciences came to have significance in the sphere of political-social

---

[8] In 1877, the Círculo Literario published a magazine in which Julio Méndez translated Louis Dumont's book, *Haeckel y la Teoría de la Evolución en Alemania*, an analysis of Haeckel's doctrines based on a lecture series which he gave in Jena in 1867.

[9] Carlos Ponce Sanginés, "La misión científica francesa Créqui-Montfort," *Kollasuyo, Revista de Estudios Bolivianos*, no. 71 (1970), 104-126.

[10] Demelas, "Darwinismo a la criolla," pp. 64-71.

[11] Francovich, *Filosofía en Bolivia*, pp. 217-222.

thought in such a way that terms like evolution, progress, and civilization were all conflated. In a similar fashion, liberalism and democracy also were conjoined in order to respond to problems of national self-definition. However, knowledge of the scientific and philosophical debates of the day did not imply any mechanical acceptance of supposedly European theories when it came to understanding the country's social development. If there were indeed academics who tried to apply the principles of struggle for existence and survival of the fittest to their own society, the majority of Bolivian leaders understood the social variants of Darwinism only as a reference to progress.[12] Consequently, I think it fair to say that European thought (in the form of social Darwinism) was not responsible for Bolivian practices of social exclusion based on ethnic criteria. It only acted as the framework within which successive measures of citizenship exclusion would be made to fit. Why did this happen?

It is possible that in a multi-ethnic society like Bolivia social Darwinism might have been invoked to justify white racial domination on the grounds of its having advanced the human species. However, historical events linked to the War of the Pacific (1879-1883) prevented this interpretation from prevailing. The defeat by Chile led Bolivians to develop a collective sense of initiating a new stage in history in which they were given the opportunity to design a future Bolivia. If Bolivians admitted the principle of "survival of the fittest," they might thereby negate the possibility of continuing as a nation. The sociological application of Darwinism implied a promise of progress because human evolution did not permit any step backwards. The dominant group would always be the best. Were this true, the incapacity demonstrated by the defeat would justify the right of the victorious nations to divide up Bolivian territory—a scenario referred to at the time as the "Polandization" (*polonización*) of Bolivia.[13] Faced with this alternative, the political leadership felt it necessary to deny in the name of prevailing natural laws the "sinister ranking of races, customs and climates" as "parallelisms maliciously used to distort and falsify the measure of humanity."[14] In other words, in the decade of the 1880s the principle of the survival of the fittest, which would have implied recognizing that Bolivia had lost the war due to racial inadequacy, could not be accepted without endangering national continuity. So entrance into a social Darwinian "paradise" would have been in contradiction with the fact that the national rebuilding effort was based on the installation of democracy understood as electoral power. However, in the following decades, as the danger of war receded and the political party system stabilized, Bolivia's national narrative returned to its former lines, this time sustained by the rhetoric of degeneration disseminated by Le Bon, Vacher de

[12] In this respect, Demelas indicates that Bolivians adopted social Darwinism because it represented a theory of social and historical progress, as well as providing a symbol of opposition to the political and social conservatism of the Church.

[13] *El Comercio* (La Paz), September 3, 1881.

[14] *El manifiesto de Don Aniceto ante su propia conciencia y ante la conciencia de los pueblos de Bolivia, América y de Europa* (Sucre, Colón, 1881), p. 2.

Lapouge, and Max Nordau.[15] Once again, Bolivian society was pictured as containing an ethnic component which undercut everything and impeded democracy. This was the Indian, although Indianness did not in itself provoke the vogue of social Darwinism.

If at first the problem of what to do with the Indian seemed to be at the center of the debate over the nation's future, the real target was the mestizo. In the decade of the 1870s some authors had opted for a mechanical interpretation of biological ideas, claiming that the ultimate solution to Bolivia's ethnic problem lay in distilling out the pure white race via immigration of Europeans and extermination of the Indians.[16] However, the following decades demonstrated that Bolivia had failed to attract European immigrants and that the indigenous population still constituted a massive presence in the Altiplano. Consequently, dreams of purification could not be realized and the Indians remained as the only available labor force. So attention then turned to neutralizing the Indians' capacity for action through the principles of liberalism which considered that any corporative residue, such as indigenous communities, should disappear to make way for the birth of a new society. Governmental measures advocating the sale of communal lands to convert the Indians into smallholders represented the objective of Bolivian leaders to transform the Indians into apolitical, productive, docile workers.[17] This policy was made explicit in many of the literary texts of this period, wherein Indians are portrayed as a victims led by abuse and ignorance to commit brutal crimes. These works suggested, therefore, that the Indian would have to be protected, kept isolated, and inculcated with the idea that the rural sector was his place in life.[18] Yet this "essentialization" of Indian status implied, above all, the desire to keep Indians from changing into mestizos.

---

[15] Demelas, "Darwinismo a la criolla," p. 64.

[16] "The Indian who because of these difficulties and his self-effacing and distrustful nature today lives in isolation from us, without participating in our poitical issues or social organization, will either become civilized or will disappear, absorbed by the white race, because it is a law of nature that in the struggle between races who fight each other, the weaker will succumb to the stronger;" José Vicente Dorado, *Impugnación de las ideas federales en Bolivia* (Sucre, Pedro España, 1877) p. 17. Also: Gabriel René Moreno, *Nicomedes Antelo, Notas biográficas y bibliográficas*, (Santiago de Chile, Cervantes, 1901), pp. 117-179; "Nicomedes Antelo. Fragmentos de una biografia," *Obras completas de Humberto Vazquez-Machicado y José Vazquez-Machicado*, (La Paz, Don Bosco, 1988), V: 367-403.

[17] Archivo Departmental de la Paz, *Posesión y venta de tierras de propiedad indígena. Extracto de opiniones valiosas del Redactor de la Convención Nacional de 1880, emitidas por los Honorables Convencionales de ese año. Sobre el derecho que tiene el Estado de vender las tierras de comunidad poseidas por los indígenas* (La Paz, 1880).

[18] Marta Irurozqui, "El negocio de la política. Indios y mestizos en el discurso de las elites bolivianas," in Jorge Pinto, ed., *Del discurso colonial al proindigenismo. Ensayos de Historia Latinoamericana* (Temuco, Universidad de la Frontera, 1996), pp. 117-140; "Qué hacer con el indio? Análisis de las obras de Franz Tamayo y Alcides Arguedas," *Revista de Indias*, nos. 195-196: 559-587; and "La pugna por el indio. La Iglesia y los liberales en Bolivia, 1900-1920," in Gabriela Ramos, ed., *La venida del reino. Religión, evangelización y cultura en América (siglos XVI-XX)* (Cuzco, CBC, 1994), pp. 377-401.

Although the effort to rationalize and order the Bolivian population based on liberal principles—order, law, peace, and authority—was initiated in 1884 by the conservative parties, it was the Liberal Party which made these initiatives concrete after its victory in the 1899 *Guerra Federal.* The era's literary and political endeavors record the progressive stigmatization of the mestizo and his conversion into the degraded *cholo*. Consequently, this paper examines changing conceptions of citizenship through an analysis of these literary and political texts. Here we can distinguish two stages which affected the definition of who could be citizens and that, therefore, encode different conceptions of the worth of the mestizo population. The first phase corresponds to the conservative period (1880-99) and receives full expression in Nataniel Aguirre's novel, *Juan de la Rosa*, where the educated and rational (*letrado*) mestizo personified in the child protagonist, Juancito, represented the citizen of the future. The second phase refers to the liberal period (1900-1920) and can be examined through works by Paredes, Pedregal, Salmón Ballivan, Finot, Chirveches, Navarro, Canelas, Tamayo and Arguedas, in which the mestizo is always a corrupt social climber who does not scruple at exploiting the indigenous population as landlord, priest or administrator. Moreover, through access to the political system this figure even places his personal ambitions before the interests of the nation. This situation prompts several questions: Why did the conception of mestizo as *letrado* change to mestizo as *cholo*? What lay behind the satirization of the politically active cholos and the threat of a bleak future for Bolivia so long as these practices continued? Why did these authors make Bolivia´s public well-being depend on the ethnic composition of its leaders? These questions really boil down to one: What is the meaning of anti-cholo prejudice? In posing these questions, I do not attempt to reconstruct the political behavior of the popular sectors encompassed by the term cholo. Rather, I propose to highlight, through political and literary texts, the connotations of this prejudice and to see how the elite manipulated it in order to curtail, limit, or control possibilities for popular political participation. At the same time, however, the elite's discussion of Indian and mestizo participation in public life reveals as much about elite values (fears, resentments, and dislikes) and intra-elite divisions as about the actual situation of the classes they sought to control.

1. THE GOOD MESTIZO CITIZEN

In 1890, Benjamín Fernández lamented that "the Democratic principles that illuminated the Republic at the moment of its foundation have been eclipsed more than a thousand times in the smoke of civil discord," giving rise to "fifty years of civil wars, barracks revolts, and demagogy." However, Fernández also asserted that the solution to this chaos was to be found in positivism, understanding this term as a synonym for liberalism and democracy, as an inevitable natural law.[19] Gabriel René Moreno also thought that all humankind was oriented towards progress and that the democratic

[19] Francovich, *Filosofía en Bolivia*, p. 198.

system adopted by the American republics anticipated the best of all possible worlds. Social harmony could be reached through education and laws favoring the development of trade, which would help combat the ignorance and poverty of the country's inhabitants—the primary causes of national backwardness. Yet, if democracy was the political system that could bring all of this about, one would still have to be quite careful with it because it furthered the practice of miscegenation (*mestizaje*). René Moreno´s proposal to prevent this factor from leading to disorder and social chaos consisted in the hope of a massive arrival of whites to drown the Indian presence. The result would be one race whose strongest tie would be a shared religious conviction, Christianity.[20]

The argument for stability through the adoption of democracy also appears in the historical novel, *Juan de la Rosa,* published in 1885 by Nataniel Aguirre. Through his protagonist, Juancito, Aguirre recounted the story of Bolivian independence placing special emphasis on the patriotic conduct of the residents of Cochabamba. Despite the local scale of the plot, Aguirre constructed a history of national scope with which he attempted to awaken the patriotic spirit lost due to successive military defeats. Through awareness of this history, Bolivians would be able to overcome the defeat by Chile and initiate a national rebirth which would return the nation to the upward track enjoyed by its neighbors. To ensure this reconversion into a "fatherland filled with unrealized promise,"[21] the author insisted that the citizen of this future nation would not be a soldier who might compromise the republican spirit through caudillo-like ambitions, but rather an agent of civilization: "You need study more than ever, Juancito. Because later on you must serve your defeated fatherland in full awareness of your obligations as a citizen and a man."[22]

Aguirre's casting of Juancito as the prototype of a new Bolivian citizen is indicative of the use of novels to emphasize the importance of reading and autodidactism in the formation of citizens, establishing a link between citizenship and teaching. Behind this proposal also lay the desire of intellectuals and politicians like Nataniel Aguirre to participate more actively in designing the nation, since they considered that intellectual knowledge was the key to achieving flourishing democratic republics. In sum, in opposition to the militarism backed by caudillos ever since Independence and which had led to an international war, the post-war climate afforded Bolivia an opportunity to construct a solid national identity built on a combination of nationalism and civic-mindedness (*civismo*). The way to ensure the correct functioning of both components

---

[20] Gabriel René Moreno, *Bolivia y Argentina. Notas biográficas* (Santiago de Chile, Cervantes, 1901), pp. 5-20; *Biblioteca boliviana. Catálogo de la sección libros y folletos* (Santiago de Chile, Gutemberg, 1979); Ramiro Condarco Morales, *Grandeza y soledad de Moreno* (La Paz, Talleres Gráficos Bolivianos, 1979), pp. 197-293.

[21] Nataniel Aguirre, *Juan de la Rosa* (La Paz, Juventud), p. 163 (original ed., 1885).

[22] *Ibid.*, p. 129.

consisted in specifying the precise capacities and qualities of future citizens, an end achieved through a novel ethnic categorization.

Among the multiple revelations contained in the retelling of past events, one of the most important proved to be that which referred to those groups who had made Independence possible. This aspect is fundamental because in the novel only those who had demonstrated patriotic capacity and conduct were recognized as suitable leaders for the future nation. Through a description in ethnic terms of the group conduct of different social actors, Nataniel Aguirre determined which citizens of Bolivia would be reborn after the War of the Pacific. To recognize them, Aguirre established a direct relationship between past actions and present public and political responsibilities. Not only did he try to place each historical subject in his proper niche by virtue of his deeds, but his subjects' future was projected in terms of responsibilities and actions undertaken in the past. This section's underlying argument was that only those who had made Independence possible had demonstrated the qualities necessary to participate in rebuilding the nation. Not all Bolivians had contributed equally. Citizenship, understood as the capacity to participate actively in the national design, belonged solely to the educated elite (*letrados*), those individuals who had cultivated their intelligence and who could think in terms of the future thanks to the knowledge they possessed of the past. To be a *letrado* was not synonymous with being literate. It referred instead to an individual who, through dedication to a proper apprenticeship, ended up possessing the art of rational thought.[23] Citizenship was understood, then, neither as a right nor as a privilege, but rather as a quality of individual *letrados*.[24] Let us see now who Nataniel Aguirre identified in this category and what criteria he applied to it.

If in *Juan de la Rosa,* independence appears as a collective achievement,[25] not all participants were accorded the same responsibility for its success. Its main architects were criollos[26] and *mestizos letrados,* while the indigenous population had an auxiliary role at best. Certainly, at times an abstract *pueblo* had aided them, but not uniformly so, and the activities of its members in the liberation of the fatherland were of very unequal value. There were clear differences in the intellectual capacity of mestizos. The majority—*la multitud*—were morally weak, lacking in ethics and discipline, and tended towards an aggressive fervor which led them to stone the houses of the Spaniards (*chapetones*) or those suspected of being such, shouting "Death to them!"—only to turn tail and flee from battle later on. In contrast was a minority with the valor and civic virtue to face history and create a nation. They were the moral force which could guide

[23] "Education, sustenance of the soul, an innner light added to that of conciiousness to make man kind more and more the king of creation, can only be obtained by a few personas and in so parsimonious a manner that it seems a joke," *ibid.*, p. 36.

[24] "Later on I learned that, poor as we were, surviving by the work of my mother, and taught to read by a diligent teacher, we felt, with respect to material means for cultivating our minds, a thousand times more fortunaate than the great mass of people composed of Indians and *mestizos*" (*ibid.*, p. 16).

[25] *Ibid.*, p. 249.

[26] *Ibid.*, pp. 28-30; also 142, 18-19, 23.

and channel the multitude's energy. That is to say, those individual *letrados*—the literate minority—were responsible for shaping the citizens who would create the future nation.[27] These mestizos were called *letrados* not only because they could read, but also because knowledge of writing permitted them to rebel and display patriotic behavior, differentiating them from an undisciplined *pueblo* all too susceptible to fanaticism, depravity, or indolence. This was the case with Juancito's distinguished relative, Alejo Calatayud, "a young man of 25 years, with mixed blood like them, a silversmith, exceptionally well instructed in reading and writing by his father, or perhaps like yourself, by some kindly friar," a leader of the 1730 rebellion, that "heroic and premature endeavor."[28] Alejo Calatayud was accused of daring to "carry in his hand the staff that did not pertain" to those of his class and of "not inclining his head before the favored vassals of our lord, the king." However, the fact that Calatayud possessed a "profound feeling for human equality" led him to explain that he behaved in this manner because he was "just as much a man as they themselves were" and that he had "the strength to protect my unfortunate brothers"[29] led Aguirre to consider him an early patriot. In doing so, the author portrays Calatayud as different from the rest of his caste, as one of the few mestizos who could transcend his origin to become part of the elite. If he himself did not achieve this due to his premature death, his descendant Juancito could indeed manage to become a leader thanks to his intellectual talent.

Presented with this kind of social classification, we might now ask why the novelist linked Independence sequentially with the mestizo rebellions but failed to consider the Indian uprisings of 1780 as an antecedent? It is important to underscore this theme because in the process of national refounding promoted in the novel the indigenous population remains on the margin and is not recognized as having any right to citizenship nor to participation in the future nation. Why does this happen? In an essay on Simón Bolivar, written a contest organized by President Narciso Campero on the occasion of the Liberator's centennial, and published in 1883, Aguirre mentions the rebellious activities of José Gabriel Condorcanqui and Tomás Catari, but does not interpret the uprising of Indians "from Cuzco to Jujuy" as a precursor to Independence. The reason was that, although the Indian leaders attempted "to make criollos and mestizos participants in the great uprising," it proved impossible for them to "control the mutual hatred of races angered and divided by the colonial regime."[30] In *Juan de la Rosa* he reiterates this argument:

[27] Laura Gotkowitz, "Las heroinas de la Coronilla," unpub. ms., pp. 5-13.
[28] Aguirre, *Juan de la Rosa*, pp. 38-39.
[29] *Ibid*., p. 41.
[30] Nataniel Aguirre, *El Libertador* (La Paz, Camarlinghi, 1973), p. 23.

> I will not tire the reader with even the briefest mention of the bloody convulsions in which the indigenous race crazily attempted to recover its independence, ensuring their defeat by proclaiming a war between the races.[31]

The indigenous population's "race war" could not be reconciled with the policies of other social sectors, since they were conceptualized in terms of a break with the *criollo-mestizo* commonwealth. The indigenous program did not lead to a union of Americans but rather to ethnic disunity. Therefore, the Indian movement played no role in Independence because it contradicted the principles of equality and fraternity proclaimed by the insurgents. As a result, the indigenous population was excluded from the design of the future nation, having demonstrated its incapacity to participate in social uniformity and promoted their own segregation. In the novel, Independence symbolized unity and conciliation:

These European brothers of ours whom you vulgarly call *chapetones*, far from enduring insults, should be the primary object of your affection. Now is the time to exalt the American character, to never threaten your neighbor nor take revenge for personal injuries. Display the nobility of your souls and the generosity of your hearts. Do not stain your hands with the blood of your brothers. Forswear all rancor, and at the same time as you depart to wage a just war against your enemies, bring the sweetest peace to your strong and worthy country.[32]

Nataniel Aguirre insisted that one should not "kill anybody while creating a nation."[33] His proposal for national rebirth after the War of the Pacific was a call to all Bolivians to understand that the success of the republican project rested on a willingness to seek consensus and integration. If he did not admit any indigenous political interference because he considered it destructive to the nation, Aguirre did advocate incorporating the Indians as citizens so long as they assumed a secondary role and acted under the tutelage of those who understood the nation as a confraternity. This could only occur if the tribute "which gave rise to their perhaps incurable degradation" were abolished along with the communal privileges which were responsible for the "the greatest degradation of the Indians labeled outsiders (*forasteros*), the laziness of the community members, and the general impoverishment of the country."[34] The author supported a kind of guided assimilation, because he recognized that Indians had participated in the independence wars, if only as simple spectators who succored the heroes in the hour of death,[35] or as pawns mobilized by the landowners.[36] The Indian population would be integrated into the new nation, but would not impact its cultural dynamics for three reasons: 1) because accumulated hatred would impede their sharing political space with the rest of Bolivians and would in the long run incur their

[31] Aguirre, *Juan de la Rosa*, p. 38
[32] *Ibid.*, p. 50.
[33] *Ibid.*, p. 32
[34] *Ibid.*, p. 91.
[35] *Ibid.*, p. 126.
[36] *Ibid.*, pp. 75 and 140-41.

destruction; 2) because the Indians had developed habits of indifference and witlessness due to their suffering as a "poor conquered race" reduced to "the condition of beasts of burden," and as a result they lacked the will and historical passion necessary to be promoters of nationhood; and 3) because they had lost mastery of their own language, leaving them with a halting command of Spanish which would not permit them to become *letrados*.[37]

2. CHOLOS MAKE BAD CITIZENS

The division of the Bolivian population into *multitud* and *mestizos letrados* in Aguirre's novel is a proposal for the type of ideal social co-optation that the elites should undertake. Aguirre establishes a model of the new citizen to which the legitimate or natural children of an elite figure and a chola should conform. Juancito fulfills these characteristics by being the son of the *mestiza* Rosita, granddaughter of Alejo Calatayud, and of the heir to an aristocratic Spanish family. The characteristics of his origin conjoined with his mental abilities make him the perfect candidate to regenerate the elite. Later novels consolidated this type of mestizo, not as a moral or learned person, but rather as a caudillo who takes advantage of the servile and social-climbing character of cholos and who exploits and mistreats Indians. Caudillos monopolize local power thanks to their participation in both the elite and popular realms. Their patronymics allow them to maintain politically and economically advantageous connections in the capital while their plebeian maternal antecedents permit easy access to the underworld of the bars (*chicherías*) and make it easy for them to serve as ward bosses.

The degree of mestizo social mobility was, if not high, still quite real as is made apparent by the change in literary qualities ascribed to this new type of mestizo. These mestizos always emerge victorious in elections against the aristocratic candidates who are the shining protagonists of these novels. Possibly it was this kind of electoral result which spurred the discourse among the elite about the lack of education of the Bolivian plebe, composed as it was of illiterates, ignoramuses, and unthinking folk disposed to let themselves be corrupted by unscrupulous caudillos. This narrative over the necessary electoral exclusion of the plebe due to its predilection for fraud has three distinct components. First, the elite needed to regenerate itself; second, the danger of mestizo upward mobility obliged the elite to generate negative discourses about the public and political capabilities of the general population; and third, they had to find a way to discipline the popular sectors in order to maintain the white population's national supremacy. The response to these concerns consisted in invalidating *lo popular* through racial arguments which ascribed Bolivia's backwardness to Indians and mestizos. This ethnic reductionism acquired a public and academic dimension through historical and

[37] *Ibid.*, pp. 127, 61, 25.

sociological studies which appeared after the Federal War of 1899. Discrimination and ethnic disdain were not new, but their systemization using pseudo-scientific authority was indeed novel. These studies linked race and geography in a cause-and-effect cycle with the military caudillismo of the Independence wars and the degeneration of the republic and its inhabitants.

Although the combination of these elements produced a discourse in which the indigenous population played a central role, beneath the discussion lay a more focused and ambitious program: the academic justification of anti-mestizo prejudice. The seeming preoccupation with the redemption of Indians masked a condemnation of the mestizo in his multiple guises of smallholder, hacendado, and local authority, all abusing the Indian who then moved to the city, became a cholo and, as a cholo, fell victim to urban demagogues and malevolent caudillos. This raises the question of why the mestizo ended up as the sole cause of all of Bolivia´s ills. Although one could put forward various explanations, this paper will only underscore those relative to mestizo political participation. The variety of literary representations of mestizos is a reflection of their social instability. Because mestizos were, above, non-Indians and thus had access to public privileges, it made no sense to either to "keep them in their place" or to tutor them. As voters, the mestizo sector had the capacity to help one elite faction attain power or even to gain access to it themselves. This ability to legitimize an electoral result, and to thereby confirm that Bolivia was a scrupulously democratic country that satisfied the objectives of liberal policy, made the elite dependent on mestizo participation to help define and negotiate their internal hegemony without excessive cost. The two-fold nature of this participation, endorsing oligarchic power and underscoring the active presence of the general public in civic life, forced the elite to find control mechanisms and ways to invalidate mestizo participation.[38] As previously mentioned, these tended to be of an ethnic nature and permitted the consolidation of stereotypes about the cholo which still retain wide public acceptance today. Let us now look in detail at some of their connotations.

In the wake of the Independence war, militarism, understood as *caudillismo*, with its conscription, forced donations, and destruction corrupted provincial habits, bastardizing the character of the rural population and making them "evil and lazy." [39] The subsequent rebirth of regionalism produced "destructive mestizos" who with their abuses forced the most important citizens, "wealthy families and intelligent people," to emigrate to the cities "sure to find there respectable social conditions, conveniences, and a future for their descendants,"[40] while the "worse social elements" stayed behind, caused "individual guarantees" to disappear from the villages, and forced youths into "partisan

---

[38] José Salmón Ballivian, *Ideario aymara* (La Paz, Escuela Tipográfica Salesiana, 1926), p. 22.

[39] Manuel Rigoberto Paredes *La Altiplanicie: Anotaciones etnográficas, geográficas y sociales de la Comunidad Aymara* (La Paz, Isla, 1955), pp. 199-200; original ed., 1911.

[40] Rigoberto Paredes, *Provincia Inquisivi. Estudios geográficos, estadísticos y sociales* (La Paz, J. M. Gamarra, 1906), p. 80; *La altiplanicie*, pp. 115, 119, 130.

armies."[41] In this way, the rural sector lost its most capable farmers and its best workers, reducing the country's agricultural wealth and its prospects for progress without allowing for any possible replacement of population by "any other higher race."[42] This void left by the traditional *hacendados* resulted in "the mestizo race which succeeded those hard-working and enterprising proprietors" neglected the tillage of the soil and abandoned it to the aborigine, "resistant to all progress." However, agriculture's sad state could not be blamed entirely on the Indian because the new mestizo hacendados had permitted a spirit of indolence to take hold by restricting possibilities for Indian enrichment or for obtaining "abundant returns, not to mention robbing them of their fertile fields or carrying off their crops at the lowest prices."[43] Creating even greater disillusionment, most provincial communities found themselves divided into factions filled with fierce hatred for one another,[44] because their mostly mestizo inhabitants had inherited the Indian feelings of "localist exclusivism" so alien to national solidarity and the idea of belonging to the Republic of Bolivia.[45]

While these developments were unfolding in the rural areas, the urban cholo population was growing. This cholo sector, which had originated in Indian immigration, fell easy prey to demagogues due its lack of education.[46] Demagogues used cholos as clients in the political struggles accompanying their drive to power. These bosses made the cholo population see them as the "pure source of virtue and abnegation. . . .awakening in them the vague notion of their value as a group and the still confused awareness of their power," as a result of which the cholos were led to "the facile notion circulated by all the demagogues of a primitive egalitarianism, wordy and undisciplined, according to which a mason or a rustic wagon-driver were worth as much as an inventor, a scholar or a student."[47] The result was the appearance "of an ignorant mob" which wanted to "satiate itself with the theft of the goods of those who were said to be their betters,"[48] and which lacked interest in grand ideals, being only interested in obtaining "social glitter, political power, and the ostentation of titles or riches." In spite of these defects, the cholo did have the positive characteristic of rejecting any illegal imposition by the authorities, but "the same authorities, far from cultivating those good qualities and channeling them in a way that would be beneficial for the country, strove to obliterate them and to vilify those who displayed them." Consequently, a bad caudillo could cause "more harm to the country than a deadly epidemic, which attacks the bodies of citizens, while he destroys their souls. The Bolivian people owe their corruption to

---

[41] Paredes, *La Altiplanicie*, p. 182.
[42] *Ibid.*, p. 182.
[43] *Ibid.*, pp. 134, 205.
[44] Paredes, *Provincia Inquisivi*, p. 83.
[45] Paredes, *La Altiplanicie*, pp. 106-109.
[46] *Ibid.*, p. 97.
[47] Alcides Arguedas, *Historia general de Bolivia, 1809-1921. El proceso de la nacionalidad*, (La Paz, Puerta del Sol, 1922), pp. 52-53, 512.
[48] Arguedas, *Pisagua* (La Paz, Biblioteca Popular Boliviana Ultima Hora, 1978), p. 129; original ed., 1903.

their caudillos."[49] In addition, the perseverance of caudillo-style practices due to ignorance, sentimentalism, audacity and fanaticism on the part of the cholos undermined the system of political parties.[50] In their craving for social betterment, the cholos gave their support to whomever could intoxicate them with "captious whiffs of ideas, which we [the elite] don't understand, but we know how to use them to best advantage: democracy, equality, socialism, suffrage, all of which are more dangerous for them than brandy and *chicha.*"[51]

In sum, the intellectual and political discourse of the era depicted the political system as undermined by two factors. The first was the incapacity of lower-class voters and the other, the carelessness of the social elite[52] who with "their ineptitude, negligence, and idleness" had permitted "the disproportionate abundance of the mestizo element and the pernicious predominance of the common people." This action converted them[53] and the white demagogues, who distracted the mestizos from manual labor, into "cholos more useless than all the cholos together."[54] With typologies like these, political mistakes assumed a personal rather than institutional character which affected all those who favored the ascendancy of the people. However, although theoretically all were guilty, the attitude of the creole-mestizo community received more benign treatment, since, given the physical and moral misery of the Bolivian people, one could understand their scruples about and disinterest in political participation. If one wished to combat the political absence of the "white people," it would require excluding from public life "the insensitivity and low instincts of the masses" which weakened democracy, preventing its recognition as the most appropriate ideology for modernization.[55] Given that it was absurd "to pretend that democracy and its accessories" would produce results in areas where they were not understood, the solution was to look for more restrictive ideologies with which to attain progress and civilization:

> One hears terms like republic, national interest, the rights of citizens, civic duty, and one hundred more, just as unintelligible and ostentatious. We all have to vote for them under the penalty of I don't know what, but, for whom? For whomever the cacique or subprefect says or for whomever, by means of the modest sum of five pesos, four reales, a bottle of brandy and four syrupy lies, has become the owner of our citizen consciousness, of

---

[49] Arguedas, *Historia general*, pp. 52, 188-89.

[50] Paredes, *Política Parlamentaria en Bolivia. Estudio de psicología colectiva* (La Paz, Cerid, 1992), pp. 37-40.

[51] Juan Francisco Pedregal, *La máscara de estuco* (La Paz, Arno, 1924), p. 162.

[52] Arguedas, *Historia general*, p. 58.

[53] Paredes, *Política Parlamentaria* , pp. 38-39.

[54] Pedregal, *La máscara* , p. 174.

[55] "Peoples have deeply rooted political vices which isolated disturbances cannot destroy, and our people have been brutalized by a servile and rapacious proselytism and rendered unable to understand republican principles or to practice them by their own will and perseverance. How can we believe that such principles might be suitable for the blunted brains of the great mass of laborers and artisans, who constitute three-quarters of the population and who find themselves in a state of ignorance, passivity and intellectual dullness analogous to the imbecile?" (Paredes, *Política Parlamentaria*, p. 134).

> our civic duty, of our democratic rights, of our sovereignty and of whatever else we own with inalienable and imprescriptible title in the name of democracy.[56]

In Bolivia everything has been weakened and destroyed "by the abuse of officials or political parties and by the passivity of the public in putting up with these misshapen practices," leaving the country governed neither by its institutions nor by justice. This occurred because the country "in race, culture, institutions, and customs" continued to be "a State under construction." As such, the country should wait until the "mixture of different races with an embryonic culture" would produce the definitive Bolivian type.[57] While this took place, international political models would have to remain in suspension without this implying any Bolivian inferiority but simply that the country had to find a national solution with its own characteristics.

The variants and complexities of the stereotype of the corrupt and corrupting mestizo in a region dominated by racial legacies and passions appear even more clearly in accounts of political parties and electoral campaigns. To make this evident, this article surveys the novels which focus on the confrontation between conservatives, liberals, and republicans during the period 1900 to 1926. The pivotal idea is the political menace represented by the "cholo" population that, due to its bad habits and inheritance, weakened the party system, keeping it from serving as an adequate channel for national modernization. The regime of parliamentary democracy only served to elevate social inferiors and hindered Bolivia´s national transformation. However, this stance did not imply a true rejection of democracy and parliamentarianism, but rather of race.

In *La candidatura de Rojas* (1908), Armando Chirveches recreated and personified the stereotypes of the scholarly essays. Interested in portraying the incessant recurrence of political corruption, Chirveches reconstructs in a satirical tone the steps that an individual takes to obtain the social recognition necessary to let him live with ease. The book deals with an individual belonging to a well-to-do family but which lacks the wherewithal to guarantee him the permanent enjoyment of his privileges. The solution is to make himself a congressional representative since this activity permitted, to whomever exercised it, the time necessary for whatever he desired as well as the pleasure of a regular stipend, applause, public attention, and "great rewards for spending a few hours seated in an Armchair covered with Cordovan leather, in an elegant site where silk velvet abounds among tall column with gilded cornices, where can spout whatever nonsense with the air of a scholar."[58] The first and most essential element in obtaining such a political office consisted in securing family influence.[59] A second condition was choosing the party with which to affiliate. Here, there only existed

---

[56] Pedregal, *La máscara*, p. 164.
[57] Paredes, *Política Parlamentaria*, pp. 96-97.
[58] Armando Chirveches, *La candidatura de Rojas* (La Paz, Urquizo, 1988), p. 18; original ed., 1908.
[59] *Ibid.*, p. 12.

two options, join the government or go over to the opposition. In no instance was presenting oneself as an independent an advantage since the circulation of elites was regulated by a two-party system which repeated the dichotomy: elite in power versus the dispossessed elite.[60] The choice of party did not depend on ideological criteria but rather on which positions were open. One would choose to be conservative or liberal depending on whether the heads of one party or the other needed candidates and which party could ensure the support of the provincial district which one sought to represent in Congress.[61] For this reason, the novel's protagonist, Enrique Rojas y Castilla, opts for the opposition once the Minister of Government has informed him of his inability to provide that support, given a prior commitment. This situation makes clear that "even though constitutions may stipulate that a citizen is an individual fit to elect and be elected, in fact the only ones elected are those who are imposed by the governing classes, by the moneyed aristocracy and by the aristocracy of power."[62]

With the party decided upon, then one had to organize the election campaign in which it was expedient to promise that "it won't cost a lot."[63] However, the voters had to be convinced not only with promises but also with money which tended to be disbursed by the relatives of the candidate who expected to benefit from his election. As Enrique Rojas ran as an opposition rather than an official candidate, his integrity was taken as a given and he was considered "a statesman who participated in suffrage, that grandiose cupola of liberty, that tournament of law, and with his heart wide open promised his electors railways, highways, and telegraphs, in return for their votes" to rebuild "their freedom, so miserably mocked by the mandarins of the province."[64] In this task, Rojas received the help of a paper entitled *La Voz del Pueblo,* in whose pages the candidate could combat his adversary, Manuel María Garabito.

Since his opponent's cronies occupied all of the province's most powerful public posts, Enrique Rojas confronted the entrenched power base of a family which had carried out all kinds of stratagems to keep the government from intervening in their territory and which also served to defeat all political adversaries. The origin of their hegemony stemmed from the caudillo era, thanks to whose instability and wars the family's founder had managed to enrich himself and to take over the region, placing all his relatives in posts which might extend the family's influence. The patriarchs' conduct was at all times dishonest and abusive, especially with the indigenous population which saw itself degraded by the land usurpations to which they were subjected by the Garabito clan. In this electoral contest, the Garabito candidate was always the candidate

---

[60] Alcides Arguedas, *La danza de las sombras. Apuntes sobre cosas, gentes, y gentezuelas de la América española* (Barcelona, López Robert, 1934), p. 90.

[61] The politicians mainly opted for candidates favored by "the lower classes" and who would be popular among artisans: "Cosas de nuestros caciques. Saben éstos que cuenta con simpatías entre los artesanos y le han ofrecido la diputación;" Alcides Arguedas, *Vida criolla* (La Paz, Juventud, 1981), p. 22.

[62] Chirveches, *La candidatura de Rojas*, p. 16.

[63] *Ibid.*, pp. 56-57.

[64] *Ibid.*, p. 78.

of the popular faction which sought votes by distributing money and alcohol among the mestizos. In contrast, Enrique Rojas decided to represent the area's aristocratic element, that is to say, the large landed interests—guardians of moral values, who would take their renters and dependents to the polls. Only at this moment was it necessary and licit to consider the Indian-peons as citizens, for which they had only to demonstrate how to write two names, their own and that of the candidate they supported. In this way, the responsibility for the local political process depended on "individuals incapable of voting for anyone else nor of selling their vote, since they scarcely knew how to trace the name of the candidate imposed by the boss." The process also depended on individuals who could be bought with alcohol since "in South America generally and in Bolivia, in particular, the best voter is alcohol, so that the representatives of the people only represent, in many cases, whiskey made by some German firm." After the voting came the vote count which tended to be carried out in a violent atmosphere, often leading to injuries among the contending groups: " Truncheons were hoisted, revolvers aimed, one shot blew away one of the hives that bees had so laboriously built in the church tower, and there erupted an unsociable battle of sticks and blows that would have end disastrously" had it not started to rain.[65] The final result was a victory for the candidate supported by the government.

The official candidate owed his triumph to the particular "carrots" he was able to dispense as well as to the "stick" provided by gangs of thugs. More generally, he depended on the entire electoral paraphernalia which he could muster. In *Vida Criolla* (1905), Alcides Arguedas describes one political procession in La Paz and the social mobilization that it provoked. At the head some "wretched children" marched, followed by Indian dupes dressed in their best holiday clothes while group leaders made "the flags unfurled to brighten up the solemn days of parish holidays or any other unforgettable event, tremble in their hands." After the Indians, "several meters away, some cholos carried the banners of white cloth that spread across the breadth of the street and on which, in black letters, supporters" had painted laudatory descriptions of the candidate. Subsequently came the artisan guild associations whose members "sweating and hoarse, shouted "*vivas*" at the caudillo, even though many couldn't even recognize him. They walked with their faces swollen, with troubled gaze, heads down as if bent by the weight of the work not yet assigned them by their holy law of redemption and so they went around with the tired air, depressed and unhappy, of those condemned to live poorly fed, without air, hidden from the sun, and locked in a perpetual carnal and alcoholic orgy." After the guild societies came the literary and scientific ones, also with their garlanded standards, followed by mule-drawn coaches carrying the party leaders: "There you could see politicians of all stripes, ages and opinions, animated by the common desire to please the candidate, earn his confidence and, with it, a modest little job in the administration where he might work for the sacred destiny of the fatherland."

---

[65] *Ibid.*, pp. 16, 183-188.

Finally, the caudillo arrived and "reclining on the velvet upholstery of the car put at his disposition by one of his wealthy backers, he looked at the sad and apparently disdainful crowd," while at his side circulated a proud group of horsemen who received the floral wreaths offered to the politician from the balconies.[66] However, the candidate's success did not reside solely in money and graft, but also and principally in knowing how to "get along with his supporters" and in compensating the efforts of the voters and electoral agents. For this it was necessary not to just get involved in politics but rather to know which "wall" to lean on, as Demetrio Canelas relates in *Aguas estancadas*:

> There wasn't even one of them who wasn't well placed. To whom did Covarrubias, that poor lad, who could find no work, even as a porter in a mercantile house, owe his career? To General Reyes? Who had freed Vidal from prison, to install him as a secretary of mission? General Reyes? Who advanced so many others, in spite of newspaper stories? General Reyes. This is what stubborn people do not want to understand.[67]

In *Los Cívicos* (1918) by Gustavo A. Navarro (Tristán Maroff), the election day narrative reveals greater drama by focusing on the brutality of the government's partisans "in their semi-barbaric bullying." First the mounted police entered the square; then came the judges with their "bandit and cretin profiles" headed by their president, "a red-faced cholo, with slanty little eyes, a roguish mouth and five hairs for a beard." After these followed a crowd which, "in a paroxysm of passion and alcohol, attacked the other side sated their fury." From a street entrance appeared a squad of soldiers, "with bayonets fixed, even more drunk with rage than the civilians" and who, instead of imposing peace, "wet their bayonets with the blood of the citizenry." With the opposition defeated in this way, "the wretched men who had just attacked the timid and defenseless citizens, proceeded to the voting booths and cast one, two, three—up to ten or twenty ballots—for their candidate, Peña." Each soldier, "his rifle in the crook of his arm, placed his shaky signature on twenty ballots" in exchange for a bank note and "a glass of liquor which they sipped with delight." After them came the prisoners who knew how to write, taken out of jail to vote up to ten times. The warden, "with his chola flesh reeking of jail and alcohol," obtained their services in return for a reduction in sentence. With the voting ended, the congressman, "emaciated, his back misshapen, ape-like profile, contorted" and with eyes that revealed "plebeian malice and a deceptive perversity," was intensely acclaimed by the crowds and came out onto the balcony to salute his subjects. The "pueblo" was no longer in attendance: "it had fled, some—fearful—to their houses, guarded by their wailing wives and children; some more brave, there, in the depths of jail, disheartened, bodies reduced to scraps, blending into the mounds of cinders and mud." At night, the abuses and outrages continued to the point where "ten masked men with Indian faces, savage, wearing military capes" attacked some drunken workers who had dared to shout hurrahs for the defeated

---

[66] Arguedas, *Vida de Rojas*, pp. 57-59

[67] Demetrio Canelas, *Aguas estancadas. Fragmentos de la vida boliviana* (Cochabamba, Canelas, 1965, p. 93; original ed., 1907.

candidate. It represented the triumph of the "barbarian- and trashocracy" (*barbarocracia y canallacracia*), imposed by a liberal president in whose face "one could see the mestizo. . . .the fatal miscegenation of the African Negro, perverse and buffoonish, loquacious and tyrannical, with the passivity of the base and servile Indian."[68]

The triumph of the *cholada* did not always occur through elections nor did it always stem from the voters' bad habits. Cholo success also came about due to elites who in their anxious search for higher status and more graft sought to marry their children to rich mestizo politicians. This unscrupulous attitude with respect to race hindered the transformation of Bolivia into a civilized nation recognized as such abroad, rather than as a country of Indians. Armando Chirveches dealt with the theme of parental ethnic responsibilities in two novels, *Celeste* (1905) and *La virgen del lago* (1920), each with a different solution. In one, the cholo triumphs; in the other, love emerges victorious. Both tell the tale of romance between a beautiful maiden and a virtuous young man of noble birth but little wealth. Alas, a rich mestizo intervenes, determined to wed the female protagonist. Although the "white" couples possess the same physical characteristics in both stories, the two mestizo suitors do not. In *Celeste*, Don Práxedes Urcullo appears with "anthropological traits that would have led a modern psychiatrist to classify him as a born criminal or immoral madman." Senator-for-life and owner of a large fortune, he was "the product of a bastard conjoining of races; he had the blood of *conquistadors*, of Indians and of slaves."[69] In *La virgen del lago,* Abelardo Topa is an elegant mestizo who suffers from "that hyperesthetic vanity of mestizos, from an unhealthy morbidity and innate lack of confidence," despite presenting "certain qualities of their race: a practical spirit, pretense, reserve, the art of using their money wisely, and the habit of looking out for their interests."[70] Despite logic, it is the first suitor who enters into marriage thanks to a ministerial appointment, while the other has to resign himself to seeing the young lovers wed. The difference between the two cases resides not only in the young woman's maturity but also in the parents' behavior. In *La virgen del lago,* the parents are aware of racial considerations:

> When you wed, do so with a well-born man, who is neither a *quidam* nor a rogue, yet I prefer to marry you to a rogue than with an Indian. Do not destroy, do not throw away your race.[71]

In contrast, the parents in *Celeste* are plutocratic, frivolous, and egotistical:

> She didn't worry herself about the species. That is something that parents don't consider when they marry off their children. And nevertheless, which shameful inheritance would the descendants of that red-headed mestizo contribute, by a rare caprice, but whose ignoble

[68] Gustavo A. Navarro, *Los cívicos. Novela de lucha y dolor* (La Paz, Arnó, 1918), pp. 29-70.
[69] Armando Chirveches, *Celeste* (La Paz, Urquizo, 1976), pp. 53, 56; original ed., 1905.
[70] Chirveches, *La virgen del lago*, (La Paz, Libreros, 1920), p. 165.
[71] *Ibid.*, p. 108.

> features, whose ordinary hair unequivocally suggested mixed blood and an obscure origin? He was a child of the crowd, just like those unknown little boys who call the first person who passes them in the street father.[72]

Celeste does not recount the negative consequences for the protagonist of marrying a cholo, but *El cholo Portales* (1926) by Enrique Finot does follow up on this idea. In Finot´s work, Señorita Velez dies due to the psychological abuse and miserliness of her spouse. Moreover, Chirveches' two books depict the cholo as a secondary character who serves to cast a pall over the happiness of the young lovers through his stratagems. In contrast, Finot gives a starring role to the cholo. His novel describes the means by which Evangelista Portales climbs up the social ladder to become a possible president of Bolivia. The character combines all of the attributes of the cholo mentioned up to now. He is fawning towards his superiors, despotic with his subordinates, miserly, avaricious, crooked, unscrupulous, not civic-minded, incapable of filial loyalty, an ingrate, and a traitor. In addition, he follows an upward trajectory similar to that previously described. The son of a chola and a priest, he receives protection from childhood on from an eminent lawyer for whom his progenitor worked as a cook. The lawyer even finances his university studies. Thanks to the friendships he acquires there, Evangelista manages to marry an elite woman who helps place him socially, ignoring the wise recommendations of his tutor who in the novel embodies white-creole honesty as it succumbs to the unbridled ambition of the cholos. Evangelista's behavior exemplifies one particular approach to the validation of social status: he rejects *los de abajo* and craves the social and cultural goods of *los de arriba.* His flaw stems from having expectations above his station and from successfully ascending the social ladder by understanding politics as a business and as a leveling activity which corrupts and perverts everything.[73] By using his university training for personal gain, Evangelista goes against the essence and destiny of the *mestizo letrado* described by Nataniel Aguirre, demonstrating that despite the appearance of material progress Bolivia remains "an Indian encampment, manipulated by some mestizos who are overreaching and more or less literate (*letrados*)."[74] This leads the author, through the character of Dr. Pérez Benavente, to deny the results of the anthropological and psychological research which made "the equality of the human species fashionable" and to assert that the country's moral climate had not improved because the white race was not yet dominant. The white race was the only one capable of "understanding, loving, and practicing democracy," in contrast to the mestizo masses who had ruined everything with caudillismo, anarchy, and their craving for power.[75] This polarity was translated into an immense race war that would determine the future of Bolivia:

> What will happen on the day when Indians and mestizos, in full awareness of their power, strike out at us, thirsty for revenge? On that day, there will take place on a large

[72] Chirveches, *Celeste*, p. 41.
[73] Enrique Finot, *El cholo Portales*, (La Paz, Juventud, 1977), pp. 43-45; original ed., 1926.
[74] *Ibid.*, p. 85.
[75] *Ibid.*, pp. 85-90.

> scale what is already happening on a small one: people of color, having seized the government, will throw us out of the territory that once belonged to their ancestors. . . . The spirit of self-preservation obliges us to be forewarned: this is a duty of the dominant race, so long as the dominated race constitutes a threat.[76]

The novel ends with honest politicians in exile, betrayed yet again by the evil arts of Evangelista Portales in his unchecked climb towards the presidency. This ending confirms race as the fundamental factor in Bolivia´s decline. Indeed, this conclusion served as a cautionary tale for the *blancos* not to go too far in charity towards their inferiors. If they did not take care, the "well born" could end up like those expelled to Antofagasta, losing their privileges and status. If they failed to support segregation in the name of the national good, caudillismo might return, bringing with it a kind of social mobility which would prove difficult to control. Admittedly, Finot´s novel was a clear attack on the Republican Party and its president, Bautista Saavedra. However, the novel's importance does not reside so much in its caricature of Bolivian politics as in the ethnic reductionism by which the political environment is understood. The stereotype of the cholo was a call for solidarity among whites and a reproof of any gesture that might call into question the advantages of a caste-based society.

3. CONCLUSION

The contrast between the novel of Nataniel Aguirre and the writings from the first decades of the twentieth century sheds light on the steady transformation of the mestizo into a cholo and the stigmatization of the latter. The first text depicts the ideal mestizo who could construct and represent a nation while the later works reveal a mestizo of flesh and blood incapable of patriotism. The novel *Juan de la Rosa* suggested an opportunity for future public participation denied in the other writings which detail the failure of this opportunity. This failure came about because the individuals who were to fulfill the opportunity seen by Aguirre did not truly possess qualities of good citizenship due to their ethnic origin. Race represented an unavoidable fate which condemned the mestizo to existence as a cholo. In part, this transformation responded to a redesigning of power relationships. However, there also existed a desire on the part of the elite to monopolize *lo mestizo* as the core of their future national identity, leaving *lo cholo* to those who were relegated to inferior social status. The necessity of providing a scientific patina to this process through essays and novels showed that the cholos were not disposed to assume a negative identity which denied or postponed their social and political existence. For Rossana Barragán the ambiguity, neutrality and positive identity acquired by the term mestizo converted it into "an escape valve, an intermediate place," and a contested space because it permitted self-identification for all social groups.[77] If

---

[76] *Ibid.*, p. 100.

[77] Rossana Barragán, *Los múltiples rostros y disputas por el ser mestizo* (La Paz, 1996), pp. 99-101.

this is true, the novels and essays mentioned previously contributed to creating opinions and perceptions around anti-cholo prejudice whereby to avoid reversing the mestizo-to-cholo transition and winding up with a cholo-to-mestizo one.

The negative theorization of ethnicity possessed two dimensions. One expressed a personal project centered on the self-image of the authors studied as well as their political stances at the time of writing. The second dimension refers to a preoccupation of the elite as a group: their self-image, international reputation, and the foundations of their social preeminence. Let us look now at both aspects.

The treatment of the mestizo changes significantly from Nataniel Aguirre's novel to the writings of the early twentieth century. Even so, all of these works probably do share a political desire to construct a sense of nationhood using literature. Their critiques were not simple expressions of disgruntlement but rather carried an implicit desire to decide Bolivia´s future. In this sense, one can distinguish two levels. First, the censure of the mestizo by these authors, as well as sympathy for his victims, implied an effort on their part to monopolize solutions to Bolivia's root problem: race. Only they could resolve this cancer as the diagnosticians of the country's illness and as the ones best-suited to explain it scientifically and to disseminate this information through their writings. By offering themselves as the country's saviors, they found a way to make themselves indispensable to the nation's destiny and as such to achieve responsible political positions which would permit them to participate in the process of national self-definition. In addition, to the extent that this group succeeded in influencing and creating opinion, they also reinforced the link between citizenship and education. Only those who possessed discipline, ethics, and honor could be true citizens, capable of writing History and, as such, building the nation. The intellectuals converted themselves, then, into leaders with the necessary moral force to guide and channel the energy of the masses. This justified defining citizenship as an exclusive quality of individual *letrados*. Inasmuch as being a *letrado* was synonymous with possessing the quality of rational thought, Aguirre's novel synthesized that which the authors of the later texts believed was their obligation to put into practice. Aguirre indicated that only the intellectual elite had the legitimate capacity to direct Bolivia's destiny. These writers' moralism and hypercriticism of their political and social surroundings was the way they declared their conviction that national self-definition was their obligation.

Anti-cholo prejudice revealed the intellectual elite's concrete public ambitions. At the same time, though, the dynamic, selective, cumulative, and consensual elaboration of this prejudice also had other objectives which went beyond the particular needs of this group. This is not the place to enter into a discussion of the specific factors which nurtured the stereotype and which might explain its idiosyncratic development in greater detail. Rather, at a more general level, one can say that the meaning of the stereotype indicated a dynamic of internal rise and decline among the elite with the possibility of regeneration through links to racially inferior social segments as a fundamental factor of elite existence. Problems of elite status and preeminence, both rising and falling, permeated the values of the entire society, contributing to the strengthening of ethnic segregation as a way to preserve the elite's privileges and identity.

The constant use of ethnic terminology to express national identity also eased the elite's sense of guilt for not having done enough to combat the country's underdevelopment and, at the same time, obviated their political responsibility. Indeed, by showing that the blame for Bolivia´s problems was not theirs, the elites made their resolution the obligation of the lower strata. Politics had differential meaning across the social spectrum, because the ethnic deficiencies of the Indians and cholos relieved the elite from all social responsibility. The discourse of the privileged indicated appreciation of the artisan vote and an exaltation of their virtues as workers, but at the same time denounced the political immaturity that their ethnic origin caused. This diminished and denigrated the civic importance of the popular classes in addition to making subalterns responsible for the flawed mechanics of the political party system. This disqualification could also be extended to the elite faction which rode the mestizo vote to power. In other words, undervaluing mestizo political activity meant fortifying one of the contradictions of elite politics. Political "outs" could cast doubt on the legitimate access to the government of the electoral victors but excuse their own underhanded activities as owing to the voters' lack of education. These voters, mainly hacienda peons and urban mestizo artisans, lacked the qualities of true citizens due to inborn racial traits and, therefore, bore the blame for the party system's perpetuation of caudillismo. They made it impossible for Bolivia to become a civilized nation. In sum, the impossibility of doing without the mestizo´s electoral participation and clientage led the Bolivian elite to make the mestizo responsible for the problems generated by intra-elite competition. This process provided ammunition for the various partisan bands to push for subjective and self-interested legislation to ensure that future voters would be racially improved.

The process of ethnic definition exposed the contradictions of the elites. For strength in the future they needed to replenish themselves in a manner which could incorporate the popular sectors without breaking down their own equilibrium, domination, and group cohesion. From 1826 until 1952, only registered literate males could vote. However, these restrictions on political participation did not mean that the population lacked their own notions about citizenship and the State and even less that they did not participate actively in electoral contests. Thanks to fraud and patron-client relationships, the registration systems included less-formalized means for popular participation, external to the mechanisms for popular representation and promoted by the needs of intra-elite competition. This consensual illegality strengthened the efforts of the disenfranchised to gain visibility as citizens. Through demands on the State to spread primary education or efforts to demonstrate that official positions were not mere sinecures, subaltern sectors sought ways to exercise suffrage and receive recognition of their voting status.[78]

Faced with this attitude, the elite needed formulas which, while permitting formal equality among individuals, would assure their inequality with respect to power and the privileges which power brings. Race and its degeneration represented one of these formulas, which not only served to diminish the capacity of many voters at least at the level of discourse, but also to question the legitimacy of the elites elevated by these voters. The racial context of debates over Indians and mestizos provided a way to maintain and strengthen the hypothetical differences between the various social actors. To recognize ethnic equality was the equivalent of consenting to the erosion of the upper class's privileges, especially those of declining elites. Consequently, the elite strove to demonstrate a "patrician" origin, which would impede their displacement from the public sphere through contradictory racial discourses which both exalted and denigrated the popular sectors. In addition, the description of popular political structures as inadequate, ignorant, and corrupt shows how among the mechanisms for controlling social mobility figured the creation of a lurking cholo menace eager to put an end to traditional social relations, imposing barbarism and international disdain. Fear acted as a guarantee of segregation and as a limit to popular excesses which threatened to subvert the social order. This fear also contributed to making the lower class sectors share the stigmatization of everything related to the Indian and the cholo, obliging these groups to interpret public life in divisive ethnic terms and to internalize their supposed political incapacity. Such consequences made this population more apt to seek authoritarian and patron-client solutions which in the long run reinforced the stereotype of the cholo as a sustainer of caudillos, giving rise in turn to new "paradises" of segregation.

Centro de Estudios Históricos, Madrid

---

[78] Marta Irurozqui, "¿Qué vienen los mazorqueros ? Usos y abusos del fraude y la violencia electorales en Bolivia, 1880-1930," in Hilda Sábato, *La expansión de la ciudadanía en América Latina* (Mexico, FCE, 1997); *idem*, "Ebrios, vagos, y analfabetos. El sufragio censitario en Bolivia, 1826-1952," *Revista de Indias*, no. 208 (1996), pp. 697-742; Rossana Barragán, "Miradas indiscretas a la Patria potestad: articulación social y conflictos de género en la ciudad de La Paz. Siglos XVIII-XIX" (preliminary version, 1996).

THOMAS F. GLICK AND MARK G. HENDERSON

THE SCIENTIFIC AND POPULAR RECEPTIONS OF DARWIN, FREUD, AND EINSTEIN

Toward An Analytical History of the Diffusion of Scientific Ideas

The reception of scientific ideas, especially fundamental ones such as those proposed by Darwin, Freud, and Einstein—when analyzed via an affective and comparative taxonomy—can be seen to take place within a field of certain obvious variables. These variables can be categorized along the following demarcations (some will apply more to scientific than popular reception): professional-disciplinary cultures, generational factors, philosophies of science, trans-national diffusion, religion, level of education, political ideology, wealth of a nation, and the imagined personas of scientists.

RESISTANCE TO AND APPROPRIATION OF SCIENTIFIC IDEAS

Beneath the particular events of reception framed by these categories *qua* variables, there exists a dynamic, intentional logic. This logic may be divided into two kinds of intentionality:[1] a response of anxiety which causes resistance on the one hand, and a response of adaptation which allows appropriation on the other. In relation to the reception of novel, fundamental ideas such as Darwinian evolution, Freudian psychoanalysis, and Einsteinian relativity, ideas which arrive to question key metaphysical values of the Western self, resistance and appropriation—depending on what one has to gain or lose—become normative responses.[2] The local arguments

[1] Here intentionality means the idea of deliberative psychology. The reception of fundamental scientific ideas, in a fine-grained history of primary diffusion, occurs in relation to a psychological deliberation. This deliberation is not solely about the truth of the idea but about how that truth will affect one's life at personal and professional levels.

[2] We are thinking, here, of diffusion in primarily the West. On normative judgment, see Allan Gibbard, *Wise Choices, Apt Feelings: A Theory of Normative Judgment* (Cambridge, Mass., Harvard University Press, 1990); on rational deliberation and intensity of feeling, see John Rawls, *A Theory of Justice* (Cambridge, Mass., Harvard University Press, 1971); on strong and weak valuation of goods, see Elizabeth Anderson,

Thomas F. Glick, Rosaura Ruiz, and Miguel Ángel Puig-Samper (eds.), The Reception of Darwinism in the Iberian World, 229-238.

attached to the course of any particular reception will be rational in a normative sense, not rational in any positivistic or empirical sense of science taken up merely as science.[3]

Anxiety which causes resistance or adaptation which allows appropriation, may be present at both the popular and scientific levels in relation to the reception of a particular idea. Resistance can take place as a defense of a generally-held worldview or, in more narrow terms, of an ideology. For example, resistance to Darwinian natural selection can take place at a generational, disciplinary level in science because professionally one may have something to lose in adopting a new idea, while at a religious level a resistance based on an anxiety of loss may occur because natural selection remains a less than fulfilling metaphysical explanation of man's existence.[4] Besides such resistances, however, there are also appropriations of scientific ideas, wherein through the act of reception, the individual or group has a chance to profit or gain in some way. Therefore, broadly, one may conceive of receptions of scientific ideas in two large frames of reference: 1) an idea is perceived as engendering a loss and so is resisted, and 2) an idea is perceived as engendering a gain and so is appropriated.[5] These are broad frames of psychological deliberation which can be divided into descriptive ratios of reception.

In order to theorize a bit of the normative logic for why a particular idea might be expected to be—in varying degrees—either accepted or rejected by a particular group or individual, Darwin's case is illustrative. With respect to evolution by natural selection, the Christian worldview of a divine creation remains in direct antithesis to Darwin's evolution—creation—without design. Here one can see a simple case of an antithetical reception based on a clash of worldviews. The reception outcome is dialectically connected to extant interests of particular people within a particular ideological group. Again in the case of Darwin, the Huxley (and Hooker) vs. Wilberforce (and Owen) encounter signifies, in some sense, the old *versus* the new generation of naturalists in England. This antithesis then is also formed, in a finer manner, upon the clash of ideological footholds or strongholds. At the level of not wishing to give up an ideological status-quo, it is perhaps a more subtle version of the first example. Part of the resistance to an idea like natural selection, therefore, is not only rooted in cultural/religious ideologies, but also in scientific/disciplinary ideologies.[6] In a wider frame, the popular rejection of Darwin in England between 1859-63 lies, perhaps most basically, in the fact that the Western notion of *self* in

---

Value in Ethics and Economics (Cambridge, Mass., Harvard University Press, 1993); for a sociological view, George C. Homans, *Sentiments and Activities* (New York, Free Press, 1962).

[3] This is true even of a thetic reception. Arguments from the point of view of a particular philosophy of science on a scientist's part will be bent first by normative concerns—by a rational economy of loss and gain which remains metascientific. Fundamental ideas first face this economy in the process of determining whether or not they are true.

[4] Though organized religion also, as a discipline, will have much to lose in letting someone else define a fundamental truth which they have traditionally defined.

[5] The anxiety of the first can be sublimated into the adaptation of the second (in a corrective reception).

[6] This tells us something about the sociology of science via reception analysis. That is, that the reception of an idea is not solely based on things like the availability of data or the accuracy of predictions.

Enlightenment and Christian terms was still based on a rational and autonomous conception of conduct and identity. Those who have closely framed their identity in such a Western image of the self will have a hard time, in purely a psychological sense, making the transition from transcendental and rational grounds for the self to that of an organic-based metaphysic.[7] This transition breeds a certain anxiety about the ontology of the self, and a certain amount of resistance may be viewed as a normative response.[8]

However normatively figured, there are distinctions to be made between popular and scientific receptions. Within the scientific reception of an idea there naturally remains the question of the idea's scientific veracity—does the idea convey a "true" description of "a real"? The question of truth becomes the one question to be satisfied in a scientific reception. But in answering this question, there remains a professional anxiety over losing an ideology under another's gain, and therefore resistance comes to take the form of antithetical practice—as was noted above at the generational level with Huxley and Wilberforce-Owen. Scientific receptions are, therefore, fought out in a professional economy, one connected to recognition and eponymy—the valuation and devaluation of ideas. The reception of fundamental ideas will do a lot to disturb the distribution of eponymous capital.[9] However, at the popular level, the nature of the reception of a fundamental scientific idea (as a validation of materialism) in relation to resistance, remains rooted more in the fear of realizing a decentered, deanthropomorphized self.

A MODEL OF PRIMARY DIFFUSION

The intentional logic, a deliberative psychology, behind acts of reception—conceived as the resistance which will arise from anxiety related to perceived loss, and the appropriation which will arise from adaptation related to perceived gain—can be described in four ratios of reception. These ratios represent the tendency in reception to divest or invest in the truth of a scientific idea. The receptions of Darwin, Freud, and Einstein can be mapped out via ratios that take the following form: antithetical, thetic, corrective, and extensional. These categories do not hold clearly or cleanly, with mutual

[7] Indeed, Darwin's metaphysic, rooted in the organic, shows why Condorcet's rational science, rooted in a Newtonian conception of the inorganic, would never work as a model for unlimited social progress. Darwin's antithetical reception, in part, at both popular and scientific levels, goes back to this kind of social philosophy which came out of the enlightenment. However, as universal science, Darwinian evolution follows the influence of Newton's program.

[8] Darwin's own full slide into materialism left him an anxious, solipsistic, existential self, one in need of great professional validation for any identity-ground.

[9] Work here is to be done on how the credit system of science, at differing times, affects the reception of ideas. For example, reception in the Latin West before the 15th century will look different than, say, reception in the 17th century. This will be true not only because of institutional changes but because of changes in the individual, in subjectivity, ones directly related to the ideas of property, authorship, originality, and the recognition of eponymous credit.

exclusion, in each case. There remains some overlapping.[10] Nevertheless they may be useful in appreciating general trends and in viewing an analytic and normative structure inherent in receptions taken up according to an intentional logic.

The *thetic* is an attempt to take up the original content of the scientific idea within a given scientific discipline—the young have the most to gain from this.[11] The *antithetical* takes form via popular and scientific resistance which is based on maintaining an ideological status-quo. A loss is perceived in the framework of an individual (popular reception) or professional (scientific reception) economy of the self. Institutionalized world-views have the most to lose. The *corrective* ratio takes form as a quasi-scientific limitation or revision of an idea's original thetic content, and as a combination of resistance and appropriation, appears to occur—in primary diffusion—mainly in the scientific discipline(s) of a country other than that of origin. Cultural or disciplinary congruence is its aim. The *extensional* takes form via cultural/ideological appropriations in political and social realms, including more narrow literary and artistic appropriations, and in a general popular reception via mass communication, with popular reception seen here as a cultural extension.

Ideally conceived, all but the thetic ratio involve some degree of misreading which is dialectically connected to particular goals of achieving a gain or avoiding a loss—personally or professionally.[12] The thetic should be, initially, a valid scientific reading of the idea (as intended by its originator), afterwards often transcending into the corrective as part of the process whereby the receiving individual works through the paradigm—this is a patterned response, particularly in the case of Freud. The antithetical is a hard movement of defense at both popular and scientific levels; the corrective comes via the scientific appropriation *qua* resistance of an idea, whereby it is transformed (or limited) to fit, usually, another national scientific culture. The

---

[10] This is a diachronic function of the cognitive act of taking up the components of an idea in real time. As a reception goes through time, however short, the categories begin to exhibit a dialectic related to a loss/gain economy. In this, an idea, or some of its component parts, may be implicated in more than one category depending on the changing social and geographical subject positions of individuals and groups. Because of time, cognition, and position, there will be hybrid categorizations, such "corrective extensional," based on the basic four modes.

[11] A thetic reception may not be achieved, while the intention remains to take the idea up thetically. There is no intentional resistance or alteration. It is an effort to capitalize on the held-to-be-true. Moreover a thetic reception may be either "active" or "passive" with respect to participation in its research program.

[12] What we mean here by misreading is profitable misinterpretation. On misreading, see Harold Bloom, *The Anxiety of Influence* (New York, Oxford University Press, 1973), pp. 42-44; and *A Map of Misreading* (New York, Oxford University Press, 1975), pp. 69-70; on the history of misinterpretation in science, see I. Bernard Cohen, *The Newtonian Revolution* (Cambridge and New York, Cambridge University Press, 1980), pp. 194, 336 notes 2 and 7. This point caused some consternation when this paper was presented orally at the November 1997 colloquium at Cancún, because our translation, *mala lectura*, is a much stronger term than English "misreading" and implies incompetence or carelessness on the part of the reader, which was not our intent. Although in this essay we use some of the concepts of post-modernist criticism, we understand that the post-modernist critique of science—*sensu lato*—is antithetic, and we do not endorse it.

extensional is a positive appropriation, an inventive gain for new directions in literature, art, politics, and social or popular movements.[13]

Darwin

In the thetic reception of Darwin, there are, of course, Lyell, Hooker, and Huxley. Yet, within the great support this threesome gave Darwin, there remained an element of rhetorical reception, not a full-blown immediate acceptance of the natural selection analytic. Hooker became a kind of convert in July-August 1858, Huxley wanted empirical proof but "converted" in 1868, and Lyell was never comfortable with the idea of evolution by natural selection—and what this implied about the place of man in nature. Darwin did have young naturalists convert thetically —and did a lot of lobbying. A. R. Wallace, as one would suspect, was with Darwin from the beginning, albeit not entirely later on. In Germany, the younger generation of naturalists accepted Darwin in a materialist reaction against the teleological, Romantic Naturphilosophie of an older generation of biologists. In Italy, there was a positive reaction, and in Spain a rhetorical reception with no real discussion of natural selection.

In Darwin's case, as noted above, there was an antithetical Christian reception of evolution by natural selection—liberal Christians preferred a Spencerian progressivism. Biblical literalism could only be antithetical to Darwin's view. European Catholics saw Darwin as an atheist and antiauthoritarian—bad for social morals. Evolutionism in social discourse generally was also non-Darwinian, not undertaken in the language of natural selection. In the United States, an antithetical scientific reception occurred in the generation of Agassiz's students: Cope, Hyatt, and Packard were neo-Lamarckians who employed recapitulationist arguments. But they were also empiricists and may have held *theory* against Darwin, seeing it as open too wide to the perils of deduction—as in the case of Huxley. In the antithetical category, the French also seemed to be able to assimilate evolutionary thought without Darwinian views—they had no field tradition, a laboratory culture, and its paleontologists were Lamarckian. Their antithesis to Darwin may have been more nationally grounded, in the intellectual heritage of Lamarck.

A corrective reception of Darwin was made via the revision which took place in the translation of Darwin by Clemence Royer, a Lamarckian. In this translation, Darwin's "selection" became *élection*, and Darwinian evolution by natural selection then reads as a kind of "conscious use" theory, which gives a type of teleological spin to Darwin's theory of evolution. This is a fundamental miswriting of the original thetic. The correction by Royer as resistance-appropriation, can be viewed as a kind of personal-national anxiety of influence reaction—a kind of defense that takes the novel quality out of Darwin. In Russia, a correction also takes place, in this case via the rejection of the Malthusian component to natural selection—that the analytic of population pressure

---

[13] The case studies which follow are intended to be illustrative of our model and are drawn from standard reception studies. In general, only Latin American and Spanish references are footnoted.

gives merely conflict—as Kropotkin uses the mutual aid thesis and sees cooperation instead of competition as the best-chance means of survival. M. M. Filippov sees the struggle for existence fundamentally between environment and individual, not between individuals. Russians see Darwin caught up in conventional metaphors. The Catholics St. George Mivart, Ceferino González, Edouard Leroy, and John Zahm, make a philosophical and biological argument for Darwinian evolution framed within the direct intervention of God in the process.[14] Later on, Pierre Teilhard de Chardin may also represent a corrective reception of Darwinism, as he appears to be a non-teleological quasi-Lamarckian Darwinist with his focus on environment and a "directed chance" thesis—here attempting to fit some kind of mystical teleology within materialism.[15]

In the extensional reception of Darwin, John D. Rockefeller, Andrew Carnegie, William Graham Sumner, or, to cite a Latin American example, the Uruguayan Martín Martínez, as types of social Darwinists, took up Darwinism via a certain politico-economic interestedness, applying and extending Darwin's ideas to social and economic realms. In an interesting extensional appropriation, some orthodox Calvinist Baptists took up Darwin's selection language, seeing it as a dimension of election, and hence subsuming it within an ultimate design. In China, Darwin was culturally-philosophically merged with Taoism, and political use was made by Mao of the Darwinian analytic to legitimate the idea and the fact of revolution.

Darwinism in Latin America

What is the significance of the repetition in a whole series of institutions in virtually every country of Latin America of the same debate over Darwin, almost always stereotypical, with the presentation of canned arguments that everyone already knew by rote? Debates, that is, in which all the participants knew in advance what their opponents were going to say. One answer is that these debates were ritual acts, a dramaturgic iteration of beliefs whose function was to strengthen the solidarity of different ideological groups in contention, a marking out of cognitive territory.

In such polemics, Darwin's ideas were typically presented thetically, even when the spokesmen themselves personally subscribed to corrective or extensive versions of Darwinism. For example, José Arechavaleta, paladin of Darwin in Uruguay, followed Alfred Giard, a French Lamarckian, in his teaching activities, while in his research he embraced the conjectures of Ernst Haeckel on the origins of life in primitive moneras—both corrective lines. Catholic opponents of evolution always presented the pure antithetic line, sometimes only slightly nuanced, as in Uruguay with tacit Catholic support for theistic "spiritualists" who opposed hard-line materialist Darwinism.

---

[14] Thomas F. Glick, "Spain," in Glick, ed., *The Comparative Reception of Darwinism*, 2nd ed. (Chicago, University of Chicago Press, 1988), pp. 307-345, on pp. 340-342.

[15] Teilhard's evolutionism could also be seen as corrective-extensional: he extends Darwin's thesis to cosmic-religious realms (subsumed however in a traditionally construed social and cultural nexus) and in this ends up performing a 'correction' [of the science] via a socially- or culturally-based extension.

Freud

The thetic reception of Freud in the United States takes place early, in 1909 at Clark University in the United States. Freud is taken up by the elite of American psychiatry and neurology—James Jackson Putnam, Morton White, Smith Ely Jelliffe, and A. A. Brill give Freud an early institutional foothold-stronghold. But it seems that most cases of the thetic reception of Freud soon turn corrective according to the existence of a middle class or a national-philosophical style of doing psychotherapy, or using Freud for merely utilitarian purposes. The wealth of nations and the economy of classes (middle to upper) come into play in the thetic reception of Freud. Freud is used most in large, cosmopolitan cities: New York, Paris, Los Angeles, Rio de Janeiro, Buenos Aires. The thetic reception of Freud changes the way medical histories are taken. "Wild analysts" practice outside of sanctioned Psychoanalytic Associations, taken here as a thetic but unsanctioned practice (Honorio Delgado, Peru). In Brazil Durval Marcondes starts out as a "wild analyst" but later on makes the full thetic transition by undergoing orthodox training.[16]

The antithetical reception of Freud takes form via conservatives particularly in Latin countries like Spain or France, who are against (presumably) Freudian ideas in their sexual- *cum*-moral-trajectory mode, or when Freudian diagnostics are marshaled to control sexuality, rather than liberate it, as in Brazil. Freud also experiences an antithetical reception in 1930s as the Nazis burn his books (1933), attacked as Jewish Science, and ban Freudian terms while continuing to use his ideas and methods (e.g., in the Goering Institute). In recent times, feminists have received Freud antithetically (in the United States, but not in France) and pharmacology has been a kind of philosophical and practical antithesis to the value of psychotherapy. Philosophers of science have been antithetical to Freud *qua* science.

At the corrective level, Richard Strachey's English translation of Freud can be seen as an attempt to make Freud sound like Darwin. This is a kind of national, linguistic-based correction worried about a perceived philosophy of science. In the United States, Freud is converted into a doctrine of adjustment and adaptation rooted in a philosophy of individualism—adaptation of ego to environment comes off as a kind of social conformism which is not in the Freudian text (the Nazis also participated in this type of correction). The Spanish organicist-materialist Santiago Ramón y Cajal and his followers read Freud as biology not psychology. Freud was also used as a "diagnostic tool' and in a utilitarian way in Spain, Brazil, and Peru. Analytical sessions were shortened by Delgado, who began group seminars to replace them completely; the Spanish psychiatrist José Sanchis Banús treated dozens of patients a day "psychoanalytically." In France, Jacques Lacan later stresses the decentered self, not primarily the ego-structure of Freud; the French, unlike their American counterparts,

---

[16] Alvaro Rey de Castro, "Freud y Honorio Delgado: Crónica de un desencuentro," *Hueso Humero*, 15-16 (Oct.-Mar. 1983), 5-76; Roberto Sagawa, "A psicanálise pioneira e os pioneiros da psicanálise em São Paulo," in Sérvulo Figueira, ed., *Cultura da psicanálise* (São Paulo, Brasiliense, 1985), pp. 15-34.

see no core personality. The act of the corrective reception of Freud both revises and extends Freud's original theses. Pro-Freudian feminists focus on pre-Oedipal, undifferentiated sexuality, sexuality as pure psychological construction. Neutral and conservative receptions would be corrective in some sense, related for example to bettering the social constitution of society (parallel to the eugenic appropriation of Darwin).

The extensional reception of Freud via cultural appropriation, takes form via artistic and literary uses: for example, Salvador Dalí's "paranoid-critical" method, and his famous Oedipal interpretation of Millet's *Angelus*.[17] In France, Apollinaire is influenced by Freud and André Breton devises automatic writing to give materiality to Freudian notions of the unconscious. Extensional appropriations which take different turns can be found in Nazi Germany and in a progressive climate in Spain, respectively: the Nazis appropriate Freud in order to effect a kind of political and social control, in order to help people adjust and adapt to the state in a repressive, nationalistic spin; while in Spain liberals hooked Freud up to a sexual reform movement, whose maximalists propounded ideas of divorce, abortion, free love, and the equality of the sexes along a Freudian reading of the self. Also in Spain, Freudian language is used to pass in the Cortes political reforms of a liberal nature—divorce and maternity leave laws take shape under the progressive logic that such measures will lead to healthier children (Sanchis Banús, 1931).[18] Finally, just as Darwinism was assimilated to Taoism, there were comparable appropriations of Freud colored by traditional religious beliefs, such as Girindrasekhar Bose's inventive interweaving of psychoanalysis and Hinduism.[19]

Einstein

The thetic reception of Einstein is taken up by disciplines of mathematics, astronomy, and physics. Mathematicians in Spain and Italy play a key role in thetic reception due to the mathematical nature of relativity (Italian mathematicians had developed tensorial calculus, the language of general relativity; the Spaniards were their disciples).[20] The verification of General Relativity by the solar eclipse of 1919 leant prestige to astronomers like Sir Arthur Eddington.

The antithetical reception of Einstein takes place in relation to an older generation tied to a mechanical view of the universe (which impeded them from appreciating the kinematic nature of Special Relativity), in relation to experimental physicists who were

---

[17] Salvador Dalí, *El mito trágico del "Angelus" de Millet* (Barcelona, Tusquets, 1978).

[18] Thomas F. Glick, "The Naked Science: The Reception of Psychoanalysis in Spain, 1893-1936," *Comparative Studies in Society and History*, 24 (1982), 533-571, on pp. 554-556.

[19] Ashis Nandy, *The Savage Freud and Other Essays on Possible and Retrievable Selves* (Princeton, Princeton University Press, 1995), pp. 81-144.

[20] Thomas F. Glick, "The Reception of Relativity in Spain," in Glick, ed., *The Comparative Reception of Relativity* (Dordrecht, D. Reidel, 1987), 231-263, on pp. 234-238.

against Einstein's intuitive-deductive method (theory without proof-via-data), and in relation to conservatives' misreading Einstein's statements on absolute time and space. The antithetical reception accuses Einstein of being abstract, incomprehensible, and undemocratic; criticisms are motivated by underlying ideologies of positivists, experimentalists, and American political philosophies, respectively. Relativity picks up baggage held against Modernism or, as in Germany, animus against Einstein because he was Jewish.

The corrective reception of Einstein is hard to document, perhaps, due to the mathematical basis of the ideas of special and general relativity. But in the United States, Special Relativity was "sold" to a distrustful physics community by stressing its experimental utility. There were "Aryan" physics texts published by the Nazis (Lenard's *Deutsche Physik*) which preserved relativity but dissociated it from Einstein. The high-brow press distorted ideas of relativity in interests of ideology and of retaining intellectual control.

The extensional reception of Einstein is made at popular, literary (Faulkner, Joyce, Capek), artistic (Dalí), social, and political levels. Anarchists take up Einstein to legitimize political agendas, as do Catalan nationalists, and some Italian Fascists: Mussolini invoked relativity to legitimize the subversion of the older generation and their traditional views. Relativity was used to promote perspectivism in philosophy and to legitimize extreme historical relativism (Ortega y Gasset, Spengler). Meaning is culture bound, not true or real (relativity semantically confused with relativism). Relativity helps to undermine traditional cultural constraints and values—while being not at all about relativity in that valu-able sense. At the popular level especially, relativity comes to mean relativism which, through feedback among levels of reception, helps to distort the scientific reception of the theory. In Russia, both Marxists and conservatives can appropriate relativity as an "open text" susceptible to multiple readings and agendas. In a wide sense, Relativity becomes a general mantra for modernization.

NORMATIVITY OF SCIENTIFIC RECPETION

There is a particular normativity to be found in an analytical aesthetics of reception. The thetic ratio shows that diverse disciplines were significant in the reception of the content of the original idea, sometimes from another discipline. Disciplines are the operational sites of thetic receptions; some disciplines replace the original discipline in taking on the thetic reception of an idea. Thus relativity was received by mathematicians, theoretical physicists, and theoretical astronomers, but in the many places where physics was thinly practiced (Spain and most Latin American countries) the popularization of Einstein's theories was left to engineers.[21]

One can see that generational and religious antithetical reactions—while differing in variables or categories—are in the same ratio. This may illustrate a desire to maintain a

[21] Glick, "Relativity in Spain," pp. 243-248.

given ideology; they show a certain anxiety-resistance over novel and non-traditional value-systems (all of which is predicated on an economy of loss or gain for the self, at either or both personal and professional levels). The corrective ratio illustrates national trends in the revision or re-presentation of ideas originating in an "other" country/thought-style. This may illustrate a kind of resistance at the level of "national identity" (modal personality) of particular scientific disciplines—i.e., Darwin's corrective reception in Larmarkian France, which is undertaken via an appropriation combined with resistance. In this type of transmission-as-transformation, the idea is altered. The extensional ratio shows a significant propensity to appropriate scientific ideas as a means of legitimacy for political, social, and popular movements. It also shows a high incidence of appropriation for innovations in literature and artistic movements or developments, which are directly connected to the novelty of the scientific idea and so can confer gains of originality upon those who take up such newness. Artistic creativity is connected, in this way, to scientific creativity.

In the end, one may say that the four ratios of reception illustrate a resistance-based and/or appropriation-based encounter between scientific ideas and particular groups and individuals with something to gain or lose. In particular, the three ideas considered are highly affective ones, creating much interest and excitement because in a real way they represent a definite movement away from an anthropomorphic view of the world and from a view of the centered self in a meaningful, transcendental kind of existence. With no special creation, no fully transcendental ego, and no absolute space and time, the very idea of the Western autonomous rational self is put into question (ironically the very self which is held to be responsible for such science) and a certain anxiety unfolds. However, at the same time, we see that the fundamental nature of Darwin's, Freud's, and Einstein's ideas not only makes them capable of engendering anxiety and resistance, but also capable of being appropriated for interested uses, exploited for a certain profit or capitalization of the self. This happens in both professionalized scientific and in less professionalized artistic realms. One can see normative considerations emerge out of a deliberative psychology to divide scientific reception into the four ratios. Movements of resistance and appropriation are related to an economy of loss and gain, of defending or exploiting intellectual capital in reference to a particular personal, social, or institutional position. Under such logic, the primary diffusion of scientific ideas takes place and moves into the world according to types of intentionality.[22] The normative aspect of the diffusion of scientific ideas, then, illustrates, under an analytic division of reception, the metascientific notion that fundamental science is not usually first taken up *qua* science, in terms of solely its truth value, but as a commodity tied to the position of a particular group or self.

Boston University
University of Massachusetts at Amherst

[22] The point here is intentionality vs. truth in reception studies; that a deliberative psychology matters as much as the truth of a scientific idea.

ROSAURA RUIZ AND FRANCISCO J. AYALA

DARWINISM: ITS HARD CORE

The objective of the following presentation is to contribute to the precision of some concepts in the analysis of the introduction of Darwinism, from England to other countries. The concern emerges from the variety of view points sustained in different studies, regarding what should be considered Darwinism.

In the first part of the presentation, we summarize the process of elaboration of the Theory of Natural Selection because we consider that the epistemological analysis of the concept, contributes to the understanding of it. Since the main objective of science is to effectively find theories which solve problems (Laudan, 1996), a historical-epistemological analysis, allows us to understand which problems Darwin intended to solve and how he did it. In this sense, it can be said that Darwin's principle concern was to provide a natural, plausible mechanism which could account for the great diversity of living beings, and those which have passed on; a mechanism, which at the same time would be a natural alternative to the concept of a perfect design stemming from a Creator, who accommodates organisms and their parts in the environment.

In the second part, we propose the basic points that should be considered in the analysis of the Darwinian concepts that are introduced. It is especially important to distinguish between Darwin's concepts and those of other authors who have played a central role in the introduction and diffusion of Darwinism. In Latin America, it is important to consider Ernst Haeckel, Herbert Spencer, and Francis Galton. Of these, the first is the most important in the spread of Darwinism in the nineteenth century. The others played a fundamental part in the extension of Darwin's ideas to certain fields, which Darwin himself did not have in mind when he developed the theory concerning the evolution of human society and the improvement of our species. Nobody equaled Spencer in the analysis of the implications of Darwinism, and in general, about the evolution of the development of human society. It was not in vain that British and especially North American elites, adhered to a such convenient conceptualization of the division of classes. However, the three authors had their own ideas about the causes of

Thomas F. Glick, Rosaura Ruiz, and Miguel Ángel Puig-Samper (eds.), The Reception of Darwinism in the Iberian World, 241-261.

evolution, and in many cases they distorted Darwin's ideas without clarifying which were their own and which were Darwin's.

Specifically, this problem arises in Haeckel and Galton. For example, in Haeckel's work there is no mention of the participation of chance in evolution. For him, as in Lamarck, variation and adaptation are the same. In reality, Haeckel's concept explicitly is the integration of Lamarckism with the idea of natural selection. Consequently, far from being more or less orthodox Darwinism, it is a distortion. Conversely, Galton espoused hereditary determinism. His preoccupation was for the improvement of the human species by "rejecting the undesirable and multiplying the desirable." The latter are those whose qualifications and disposition establish them in positions of leadership. For Galton, character, intelligence, and the capacity for innovation were as inheritable as height, color of eyes, and of hair. For him, the difference in success among human beings was due to the quality of the individual reproducers. He believed that Darwin was correct on this point. Galton did not understand the importance that the environment has in evolution. He did not perceive that there was any relationship between the environment and the adaptive character of variation. Consequently, he did not distinguish characters in the context of adaptation, giving it an absolute character. In a letter that Darwin sent him after reading ***Hereditary Genius*** (Galton, 1869), he manifests only a partial agreement with Galton because Darwin always thought that all men, with exception of fools, do not differ much in intelligence, only in enthusiasm and in their capacity to work; and he always considered that it was an important distinction (F. Darwin y A. Seward, 1903, Vol. I, p. 317). Galton replied that such enthusiasm and capacity were, of course, inherited characters. According to Galton, individuals are born with specific capabilities and the environment can do little to change them. On the other hand, Darwin was very impressed by the case of the Fuegians "collected" by FitzRoy, on an earlier voyage of ***The Beagle*** to America and that he taken with him on the voyage with Darwin. While they were in England, the surviving Fuegians (one died shortly after their arrival) learned English and all that the "civilized" British had shown them. In addition, in ***The Origin*** he wrote that one should not confuse habits and instinct (p. 209).

The case of Spencer is different in an important sense, because he influenced Darwin's work more than Darwin influenced Spencer's. Darwin accepted Spencer's term, "Survival of the Fittest" as a synonym of natural selection. For Spencer, evolution is a universal principle. From the solar system to living beings, human society and their products have evolved from the simple and homogeneous to the complex and heterogeneous. This process of differentiation and individualization is inevitable; it is a *law* that leads to progress in all levels of evolution. For Spencer, Darwinism, particularly its central notion of natural selection, came to validate his evolutionist thesis. In the case of humanity, social prosperity was a demonstration of an inherited superiority that guaranteed success in the fight for survival. The major difference between Darwin and Spencer is found in Spencer's defense of an inner, mysterious, unknown force that worked continuously in the production of variation. This force, which he called "Inscrutable Power" or "Unknowable", was a religious idea because such a force could be identified with God. This power created the initial diversity upon which natural selection would act mechanically (Shipman, 1994, p. 109).

On the other hand, Spencer's theory of evolution has strong roots in Lamarckism. Spencer, like Haeckel, did not accept the participation of chance in evolution. For him, all structures and all functions have a purpose, and these have to do with progress. Spencer's concept of evolution is Lamarckism; in Darwin, natural selection favors the strong in the fight for survival. The entities that interact in this battle, function as the selectors of hereditary characters that determine the most vigor. Here, natural selection is a creative force. In Spencer, the struggle has another meaning. Suffering, obligates individuals to go beyond the limits of heredity. As Bowler points out, the purpose of the fight is not to eliminate the least competent, but to force him to become competent. The misery that accompanies failure is the best way to educate the weak, so that in the future they become more clever and enterprising. The elimination of congenital stupidity is only a secondary factor, the great majority of the people have the ability, if only they put their mind to it, to adequately function in the world. The advantage of unrestricted individualism is that it forces all people to maximize their efforts and to completely exploit their initiative (Bowler, 1990, p.198). For Spencer, the changes accomplished would be converted into hereditary and only those not capable of changing would be eliminated (Spencer, 1851b).

Lastly, it is important not to forget that Darwinism has increased in precision. This precision assists historical analysis by making a greater understanding of Darwin's concepts possible. This is valid only when present-day concepts are not confused with the original ideas. It is evident in Darwin's acceptance of the inheritance of acquired characters. Neither should there be confusion concerning the asserttion that the incorporation of Mendelism as an explanation of heredity, does not affect the conception of the causes of Darwin's evolution. For this reason, we include a reflection about the origin of the central ideas of Darwinism. This allows us to demonstrate that modern explanations do not alter the original meaning of the before-mentioned concepts.

THE DISCOVERY OF NATURAL SELECTION

Once Darwin was posted to *The Beagle* as a naturalist, he became a participant in the research tradition of the period.[1] Among the most outstanding representatives of that tradition was Charles Lyell. During that time, Darwin was a convinced creationist. He thought that creation should be enough for the understanding of all of the existing phenomena. It is with this vision that he began his incursion into science.

When he returned to England in December of 1836, he began writing *The Journal of Researches* (published in 1839). It was at that time, that he received the reports on the

[1] We use the concept of "tradition of investigation" in Lauden's terms (1977, 1996). According to him, research traditions work as guides for investigation, insoso far as they: 1) indicate which assumptions can be seen as "antecedent information" that is not controversial, but worked on by all of the scientists that follow in that tradition; 2) help identify those portions of the theory that have difficulties and should be modified or corrected; 3) establish rules for the collection of data and for the proposal under investigation of the theories; 4) propose conceptual problems for whichever theory violates the ontological and epistemological foundations of the parental tradition (Laudan, 1996, pp. 83-84).

species which he collected from the taxonomists who identified them.[2] Understanding what organisms constituted which species and the task of relating them to geographical distribution, led him to conclude that there were certain matters that could not be explained by existing theories. He especially began to question Lyell's vision of living beings and to think about the possibility of evolution. In an abstract about ornithology, from journals written during the *Beagle* voyage, he wrote that if these observations have the least bit of foundation, it would be worth studying the zoology of archipelagos. For such findings would undermine the stability of species. (This entry in his diary—contemporary with his first Notes on the Transmutation of the Species—has been considered an indubitable demonstration that, in the Spring of 1837, Darwin had abandoned the traditionalist point-of-view. From that moment on, he decided to develop an explanation for evolution because the ones that existed at that time, that of Lamarck and his grandfather, did not seem correct to him.

The fact that he made these notes after his return to England, indicates that Darwin had not abandoned the traditional concept during his visit to the Galapagos. That is to say, his observations in that "laboratory of evolution" (as Ernst Haeckel called it), were not enough for him to become convinced of evolution. The abandonment of the traditional view point took place during the writing of *The Journal of Researches*, in the Spring of 1837, when the confrontation between natural theology and the observations made during the trip made clear the lack of fit between the two. On the contrary, after observing the peculiar Australian fauna at the end of the trip, it caught his attention that despite all of the differences with other environments on other continents, there were also surprising similarities. Then, he thought about the possibility of separate creations. In a diary entry dated January 1836, he states that the same hand has acted throughout the universe and, to explain the differences, a geologist might propose that there were different moments of creation, separate and distinct from one another, because the creator had rested in his labor.

NATURAL THEOLOGY IN CRISIS

Before becoming convinced of evolution, Darwin considered Lyell's version of natural theology. It is sufficient to say that when he returned to England, he began to reappraise his material, in order to write up his notes from the trip. It is important to clarify that the differences between Darwin and Lyell are particularly related to the origin of living human beings. The major point of contact stems from Lyell's geological ideas (there were differences concerning time) and his uniformitarian conceptualization. These were not rejected by Darwin. He actually incorporated them, to a great extent, into his theory. The causes of evolution are just as constant, gradual, and current, as the causes of the transformation of the earth's surface.

In regard to the geographic distribution of living beings, Darwin found the following contradictions with natural theology:

1. The differences in biological diversity between islands and continents. For example, there is much more biological diversity on continents than on islands.

---

[2]Other participants included John Gould, who identified the birds; Richard Owens who identified the living mammals; Joseph Hooker who identified the vegetables; and Leonard Jenyns who identified the fish.

2. Groups definitely excluded on islands. An example of this are amphibians, because their eggs are not resistant to salt water; or mammals (except for bats that reside on islands far away from continents).

3. There are simiarities owing to latitudinal and altitudinal geographical distribution.

4. There exist similar species in different environments, while in similar environments similar species do not develop, as natural theology supposes. This contradicts the notion of the "perfection of design" and, indeed, the idea of design itself.

Darwin considers that more than being the result of a perfect design, the patterns of geographic distribution are determined by the types of locomotion and dispersion of the organism, together with the obstacles that they encounter. The Galapagos Islands are populated by forms that are capable of crossing the ocean or that have been introduced by man. That is, Darwin saw that geographic distribution is "arbitrary. It is not permanent. This might be made very strong if we believe the Creator creates by any laws, which, I think is shown by the very facts of the Zoological character of these islands. So permanent a breath cannot reside in space before island existed. Such an influence must exist in such spots. We know that birds arrive & seeds" (Darwin, 1987, Notebook B, p. 98). On this point, Darwin agrees with Linnaeus and Lyell in one respect: each species originated in only one place, and from there, were dispersed. (Linnaeus and later Lyell espoused the idea of a central point of creation.)

Another point that Darwin immediately questions is the explanation of extinction. Lyell held that there are two factors that elucidate the extinction of species. The first is change in local circumstances, which implies the affected species' failure to adapt. This is due to the fact that they were designed with the capacity of adaptation for the precise environment that they find themselves in. Competitive exclusion provides the second reason. If two species compete for the same resource, the weakest one will disappear.

On this topic, as with the case of geographic distribution, Darwin made observations which neither of the two explanations could account for. In the Red Notebook (Herbert, ed. 1980), Darwin recorded his observations regarding the two cases that exemplify this question. These concern the relationship between what he believes to be an extinct llama *Macrauchenia patachonica*, and a present-day llama called *Lama guanicoe*, and two ostriches. The big or common ostrich is known as *American Rhea* and the small one, *Rhea darwinii* (Darwin's rhea, today called *Pterocnemia pennata*).

With foundation in his geological studies, Darwin concluded that there had not been any climate changes in the plains where the remains of *Macrauchenia* were found. As a consequence, its extinction could not be blamed on changes in circumstances. He believed that it could not be attributed to competitive exclusion either, because of the large size of the extinct llama. In retrospect, he concluded that a better explanation was that the present-day llama is a modified descendent of *Macrauchenia* which might have disappeared suddenly.[3] Darwin considered that in regard to the ostriches whose

[3]Something which is of great historical and scientific interest, is that the extinct "llama" that led Darwin to abandon the traditional point-of-view, in reality, was not a llama. When Darwin returned to England, Richard Owen informed him that the fossils were those of a gigantic camel, belonging to the order of Ruminantia. Consequently, Darwin reasoned that it could be the ancestor of the guanaco, a small camelid that roamed the plains that were then dominated by *Auchenia*. However, at the beginning of 1838, that is after Darwin had completed the analysis that had led him to conclude that the *Macrauchenia* was a direct ancestor of the llama,

populations were separated only by the Rio Negro (which was not an obstacle for them), that if you accept the notion of competition in Lyell's terms, this kind of distribution could not have happened.

Darwin thought that the temporal distribution of *Macrauchenia* and llama and the spatial distribution of the ostriches, could be better explained if their family relationships were considered. At that moment, he thought that in both cases one species could have given rise to the other in only one step. In other words, one generation produced the next. In this way, a new species resulted from the modifications that the sex cells underwent in response to environmental changes.[4]

In this first stage, he considered that the extinction of a species could be pictured as a process very similar to the death of a person. That is to say, having reached senescence, a species could be extinguished just as an individual could.

FROM PERFECT ADAPTATION TO DIFFERENTIAL ADAPTATION

Besides *Macrauchenia* (identified by Owen) and the ostriches, Darwin was especially impressed by John Gould's report on mockingbirds. This is probably because they were the ones that he had best collected and, contrary to the finches whose labels were incomplete, the origin of the mockingbirds had been clearly identified. Once the mockingbird species had been established, Darwin realized that the island species were similar to those of the continent. Interpreted from the vantage point of his new idea, this meant that the island species could be modified descendants of individual colonizers that came from the continent. This conclusion resolved the questions that he himself had raised regarding the traditional explanation of biogeographic distribution given by natural theology. Furthermore, this led Darwin to pose two additional questions. On the one hand, he gave clues respecting the gradual progression of this process, inasmuch as the species of the island were very similar to those of the continent. That is, there had not been a leap, as there had been in the case of the ostriches. On the other hand, it led him to emphasize the role of geographic isolation in the generation of new species. Due to his conception of mixed heredity, he believed that if only two varieties are separated for reproduction by geographic barriers, they could lead to the formation of two distinct species.[5]

Owen informed him that he had named the fossil *Macrauchenia*, and that it was not a camel, but more like a tapir with a camel-like neck; not even a ruminant. "Owen assigns the *Macrauchenia* to the order of pachyderms (elephants, hippopotamus, pigs, tapirs, rhinoceros, and horses)." As Darwin states in *The Origin*, "Cuvier considered the ruminants and pachyderms as the most distinct orders of mammals" (Rachootin, 1985). Rachootin classifies *Macrauchenia* as a member of Litopterna ("smooth-heel"), an order of extinct ungulates.

[4] Hodge and Kohn (1985) argue that Darwin perceived a parallelism between the sexual generation of one individual from two others and the propagation of one species from another. The construction of this analogy is possible because Darwin considered that the environment can influence the process of embryonic maturation. The variations of adaptation which accompany sexual generation in changing conditions can be incorporated into the new species' structure. Only adaptations are incorporated, not other changes such as mutilations, for the process of ontogenetic recapitulation only allows innovations which are in harmony with the previous structures.

[5] Despite the fact that after Mendel, we know that heredity is particulate, it is still considered essential that the two populations be separated for species to be formed. But the reason for this is different from Darwin's.

With the previous speculations in mind, Darwin took on the problem of variation. Clearly, without variation there is no evolution. The over-population which results from the excessive reproductive capability of a large part of the species, in relation to the number of organisms that a determined territory can sustain, is eliminated by chance.

To clarify this problem, Darwin attempted to understand what the connection was between the variations produced and the subsequent formation of permanent varieties and species. With that objective in mind, he read the work of hybridologists, horticulturists, and breeders hoping to elucidate the process whereby varieties appear and are preserved.

A fundamental step in the elaboration of his theory was Darwin's recognition that the variations are not in themselves adaptive. A monstrosity can, for example, be very unfavorable for the original form in a determined environment, but favor an organism already undergoing change. This is the notion that differentiates Lamarck's adaptation concept with Darwin's. For Lamarck, all variation is adaptive because it is the instantaneous result of the actions of the organism, whose goal is the improvement of the relationship between itself and the environment. From that moment on, for Darwin (we previously saw that he supported the idea of instantaneous adaptation), adaptation is a process that begins with the appearance of variation. Consequently, it can follow two distinct paths, depending on the origin of the variation. If it developed due to the direct action of the environment or from use or disuse of organs (Lamarck's concept of variation), such a variation is immediately adapted and incorporated into the general characteristics of the species; it could be reinforced by natural selection, but it is clearly not rejected. On the other hand, if the variation occurred spontaneously, it is not in itself an adaptation. Its character, adaptive or not, is going to depend on the relationship between the organism and the environment. If the variation improves this interaction, natural selection will favor the carrier, but if the contrary takes place, it will discard it.

One of Darwin's most important contributions to the modern idea of evolution is the understanding of this type of variation. Without this notion, evolution would not be understood as a contingent phenomenon. The fact that the variations are spontaneous does not mean that they necessarily occur by chance. Darwin considered them spontaneous because he acknowledged his ignorance of the process that produces them. Where there is contingency is whether the variations are adaptive or harmful. A single variation can turn out to be favorable in one environment, detrimental in another, and even neutral in a third one.

Despite the fact that Darwin understood the importance of variation in evolution and that he integrated it adequately into his concept of natural selection, his analysis of a variety of studies on variation under domestication and in nature (including his own studies of barnacles) did not lead to any advances in undertanding the mechanisms that produced variation and its transmission.

Although the possibility of sympatric speciation is allowed, it is recognized that an indispensable condition for the multiplication of species is reproductive isolation. This is accomplished more radically in allopatric conditions. Nevertheless, today it is accepted, as Darwin later realized, that the origin of new species can also occur in sympatric conditions.

THE MALTHUSIAN EPISODE

Darwin's reading of Malthus's essay on population has been considered by the majority of students of Darwinism as a key moment in the structuring of the Theory of Natural Selection (Hodge, 1983; Herbert, 1971; Kohn, 1980; Ospovat, 1981; Ruse, 1979; Schweber, 1977, 1980). Malthus's book contributed to Darwin's understanding of some of the ideas central to evolution.

Among the most important are the notion of intraspecific competition and population analysis. In natural history (whether in the tradition of Paley and Linnaeus or that of Lyell and De Candolle) interspecific competition had been recognized. However, it is in Malthus where Darwin understands the difficulty of the struggle among individuals of the same species motivated by similar necessities. The analysis passes from the struggle between lions and gazelles, to the war among gazelles.

The influence of Malthus was decisive. His work is an analysis of a single population of a single species, allowing for the understanding, among other points, of the consequences of the interaction between the individuals of a population. It produces the change from a typological vision to one focused on population. From that moment on (according to Ospovat) Darwin would focus on the long-term effects of the differences between individuals on the composition of a population. He abandons thinking in terms of idealized forms and begins to concentrate on the activities of individuals. At that point, the principle of natural selection was obvious. To build an exact theory of natural selection, it is necessary to see how the success or failure of an individual may affect the properties of the species, gradually altering the proportion of individuals with a certain characteristic. Ghiselin (1969) states that before reading Malthus, Darwin had the necessary components of his theory, especially because he distinguished between collective selection and the traditional, natural (individual) selection. Although he did not have the theory itself, he still thought of species and varieties as new groupings of individuals characterized by particular attributes. He had to conceive of the species as new units of interaction, composed of biological individuals, as populations more than classes.

It was only after reading Malthus, that Darwin changed his vision of adaptation. He understood its relative character: organisms are more-or-less adapted in comparison with others. From that moment on, he began altering his conception. He began re-considering adaptation not as perfect, but as a process that arises and that instantly corrects the organism, adjusting it to its surroundings. When he comprehended this, he presented his definitive vision in 1844 (in the Sketch). This was an action that gradually adapted structures and functions of form in accord with the environment. Before the Malthusian episode, Darwin believed that all changes in organisms corresponded to environmental changes. That is to say, that all variation would be an appropriate response to the environment.

Until the beginning of 1839, Darwin viewed variation as differences, not adaptations. Then, he explains the second part of the process, that of the selection of variations. Before that moment, Darwin only saw the first part of the process. In order to explain evolution, it seemed sufficient to understand the response to problems offered by the environment in terms of what was relevant to the production and conservation of variation. What was lacking was the understanding that the changes in organisms are not adaptive in-and-of-themselves, the understanding that each variation makes the organism stronger or weaker, and more-or-less capable of reproduction; in sum, the understanding of the risky character of the quality of variation. This makes the discovery of natural selection possible. The capacity of a variation to adapt or not will then be defined during the second stage of the process as the environmental agents which, although they are not the instigators of the variation, are the ones that account for pressure in favor or against such variations. In this way, Darwin introduces a difference that makes his theory incompatible with the most important one among his predecessors: in Lamarck, the environment instructs the organism to change in the required direction. Evolution is a result of a one-and-only stage, and organisms are transformed because they adapt.

In classical political economy Darwin found the effects of elimination caused by an increase in population larger than the resources available for survival. He draws an analogy between what he knew about the war between species, with the struggle for existence among the members of a population. He understands survival at an individual level as the impulse that orients the evolution of a species. This struggle produces adaptation and consequently explains it. Without the notion of the battle between individuals of the same species, the idea of war in nature on the species level does not lead to the theory of natural selection that the concept of differential adaptation requires. What Malthus demonstrates is contrary to the vision of natural harmony that Darwin, even as an evolutionist, sustained until he read Malthus's work.

Between September 28 and October 3, 1838, somewhat after reading the essay about the principles of population, Darwin ceased viewing adaptation as a slow process of a species' accommodation to the environment, that is due to its flexibility. Adaptation is converted into a question of thousands of wedges, forced into narrow cracks: "One may say that there is a force like a hundred thousand wedges trying force <into> every kind of adapted structure into the gaps <of> in the economy of Nature, or rather forming gaps by thrusting out weaker ones. The final cause of all of these wedgings, must be to sort out proper structure and adapt it to change" (Darwin, 1987, Notebook D, p. 135).

In summary, Darwin reads Malthus with the knowledge of the problem of variability, and he conjoins two essential issues: variation and the struggle for existence. From there stems the fundamental step in this whole story: organisms are different and they must fight among themselves because there are too many for the resources available. The stronger organisms triumph and pass their characteristics on to their descendants. By reading about plant and animal breeding, he learned the importance of the selection of progenitors. From there emerges the term "selection" that Darwin placed in a new conceptual context. Evidently, the word "natural" distinguishes between human selection and the one that nature carries out.

THE PRIMARY POINTS THAT SHOULD BE CONSIDERED IN THE ANALYSIS OF DARWIN'S CONCEPTIONS[6]

Natural Selection

Darwin's response to the question about the causes of organisms' design is the combination of variation and natural selection is. Natural selection can only act if variations exist that give their carriers the advantages of competition for survival and reproduction. From Darwin's point-of-view, abrupt variations generally cause such disorganization that they impede survival. Due to this, he sustains that the design of organisms that permit their adjustment to the environment, is a result of a gradual accumulation of changes that occur thanks to the constant action of natural selection.

Darwin has two problems: first, to convince people of evolution, and secondly, to establish a mechanism that accounts for the changes of living beings in history and their adaptation to the environment. The explanation is found and extends itself to the fossil register, the presence of rudimentary and atavistic organs, biogeographic distribution, extinction, and other biological problems. Darwin's theory resolves the polemic debate between the defenders of Natural Philosophy, such as Geoffroy Saint-Hilaire, and natural historians, such as Cuvier, concerning whether or not the structure of organisms is fundamentally due to the environment's need to change or the *embranchement* to which they belong.

Darwin offers an alternative answer to the problem of design. The organs of living beings seem to be designed to carry out a determined function. Before Darwin, the origin of the organisms and their extraordinary adaptations were attributed to the design of an omniscient creator. God created the birds and the bees, the trees in the forest, and best of all, man. God gave man eyes so that he could see, and gave the fish gills so that they could breathe in the water. Philosophers and theologians argued that the functional design of organisms manifests the existence of an all-powerful creator. Wherever there is a design, there is a designer; the existence of a clock, proves the existence of a watchmaker.

It is important to take this point into consideration because many investigators confuse theological, Lamarckian-type beliefs with the idea that all organs have some kind of function. In Lamarck, as in Darwin, the organs are adapted to carry out a function. The difference lies in the explanation of the causes of such an adaptation.

Darwin summarizes the idea of natural selection as "the principle by which all favorable variation, as small as it may seem, is kept." In *The Origin of the Species*, he argues this concept by using artificial selection as a justifying analogy. He writes: "Can it then, be thought improbable, seeing that variations useful to man have undoubtedly occurred, that other variations useful in some way to each to each being in the great and complex battle of life, should sometimes occur in the course of thousands of generations? If such do occur, can we doubt (remembering that many more individuals are born, than can possibly survive) that individuals having any advantage, however

---

[6]In some cases, we use current terminology in hope of a better understanding of the original concepts. We do this only when such a use does not change the meaning of Darwin's ideas.

slight, over others, will have the best chance of surviving and of procreating their kind? On the other hand, we may be feel sure that any variation in the least degree injurious, would be rigidly destroyed" (pp. 80-81). Darwin called this preservation of favorable variations and the suppression of unfavorable ones Natural Selection (p. 81). He concludes this paragraph by indicating that the variations which are neither beneficial, nor damaging (the neutral ones), would not be affected by natural selection and could be left as fluctuating elements, possibly in the same way that we consider polymorphic species.

The Argument

The central argument of Darwinism is that all of the species reproduce in larger numbers than the ones that can survive in a territory. The over-population, added to the problem of limited resources, provoke a struggle for existence. All of the organisms that carry some kind of variation which improves their possibility of taking advantage of the place that they occupy in nature's economy (niche), increases the number of their offspring.[7] The modified descendants will orient the transformations of the species in this new sense.

Darwin's theory resolves the problem of explaining the adaptive character of organisms. Darwin argues that adaptive variations appear occasionally, and that it is possible that they increase the reproductive opportunities of their carriers. Through the generations, the favorable variations will be conserved and the harmful ones will be eliminated. Natural selection does not have limits. For Darwin we cannot see the limit of this power to beautifully and slowly adapt each form to the most complex relationships of life. Darwin proposed natural selection, principally to give an account of the adaptive organization, or the "design" of living beings; it is a process which promotes or maintains adaptation. Through time, the changes of evolution and evolutionary diversification (multiplication of species), are not directly promoted by natural selection, but they often result as co-products of it.

For Darwin, natural selection was above all, the differential survival which was intimately related to reproduction. This is an important difference from Neodarwinism because here the central issue of the theory has passed on to the analysis of differential reproduction, taking as a fact that survival is not enough. The modern understanding of the principle of natural selection is formulated in genetic and statistical terms. Natural selection implies that some genes and typical genetic combinations are transmitted to following generations more frequently than their alternative forms. Such genetic units will become more common in each subsequent generation and their alternatives will become less common. Natural selection is a biased statistic in the alternative-genetic-units' relative reproductive scale.

[7]Darwin's concept of "the place in the economy of nature" or the "place in the politics of nature" is an antecedent to the concept of niche. We use it here in order to be much more concise than the original, even though we are aware of the differences, which despite everything, do not leave room to misunderstandings.

Natural Selection's Capacity to Create: Its Chance Character

Natural selection has been compared to a sieve that keeps the useful genes (which rarely appear), and allows the hazardous mutants (which occur more frequently), to pass through. Natural selection acts in this manner, but it is much more than a purely negative process because it is capable of generating novelties, and increasing the probability of genetic combinations which, otherwise, would be extremely improbable. Therefore, in this sense, natural selection is a creator. It does not create the entities over which it operates, but it produces adaptive, genetic combinations which in another form, would not have existed.

The creative role of natural selection should not be understood in the sense of absolute creation that the traditional Christian theology preaches about, concerning the Divine act through which God had created all species. More precisely, natural selection should be compared to a painter who creates a piece of art by mixing and distributing pigments on a canvas in various ways. The canvas and the pigments are not created by the artist, but the work of art is. It is conceivable that a chance combination of the paints, could result in the ordered set that is the final work of the art. However, the probability that a piece of art like "Las Meninas" by Velázquez, would result from a chance combination of paints, is infinitely small. In the same way, the combination of genetic units which carry the hereditary information responsible for the formation of the vertebrate's eyes, could not have been produced by a process of chance, as is the process of mutation. Even when we allow the millions of years of the existence of the earth to pass, the complicated anatomy of the eye, just as the functioning of the kidney, is not the result of chance, but of the strict process of natural selection.

Selection as an Accumulative Process

The deficiency in understanding natural selection as a creative process and not one of chance, has been the motive of permanent criticism. For example, Gabino Barreda, one of the first Mexicans to discuss Darwinism, argued that the human eye could not be the result of natural selection because selection cannot favor the parts of an organ and then expect that the following parts will develop, until the organ is complete. Barreda would be correct, if this in fact was what Darwinism professed. What he did not understand, and none of his peers and disciples of evolutionism from the *Asociación Metodófila* could not explain, was that an organ, such as the eye, does not newly arise each time. Rather, it is the result of the organ's evolutionary history. From the *Euglena* optic stain, to the eyes of insects, mollusks, and vertebrates, there is an accumulative history of evolution.

The same problem was proposed to Darwin. His answer was that the eye had to be studied throughout the animal scale and not only the perfected eye of the vertebrates. Today, we say that the vertebrae's eye did not suddenly appear in its perfected form. Its formation required the appropriate integration of many genetic units. Therefore, the eye could not have resulted solely by a process of chance. The ancestors of today's vertebrates, had, during half-a-billion years, a class of organs which were sensitive to light. The perception of light (that later evolved into vision), was important for the survival and successful reproduction of these organisms. Consequently, natural selection favored the genes and genetic combinations that increased the functional efficiency of the eyes. Such genetic units accumulated, gradually reaching the efficient and highly complex system of the eyes of the vertebrates. Natural selection can give an account of the origin and expansion of genetic constitutions, and therefore, of all types of organisms which never would have existed under the uncontrolled action of incidental mutations. In this sense, natural selection is a creative process, even when it does not create the primary material (the genes) over which it acts.

Natural Selection as an Opportunistic Process without a Pre-Determined Direction

There is an important aspect which makes the painter a bad analogy of natural selection. A painter usually has a preconception of what he wants to paint. He consciously modifies the painting so that it will represent what he wants it to. Natural selection does not have a project, nor does it operate according to a preconceived plan. On the contrary, it is a purely natural process which results from the interaction between physicochemical and biological (within its properties) entities. Each step of natural selection is oriented towards the improvement of the species' capacity for adaptation. However, the total sequence is not pre-determined, nor does it have a preconceived end. Furthermore, each step may alter the functional direction of the different pressures of selection. Natural selection, is simply a consequence of the differential multiplication of living beings. It has a certain appearance of having a purpose, for it is conditioned by the environment. The fact that organisms reproduce more efficiently depends on what variations that are useful in the surroundings in which they live, they posses. However, natural selection does not foresee future environments; the organisms that had survived in previous conditions, may not be able to overcome new and drastic changes in the environment.

Wallace criticized Darwin for the use of the term "natural selection" because from his point-of-view, the use of the analogy with artificial selection, lent itself to the personification of the mechanism of selection. Darwin was never in agreement with abandoning the name of his theory. He said that it was quite obviously a metaphor. A metaphor similar to Newton's when he spoke of "gravitation" as an attraction between celestial bodies. In any way, he accepted Wallace's suggestion of using Spencer's phrase: "survival of the fittest" as a synonym of his theory. For Darwin, natural selection was not directed by anyone. It is a process that is constantly performed. Predators, competitors, parasites, etc., just as the physical agents, constantly evaluate the organisms. Only those that triumph over such factors, survive and may reproduce. In evolution, the adaptive combinations are not being selected by anybody. These combinations select themselves by multiplying more efficiently than the less adaptive ones.

Moreover, natural selection does not insist on producing predetermined types of organisms, but only the organisms which are adapted to the present environment. What characteristics are selected depends on what variations are present at a given moment and place. At the same time, this depends on the chance process of mutation, just like the organisms' previous history. This history is composed of the genetic composition that they have as a consequence of their evolution. Natural selection is an opportunistic process. The variables which determine which direction it will follow are the environment, the pre-existing constitution of the organisms, and the mutations develop by chance.

The Combination of Chance and Necessity: The Central Idea In Darwin's Theory

Only the combination of the spontaneous development of variations and the actions of natural selection can explain the process of change in organisms. In Darwin's opinion, mere chance can cause a variety to differ from its progenitors in some character, and the descendants of this variety to differ once again from their parents at a higher level, but this will never give an account of the habitual and large quantity of differences that develop between the varieties of the same species, and between species of the same genus (***Origin***, p. 111).

As Darwin argues, the origin of chance variations, does not explain the adaptive character of organisms. The organism's natural selection, which is subject to the whims of genetic mutation and the challenges of the environment, is the one that best illustrates the radiations, expansions, changes from one form to another, occasional but irregular tendencies, and the extinctions that are always present. These events are not compatible with a pre-ordered plan, even if they are accomplished without an omniscient and all-powerful designer, or as the result of a necessary force which inclines the process towards definitive results. Biological evolution differs from a painting or an artifact in that it is not the result of a preconceived design, such as the one that an artist or an artisan designs.

Natural selection explains the design of organisms because the adaptive variations tend to increase the probabilities of survival and reproduction of its carriers at the expense of the less-adaptive variations. The arguments of Saint Thomas of Aquinas or William Paley against the incredibly improbable appearance of organs such as the eye, are acceptable up to a point. But not even these scholars, nor any other authors before Darwin, were capable of discerning that there is a natural process (that is, natural selection) that does not occur by chance but rather is directed and capable of "creating" or generating order. The characteristics that the organisms acquire in their evolutionary histories are not fortuitous. They are determined by the functional utility that they provide for the organisms.

Chance, however, is an integral part of the process of evolution. The mutations that produce hereditary variations that are available for selection occur by chance, independent of whether they are beneficial or hazardous to their bearers. But this fortuitous process (the same as all of the ones that have a role in the great theater of life) is counteracted by natural selection, which maintains what is useful and eliminates what is detrimental. Evolution could not take place without mutations, because there would be no variations to be differentially transmitted from one generation to another. Nevertheless, without natural selection, the process of mutation would result in disorganization and extinction, due to the fact that most mutations are disadvantageous. Together, mutations and natural selection have led the marvelous process which, beginning with microscopic organisms, gave rise to orchids, birds, and human beings.

The theory of evolution manifests chance and necessity, intricately combined in the essence of life; chance and determinism working in a natural process which has given rise to the most complex, diverse, and beautiful entities in the universe: the organisms that populate the earth. This includes human beings that think and love, are endowed with free will and creative powers, and are capable of analyzing the same process of evolution which gave them their existence. This is Darwin's fundamental discovery, that there is a creative, although not conscious process. This is the conceptual revolution that Darwin completed: that everything in nature, including the origin of living beings, can be explained by material processes, which are governed by natural laws. This is not but a fundamental vision that has forever changed the way in which human kind perceives itself and its place in the universe.

The Difference Between Natural Selection and Artificial Selection

The differences between artificial and natural selection occur, above all, because in the first, the selector is an individual who is conscious of his objectives. A cattle rancher in the field and a scientist in the laboratory, perform selection with pre-determined ends. Nonetheless, their performance, is in a large way, equivalent to that of the multiple, biotic and physical factors that direct evolution in nature. Because of this, the investigations, and in general, the practices of artificial selection are considered as evidence in favor of the theory of natural selection. The conscientious character of artificial selection, prevented Darwin from accepting the fact that it was one more form of natural selection. One of his main problems was demonstrating that in natural selection there does not precisely exist a being who deliberately directs evolution.

Due to this, even if we utilize artificial selection as a justifying analogy, Darwin emphasizes the difference between the two. The selection which creates the human species is a directional selection with bases in human interests. There is a finality, a pre-defined point of arrival that does not exist in natural selection. On the other hand, nature acts for the creatures benefit. It tries out the characteristics that it selects so that each living being is placed in the conditions of existence among which it has developed. This road may suffer changes according to the species' necessity. On the contrary, human beings never consider the precise conditions that a species requires for its life, they place mixed species (without natural ecological relationships), in the most dissimilar environmental conditions. They do not allow the males to fight for the fecundating of the females, and they do not destroy the organisms which are deficient, etc. None of this makes the selection natural (*Origin*, p. 83). Natural selection "is daily and hourly scrutinizing, throughout the world, every variation even the slightest; rejecting that which is bad, preserving and accumulating all that is good; silently and insensibly working, whenever and wherever opportunity offers, at the improvement of each organic being in relation to its organic and inorganic conditions of life. We see nothing of these slow changes in progress, until the hand of time has marked the long lapse of ages, and then so imperfect is our view into long past geological ages, that we only see that the forms of life are now different from what they formerly were" (p. 84).

Sexual Selection

In *The Origin of Species*, Darwin briefly mentions sexual selection. As we know, a topic that develops in its plenitude in *The Descent of Man and Sexual Selection*. He states that this type of selection does not depend on the struggle for existence, but of the battle between two males for the possession of the females, or a competition between males to attract them. The result is not the death of the unsuccessful competitor, but a reduction to some extent of the size of the progeny. Sexual dimorphism is another result of this type of competitiveness.

OTHER CENTRAL CONCEPTS

From our point of view, an author who assumes Darwin's central ideas concerning natural selection and chance (even though if it is about authors prior to the modern synthesis, they also accept the inheritance of acquired characteristics), should be considered a Darwinian. Nevertheless, it is important to analyze their understanding of other important topics, both for their own intrinsic interest, and also because it may signal the development of something that we could call "local Darwinism," where an author who defends natural selection and the role of chance in evolution may be recognized as a Darwinian and, at the same time, may have a series of original ideas concerning other questions. This re-elaboration or modification of Darwinism has occurred with evolutionist who were Darwin's contemporaries, as well as with present day investigators.

Variation

Although Darwin admitted his lack of knowledge of the laws that control heredity, for him it was clear that natural selection can only act if there is great variability between the individuals of a species. He asserted that unless one individual (one is enough) among progeny inherits useful variations, there is nothing on which selection can act. He recognized that there are two forms of variation: individual variations, which are the small, gradual ones, and the ones that he named "sports", or sudden changes. He accepted three causes of variations: the environment's direct action upon the organism, the use and non-use of organs, and of spontaneous variation. As a consequence of the first two, adaptive characters are necessarily produced because they are the result of the organisms' accommodation to the environment. These adaptive characteristics may be sustained by natural selection, but only the spontaneous variations, which develop without any direct relationship with the organism's requirements, would be subject to the action of natural selection.

Darwin proposed the difference between the explanation of chance as a result of ignorance and true chance. In the first case, we find the explanation of variation is the result of lack of knowledge concerning its causes. When these are unknown, they are frequently said to be owing to chance. In *The Origin*, he emphasizes: "I have hitherto sometimes spoken as if the variations —so common and multiform in organic beings under domestication, and in a lesser degree in those in a state of nature— had been due to chance. This, of course, is a wholly incorrect expression, but it serves to acknowledge plainly our ignorance of the cause of each particular variation" (*Origin*, p. 131).

Neither Darwin, nor his contemporaries knew what accounted for variation. Chance really presents itself in the relationship between a variation and the environment in which it occurs. The same modification may be positive in one medium, while negative in another, and even neutral in a third.

The Struggle for Existence

This concept is very important because Darwin was severely criticized by those who interpreted the struggle for existence in a literal sense. In Mexico, for example, such a concept provoked the rejection of some positivist who understood that this notion justified what they had just experienced, namely the civil war and the invasions. *That* struggle the emblem of liberty, order, and progress. In the same way, it invited the dissent of socialists who believed that it was cooperation, not competition, that allowed for progress.

The explanation of what Darwin understands as "the struggle for existence", is developed, especially in chapter 3 of *The Origin*. There Darwin argued that in all species there is an overproduction of descendants that, combined with evidently limited resources, incites the struggle for existence. Darwin refers to Malthus's arguments, asserting that the tendency of organisms is to increase in number at a geometric rate, while resources only augment at an arithmetical rate.

The idea of the struggle for existence only as a battle among individuals is positively a misunderstanding, Darwin includes all of the positive and negative relationships between living beings themselves (and living beings with the physical environment) in the concept. It is about a fully ecological concept that includes the impact of all of the possible relationships. For example, the impact of intra- and interspecific competition, predation, parasitism, climatic action, etc., on the efficiency of reproduction.

The following paragraph is key in the understanding of the Darwinian notion of the struggle for existence: "A struggle for existence inevitable follows from the high rate in which all organic beings tend to increase. Every being, which during its natural lifetime produces several eggs or seeds, must suffer destruction during some period of its life, and during some season or occasional year, otherwise, on the principle of geometric increase, its numbers would quickly become so inordinately great, that no country could support the product. Hence, as more individuals are produced than those that can possible survive, there must in every case be a struggle for existence, either one individual with another of the same species, or with individuals of distinct species, or with the physical conditions of life" (*Origin*, p. 63).

In the previous paragraph, Darwin states that he utilized the concept of the struggle for existence in a metaphoric sense. This includes the dependence of one being on another, the fight of a plant against drought and other physical factors, and most importantly, it not only includes the survival of the individual, but its success in leaving progeny (*ibid*., p. 62). Further on, Darwin clarifies that although all relationships are important, especially the biotic ones, it is competition, especially that occuring between species of the same kind, varieties of the same species, and the intraspecific, that provoke the most severe conflict. This is explained by the fact that organisms with the specified levels of taxonomic affinity, normally have the same necessities (*ibid.*, p. 76).

The notion of the struggle for existence is of fundamental importance for understanding Darwin's ecological conceptions. For it is here that he analyzes the form in which the growth or the decline in growth of a population affects the increase or decrease of other species. In *The Origin of Species*, Darwin is already very far from Linnaeus's notion of static balance in nature.

Divergence of Character

Divergence of character favors the increase in biological diversity due to the specialization in the exploitation of a niche in the economy of nature. His argument is that the more differentiated (in respect to other species) the structure, the descendant's constitution and habits in any species will have a higher capacity of adaptation to a diversity of locations in nature's economy. This implies a certain specialization in the exploitation of different niches, and perhaps the avoidance of entering into competition with other species. This favors both, and they'll be able to increase their number more easily. Darwin considered such a process of specialization as analogous to Milne Edward's idea of the physiological division of labor (p. 116). The energy that a system attains is put to better use, for example, in a digestive system that is specialized in the digestion of one type of nutrient, such as meat in the case of carnivores. Ecological specialization permits better use of resources. The difference between the varieties of a single species is broadened, thanks to the fact that natural selection favors this process. In this way, the combined action of the production of variations and natural selection propel an increment in the diversity of species.

Darwin uses a fictitious example to explain why natural selection favors divergence of character: this is the case of a quadruped carnivore, whose number reaches the maximum that can be tolerated in a region. Darwin imagined what would be happen if the total reproductive capacity of this species could be developed in an area that did not suffer any structural change. Under such conditions, he replies, the only possibility for the species' increase was that its modified descendants might settle in places presently occupied by other animals. For example, if some of them would be capable of feeding on new types of game (alive or dead), or inhabiting new locations, climbing trees, treading water, and some even becoming less carnivores. In conclusion: The more diverse the habits and structure of our carnivore's descendent, the more places they will be able to occupy (p. 113).

Divergence of character, by making possible the exploitation of different resources at different times, allows the organisms to evade competition. Recently there have been important studies bearing on this issue, the best-known being those of Lack (1947) and Grant and Grant (1979). They have discovered how different species of birds have become specialized predators of insects that inhabit different altitudes of trees, or that have diurnal or nocturnal habits.

This principle is of great importance for evolution because it stimulates the preservation of characters that have an equivalent adaptive value. Due to this factor, Darwin considers natural selection as a complementary process. So long as two organisms empower their carrier for the differential exploitation of the niche in similar ways, it cannot be said that one is more favorable than the other. Consequently, both remain, promoting an increase of diversity.

Speciation

In chapter IV of *The Origin*, Darwin postulates the nucleus of his theory. He explains the process by which new species are produced. In other words, he develops the theory of the origin of species. The term "speciation", in the present context, refers to the multiplication of species from an original progenitor species.

An indispensable prerequisite for the formation of two or more species from one species, is the separation of two or more populations for reproduction. Darwin was convinced that this was an explicable problem. He proposed two forms in which reproductive separation could be achieved. The first involved geographic isolation. The second entailed isolation caused by ecological preferences, in the same environment.

In the first stage of his conception of evolution, Darwin believed that geographic isolation was indispensable for the creation of species. He considers that the formation of species is very effective under these terms. The reasoning behind this, was that the separation for reproduction is achieved immediately. Here, Darwin proposed two problems to himself. First, populations on islands are usually smaller than on continents. Therefore, the amount of variation is not very large. Second, in small populations the effect of mixing hereditary characters[8] would provoke a considerable reduction in the variability upon which natural selection would act: "if an isolated area be very small, either from being surrounded by barriers, or from having very peculiar physical conditions, the total number of the individuals supported on it will necessarily be very small; and fewness of individual will greatly retard the production of new species through natural selection, by decreasing the chance of the appearance of favourable variations" (*Origin*, p. 105). The way to minimize both problems may be with the formation of species in very large areas, where reproductive separation is accomplished in different ways than those of geographic separation.

Moreover, in a large territory the structural and biological conditions are much more complex. This would allow much more environmental variability: In an extensive and open territory, there would not only be more probability of the development of favorable variations among the great number of same-species individuals that inhabit it, but the conditions of life are also much more complex. This is caused by the large number of already existent species. If one of them is modified with an improvement, the others will also have to change, or they will be exterminated. Each new form... will enter into competition with many others. In this way, new places will be formed, and

[8]Until before the re-discovery of Mendel's laws, biologists sustained that the hereditary material that stems from both progenitors was mixed. Such a mixture provoked a tendency towards homogenization of characters, and consequently, a reduction in variability.

the competition to fill them will be more difficult in a large area than in a small, isolated one (*ibid.*).

According to Darwin, the methods for preventing two populations from crossing today are known as Reproductive Isolation Mechanisms (RIM). Darwin, who labeled them as mechanisms for the acquisition of sterility, believed that they were caused by environmental conditions or behavioral habits that, in the long run, produced barriers of sterility between newly-developed species. In one territory, two variants of the same animal may continue to be distinct for a long period of time. This may occur because they do not frequent the same places, or they procreate during different periods, or the individuals of each variety prefer gathering among themselves" (*Origin*, p. 103).

In 1841, when Darwin already had his theory but had not made it public yet, Moritz Wagner published the observations he made during his trips to Asia, Africa, and America. During these journeys, he discovered that, almost invariably, close species occupied adjacent areas, separated from others by rivers, mountains, valleys, or other obstacles to their dispersion. His observations led him to sustain that geographic isolation was the only way to form two species from one.

Darwin's supposition that the amount of variability as the prime material for natural selection's action led him to look for other ways in which reproductive isolation may be accomplished. He concluded that geographic isolation was not indispensable for the multiplication of species.

Extinction

Innumerable critics pointed out to Darwin that the fossil register was evidence against his theory because it did not corroborate gradualism. Nevertheless this was exactly his primary piece of evidence for evolution especially in view of the mutual affinity of living and extinct species. In Darwin's view, the fact that species can be grouped into a small number of great classes can be quickly explained by the descent principle. Extinct species can all be classified within the still-existent groups, or in the intervals between them. It is true that extinct organic forms help fill the intervals that exist between living types, families, orders.

As we know, Darwin defended strict graduality. Consequently, he rejected the possibility that species may be able to appear in leaps. Against catastrophism, he states that because of our ignorance, we invoke catastrophes to desolate the Earth. In his view both the origin and the extinction of species are fundamentally a result of the action of natural selection. Certainly, the theory of natural selection was founded on the belief that each variety, and finally, every new species, is produced and sustained because it has some advantage over those who enter into competition with it. The consequential extinction of the least favored forms almost inevitably follows. Paleontological studies clearly demonstrate the important role that extinction has played in the history of life. This is in virtue of the appearance of new forms and the disappearance of old ones. In many cases of recent, tertiary formations, we see that the rarity of the species precedes its extinction, and we know that this has been the course of events in those animals that have been locally or totally exterminated by human actions. For Darwin, to admit that species are generally made rare before extinction, yet marvel when the species ceases to exist, is almost the same as admitting that an individual's illness is a precursor of death, while finding the illness surprising.

With the permanent flow of criticism that stem as much from the creationist tradition, as from evolutionists, concerning the absence of intermediary forms in the fossil register, Darwin acknowledges that geological research has not revealed the previous existence of infinitely numerous gradations, which are as delicate as the actual variations, and that link together almost all species, both living and extinct. However, he states that we would not expect to find in our geological formations, an infinite number of those subtle forms of transition, that according to the theory of natural selection have reunited all past and present species of the group into a large and branching chain of life. We should find only some links. Certainly we will find them, some more distant and others closer. If they are found on different levels of the same formation, many paleontologists will consider them as distinct species. He explains that the type of fossil registers found are the result of the action of natural selection. In his opinion, in the places that are completely populated, natural selection favors the varieties or species that have some advantage over other forms in the contest for life. For this reason, there would be a constant tendency in the perfected descendants of whatever species to supplant and exterminate the precursors of every generation, as well as the primitive trunk. Darwin's point is that all of the intermediate forms between the primitive state and the more recent ones tend to die out.

Progress

In Darwinism, there is an explicit rejection of the existence of "tendencies to progress". On this topic, Darwin states that even though the extinction of species and of entire groups of species is an almost inevitable consequence of the principles of natural selection (previous forms are replaced by new and improved ones), one cannot speak of a general line. One must refer to the advancement of solutions to problems offered by the environment in a determined time and space to different groups of organisms (see Ayala, 1974, 1977). Indeed, the "best" in one given time and place may have a lesser level of organization than a previous form; let us think, for example, about parasites that have lost organs, as in the case of the cestods. This has permitted them better adjustments. Darwin states that this is due to the fact that even when the organization in general might have advanced, and is still advancing in all of the world, the scale will always present many levels of perfection. The great progress of entire classes, or of determined members of each class, does not in any way necessarily lead to the extinction of groups with which they do not enter into direct competition. There are two central arguments in the opposition to the idea of progress in evolution. First, natural selection can only act in favor of the survival and reproduction of organisms or species, in a given moment, not in the long run. Although selection acts in favor of adaptive improvement, it cannot foresee what will occur in successive generations. Second, there is a multiplicity of events that play a role as important as natural selection in the process of evolution, and which do not depend on natural selection. These are events that take place without direction and that by chance may or may not favor the survival and reproduction of an organism.

In Darwin's terms, there is no tendency towards progress because each formation does not represent a new and complete act of creation, but only one incidental scene, taken by chance from a constantly changing drama. The arguments against the idea of progress in general, also demonstrate the impossibility of the reappearance of an extinct species. A species, once lost, does not have exactly the same organic and inorganic conditions of life; for even when the descendants of a species are able to adapt and fill the place of another in the economy of nature, replacing it in this manner, the two forms (the old and the new) would not be identically the same. Both would almost surely inherit altered characters from their distinct forefathers, and dissimilar organisms would have to vary in different ways.

National Autonomous University of Mexico
University of California, Irvine

REFERENCES

Ayala, F. J. and T. Dobzhansky, eds. 1974: *Studies in the Philosophy of Biology*. MacMillan, London, and University of California Press, Berkley.

Ayala, F. J. 1977: in Dobzhansky, Ayala, Stebbins, *Evolution*, ch. 16. W. H. Freeman, San Francisco.

Bowler, P. 1990: *Charles Darwin, The Man and his Influence*. Basil Blackwell, Cambridge, Mass.

Darwin, C. [1839]: *Journal of Researches into the Geology and Natural History of the Various Countries visited by H: M. S. Beagle*. Ward, Lock and Bowden, 1894.

Darwin, C. 1859: *On the Origin of Species by means of Natural Selection, or the Preservation of Favored Races in the Struggle for Life.* A facsimile reprint of the first edition, with an introduction by Ernst Mayr. Harvard University Press, Cambridge, 1964.

Darwin, C. 1987: *Notebooks on Transmutation of Species*. P. H. Barret, *et al.*, eds. Cornell University Press, Ithaca.

Darwin, F. and A. Seward, eds. DATE: *More Letters of Charles Darwin*. London, John Murray, London.

Ghiselin, M. T. 1969: *The Triumph of the Darwinian Method*. University of California Press, Berkeley.

Grant, B R. and P. R. Grant 1979: "Darwin's Finches: Population variation and sympatric speciation," *Proceedings of the National Academy of Sciences*, 76:2359-2363

Herbert, S. 1971: "Darwin, Malthus and Selection," *Journal of the History of Biology*, 4: 209-217.

Hodge, M. J. S. 1983: "The Development of Darwin's General Biological Theorizing," in *Evolution from Molecules to Men*, D. S. Bendall, ed., Cambridge University Press, London and New York, pp. 43-62.

Hodge, M. J. S. and D. Kohn, 1985, "The Immediate Origins of Natural Selection", in *The Darwinian Heritage*, D. Kohn, ed., Princeton University Press, Princeton, pp. xxx-xxx.

Kohn, D. 1980: "Theories to Work by: Rejected Theories, Reproduction and Darwin's Path to Natural Selection," *Studies in the History of Biology*, 4: 67-170.

Lack, D. 1947: *Darwin's Finches*. Cambridge University Press, Cambridge.

Laudan, L. 1977: *Progress and its Problems, Towards a Theory of Scientific Growth*, University of California Press, Berkley.

Laudan, L. 1996: *Beyond Positivism and Relativism: Theory, Method, and Evidence*, Westview Press, PLACE.

Limoges, C. 1976. *La selección natural*, Siglo XXI, Mexico City.

Malthus, T. R., [1798]: *An Essay on the Principle of Population*, London.

Ospovat, D. 1979: "Darwin after Malthus," *Journal of the History of Biology*, 12:211-223.

Ospovat, D. 1981: *The Development of Darwin's Theory*, Cambridge University Press, Cambridge.

Ruse, M., 1979: *The Darwinian Revolution*. University of Chicago Press, Chicago.

Schipman, P. 1994: *The Evolution of Racism*. Simon and Schuster, New York.

Schweber, S. S. 1977: "The Origin of The Origin Revisited," *Journal of the History of Biology*, 10:229-316.

Schweber, S. S. 1980: "Darwin and the Political Economists: Divergence of Character," *Journal of the History of Biology*, 13 (1980) 195-289.

Spencer, H. 1851a: *Social Statics: Or the Conditions Essential to Human Happiness Specified, and the First of them Developed*, J. Chapman, London.

Spencer, H. [1851b]: "The Development Hypothesis," reprinted in his *Essays*, 1983, I: 381-387.

INDEX

Boston Studies in the Philosophy of Science

Editor: Robert S. Cohen, *Boston University*

1. M.W. Wartofsky (ed.): *Proceedings of the Boston Colloquium for the Philosophy of Science, 1961/1962.* [Synthese Library 6] 1963 ISBN 90-277-0021-4
2. R.S. Cohen and M.W. Wartofsky (eds.): *Proceedings of the Boston Colloquium for the Philosophy of Science, 1962/1964.* In Honor of P. Frank. [Synthese Library 10] 1965 ISBN 90-277-9004-0
3. R.S. Cohen and M.W. Wartofsky (eds.): *Proceedings of the Boston Colloquium for the Philosophy of Science, 1964/1966.* In Memory of Norwood Russell Hanson. [Synthese Library 14] 1967 ISBN 90-277-0013-3
4. R.S. Cohen and M.W. Wartofsky (eds.): *Proceedings of the Boston Colloquium for the Philosophy of Science, 1966/1968.* [Synthese Library 18] 1969 ISBN 90-277-0014-1
5. R.S. Cohen and M.W. Wartofsky (eds.): *Proceedings of the Boston Colloquium for the Philosophy of Science, 1966/1968.* [Synthese Library 19] 1969 ISBN 90-277-0015-X
6. R.S. Cohen and R.J. Seeger (eds.): *Ernst Mach, Physicist and Philosopher.* [Synthese Library 27] 1970 ISBN 90-277-0016-8
7. M. Čapek: *Bergson and Modern Physics.* A Reinterpretation and Re-evaluation. [Synthese Library 37] 1971 ISBN 90-277-0186-5
8. R.C. Buck and R.S. Cohen (eds.): *PSA 1970.* Proceedings of the 2nd Biennial Meeting of the Philosophy and Science Association (Boston, Fall 1970). In Memory of Rudolf Carnap. [Synthese Library 39] 1971 ISBN 90-277-0187-3; Pb 90-277-0309-4
9. A.A. Zinov'ev: *Foundations of the Logical Theory of Scientific Knowledge (Complex Logic).* Translated from Russian. Revised and enlarged English Edition, with an Appendix by G.A. Smirnov, E.A. Sidorenko, A.M. Fedina and L.A. Bobrova. [Synthese Library 46] 1973 ISBN 90-277-0193-8; Pb 90-277-0324-8
10. L. Tondl: *Scientific Procedures.* A Contribution Concerning the Methodological Problems of Scientific Concepts and Scientific Explanation.Translated from Czech. [Synthese Library 47] 1973 ISBN 90-277-0147-4; Pb 90-277-0323-X
11. R.J. Seeger and R.S. Cohen (eds.): *Philosophical Foundations of Science.* Proceedings of Section L, 1969, American Association for the Advancement of Science. [Synthese Library 58] 1974 ISBN 90-277-0390-6; Pb 90-277-0376-0
12. A. Grünbaum: *Philosophical Problems of Space and Times.* 2nd enlarged ed. [Synthese Library 55] 1973 ISBN 90-277-0357-4; Pb 90-277-0358-2
13. R.S. Cohen and M.W. Wartofsky (eds.): *Logical and Epistemological Studies in Contemporary Physics.* Proceedings of the Boston Colloquium for the Philosophy of Science, 1969/72, Part I. [Synthese Library 59] 1974 ISBN 90-277-0391-4; Pb 90-277-0377-9
14. R.S. Cohen and M.W. Wartofsky (eds.): *Methodological and Historical Essays in the Natural and Social Sciences.* Proceedings of the Boston Colloquium for the Philosophy of Science, 1969/72, Part II. [Synthese Library 60] 1974 ISBN 90-277-0392-2; Pb 90-277-0378-7
15. R.S. Cohen, J.J. Stachel and M.W. Wartofsky (eds.): *For Dirk Struik.* Scientific, Historical and Political Essays in Honor of Dirk J. Struik. [Synthese Library 61] 1974 ISBN 90-277-0393-0; Pb 90-277-0379-5
16. N. Geschwind: *Selected Papers on Language and the Brains.* [Synthese Library 68] 1974 ISBN 90-277-0262-4; Pb 90-277-0263-2
17. B.G. Kuznetsov: *Reason and Being.* Translated from Russian. Edited by C.R. Fawcett and R.S. Cohen. 1987 ISBN 90-277-2181-5

Boston Studies in the Philosophy of Science

18. P. Mittelstaedt: *Philosophical Problems of Modern Physics.* Translated from the revised 4th German edition by W. Riemer and edited by R.S. Cohen. [Synthese Library 95] 1976
ISBN 90-277-0285-3; Pb 90-277-0506-2
19. H. Mehlberg: *Time, Causality, and the Quantum Theory.* Studies in the Philosophy of Science. Vol. I: *Essay on the Causal Theory of Time.* Vol. II: *Time in a Quantized Universe.* Translated from French. Edited by R.S. Cohen. 1980 Vol. I: ISBN 90-277-0721-9; Pb 90-277-1074-0
Vol. II: ISBN 90-277-1075-9; Pb 90-277-1076-7
20. K.F. Schaffner and R.S. Cohen (eds.): *PSA 1972.* Proceedings of the 3rd Biennial Meeting of the Philosophy of Science Association (Lansing, Michigan, Fall 1972). [Synthese Library 64] 1974 ISBN 90-277-0408-2; Pb 90-277-0409-0
21. R.S. Cohen and J.J. Stachel (eds.): *Selected Papers of Léon Rosenfeld.* [Synthese Library 100] 1979 ISBN 90-277-0651-4; Pb 90-277-0652-2
22. M. Čapek (ed.): *The Concepts of Space and Time.* Their Structure and Their Development. [Synthese Library 74] 1976 ISBN 90-277-0355-8; Pb 90-277-0375-2
23. M. Grene: *The Understanding of Nature.* Essays in the Philosophy of Biology. [Synthese Library 66] 1974 ISBN 90-277-0462-7; Pb 90-277-0463-5
24. D. Ihde: *Technics and Praxis.* A Philosophy of Technology. [Synthese Library 130] 1979
ISBN 90-277-0953-X; Pb 90-277-0954-8
25. J. Hintikka and U. Remes: *The Method of Analysis.* Its Geometrical Origin and Its General Significance. [Synthese Library 75] 1974 ISBN 90-277-0532-1; Pb 90-277-0543-7
26. J.E. Murdoch and E.D. Sylla (eds.): *The Cultural Context of Medieval Learning.* Proceedings of the First International Colloquium on Philosophy, Science, and Theology in the Middle Ages, 1973. [Synthese Library 76] 1975 ISBN 90-277-0560-7; Pb 90-277-0587-9
27. M. Grene and E. Mendelsohn (eds.): *Topics in the Philosophy of Biology.* [Synthese Library 84] 1976 ISBN 90-277-0595-X; Pb 90-277-0596-8
28. J. Agassi: *Science in Flux.* [Synthese Library 80] 1975
ISBN 90-277-0584-4; Pb 90-277-0612-3
29. J.J. Wiatr (ed.): *Polish Essays in the Methodology of the Social Sciences.* [Synthese Library 131] 1979 ISBN 90-277-0723-5; Pb 90-277-0956-4
30. P. Janich: *Protophysics of Time.* Constructive Foundation and History of Time Measurement. Translated from German. 1985 ISBN 90-277-0724-3
31. R.S. Cohen and M.W. Wartofsky (eds.): *Language, Logic, and Method.* 1983
ISBN 90-277-0725-1
32. R.S. Cohen, C.A. Hooker, A.C. Michalos and J.W. van Evra (eds.): *PSA 1974.* Proceedings of the 4th Biennial Meeting of the Philosophy of Science Association. [Synthese Library 101] 1976 ISBN 90-277-0647-6; Pb 90-277-0648-4
33. G. Holton and W.A. Blanpied (eds.): *Science and Its Public.* The Changing Relationship. [Synthese Library 96] 1976 ISBN 90-277-0657-3; Pb 90-277-0658-1
34. M.D. Grmek, R.S. Cohen and G. Cimino (eds.): *On Scientific Discovery.* The 1977 Erice Lectures. 1981 ISBN 90-277-1122-4; Pb 90-277-1123-2
35. S. Amsterdamski: *Between Experience and Metaphysics.* Philosophical Problems of the Evolution of Science. Translated from Polish. [Synthese Library 77] 1975
ISBN 90-277-0568-2; Pb 90-277-0580-1
36. M. Marković and G. Petrović (eds.): *Praxis.* Yugoslav Essays in the Philosophy and Methodology of the Social Sciences. [Synthese Library 134] 1979
ISBN 90-277-0727-8; Pb 90-277-0968-8

Boston Studies in the Philosophy of Science

37. H. von Helmholtz: *Epistemological Writings.* The Paul Hertz / Moritz Schlick Centenary Edition of 1921. Translated from German by M.F. Lowe. Edited with an Introduction and Bibliography by R.S. Cohen and Y. Elkana. [Synthese Library 79] 1977
ISBN 90-277-0290-X; Pb 90-277-0582-8
38. R.M. Martin: *Pragmatics, Truth and Language.* 1979
ISBN 90-277-0992-0; Pb 90-277-0993-9
39. R.S. Cohen, P.K. Feyerabend and M.W. Wartofsky (eds.): *Essays in Memory of Imre Lakatos.* [Synthese Library 99] 1976 ISBN 90-277-0654-9; Pb 90-277-0655-7
40. Not published.
41. Not published.
42. H.R. Maturana and F.J. Varela: *Autopoiesis and Cognition.* The Realization of the Living. With a Preface to "Autopoiesis' by S. Beer. 1980 ISBN 90-277-1015-5; Pb 90-277-1016-3
43. A. Kasher (ed.): *Language in Focus: Foundations, Methods and Systems.* Essays in Memory of Yehoshua Bar-Hillel. [Synthese Library 89] 1976
ISBN 90-277-0644-1; Pb 90-277-0645-X
44. T.D. Thao: *Investigations into the Origin of Language and Consciousness.* 1984
ISBN 90-277-0827-4
45. F.G.-I. Nagasaka (ed.): *Japanese Studies in the Philosophy of Science.* 1997
ISBN 0-7923-4781-1
46. P.L. Kapitza: *Experiment, Theory, Practice.* Articles and Addresses. Edited by R.S. Cohen. 1980 ISBN 90-277-1061-9; Pb 90-277-1062-7
47. M.L. Dalla Chiara (ed.): *Italian Studies in the Philosophy of Science.* 1981
ISBN 90-277-0735-9; Pb 90-277-1073-2
48. M.W. Wartofsky: *Models.* Representation and the Scientific Understanding. [Synthese Library 129] 1979 ISBN 90-277-0736-7; Pb 90-277-0947-5
49. T.D. Thao: *Phenomenology and Dialectical Materialism.* Edited by R.S. Cohen. 1986
ISBN 90-277-0737-5
50. Y. Fried and J. Agassi: *Paranoia.* A Study in Diagnosis. [Synthese Library 102] 1976
ISBN 90-277-0704-9; Pb 90-277-0705-7
51. K.H. Wolff: *Surrender and Cath.* Experience and Inquiry Today. [Synthese Library 105] 1976
ISBN 90-277-0758-8; Pb 90-277-0765-0
52. K. Kosík: *Dialectics of the Concrete.* A Study on Problems of Man and World. 1976
ISBN 90-277-0761-8; Pb 90-277-0764-2
53. N. Goodman: *The Structure of Appearance.* [Synthese Library 107] 1977
ISBN 90-277-0773-1; Pb 90-277-0774-X
54. H.A. Simon: *Models of Discovery* and Other Topics in the Methods of Science. [Synthese Library 114] 1977 ISBN 90-277-0812-6; Pb 90-277-0858-4
55. M. Lazerowitz: *The Language of Philosophy.* Freud and Wittgenstein. [Synthese Library 117] 1977 ISBN 90-277-0826-6; Pb 90-277-0862-2
56. T. Nickles (ed.): *Scientific Discovery, Logic, and Rationality.* 1980
ISBN 90-277-1069-4; Pb 90-277-1070-8
57. J. Margolis: *Persons and Mind.* The Prospects of Nonreductive Materialism. [Synthese Library 121] 1978 ISBN 90-277-0854-1; Pb 90-277-0863-0
58. G. Radnitzky and G. Andersson (eds.): *Progress and Rationality in Science.* [Synthese Library 125] 1978 ISBN 90-277-0921-1; Pb 90-277-0922-X
59. G. Radnitzky and G. Andersson (eds.): *The Structure and Development of Science.* [Synthese Library 136] 1979 ISBN 90-277-0994-7; Pb 90-277-0995-5

Boston Studies in the Philosophy of Science

60. T. Nickles (ed.): *Scientific Discovery.* Case Studies. 1980
ISBN 90-277-1092-9; Pb 90-277-1093-7
61. M.A. Finocchiaro: *Galileo and the Art of Reasoning.* Rhetorical Foundation of Logic and Scientific Method. 1980 ISBN 90-277-1094-5; Pb 90-277-1095-3
62. W.A. Wallace: *Prelude to Galileo.* Essays on Medieval and 16th-Century Sources of Galileo's Thought. 1981 ISBN 90-277-1215-8; Pb 90-277-1216-6
63. F. Rapp: *Analytical Philosophy of Technology.* Translated from German. 1981
ISBN 90-277-1221-2; Pb 90-277-1222-0
64. R.S. Cohen and M.W. Wartofsky (eds.): *Hegel and the Sciences.* 1984 ISBN 90-277-0726-X
65. J. Agassi: *Science and Society.* Studies in the Sociology of Science. 1981
ISBN 90-277-1244-1; Pb 90-277-1245-X
66. L. Tondl: *Problems of Semantics.* A Contribution to the Analysis of the Language of Science. Translated from Czech. 1981 ISBN 90-277-0148-2; Pb 90-277-0316-7
67. J. Agassi and R.S. Cohen (eds.): *Scientific Philosophy Today.* Essays in Honor of Mario Bunge. 1982 ISBN 90-277-1262-X; Pb 90-277-1263-8
68. W. Krajewski (ed.): *Polish Essays in the Philosophy of the Natural Sciences.* Translated from Polish and edited by R.S. Cohen and C.R. Fawcett. 1982
ISBN 90-277-1286-7; Pb 90-277-1287-5
69. J.H. Fetzer: *Scientific Knowledge.* Causation, Explanation and Corroboration. 1981
ISBN 90-277-1335-9; Pb 90-277-1336-7
70. S. Grossberg: *Studies of Mind and Brain.* Neural Principles of Learning, Perception, Development, Cognition, and Motor Control. 1982 ISBN 90-277-1359-6; Pb 90-277-1360-X
71. R.S. Cohen and M.W. Wartofsky (eds.): *Epistemology, Methodology, and the Social Sciences.* 1983. ISBN 90-277-1454-1
72. K. Berka: *Measurement.* Its Concepts, Theories and Problems. Translated from Czech. 1983
ISBN 90-277-1416-9
73. G.L. Pandit: *The Structure and Growth of Scientific Knowledge.* A Study in the Methodology of Epistemic Appraisal. 1983 ISBN 90-277-1434-7
74. A.A. Zinov'ev: *Logical Physics.* Translated from Russian. Edited by R.S. Cohen. 1983 [*see also* Volume 9] ISBN 90-277-0734-0
75. G-G. Granger: *Formal Thought and the Sciences of Man.* Translated from French. With and Introduction by A. Rosenberg. 1983 ISBN 90-277-1524-6
76. R.S. Cohen and L. Laudan (eds.): *Physics, Philosophy and Psychoanalysis.* Essays in Honor of Adolf Grünbaum. 1983 ISBN 90-277-1533-5
77. G. Böhme, W. van den Daele, R. Hohlfeld, W. Krohn and W. Schäfer: *Finalization in Science.* The Social Orientation of Scientific Progress. Translated from German. Edited by W. Schäfer. 1983 ISBN 90-277-1549-1
78. D. Shapere: *Reason and the Search for Knowledge.* Investigations in the Philosophy of Science. 1984 ISBN 90-277-1551-3; Pb 90-277-1641-2
79. G. Andersson (ed.): *Rationality in Science and Politics.* Translated from German. 1984
ISBN 90-277-1575-0; Pb 90-277-1953-5
80. P.T. Durbin and F. Rapp (eds.): *Philosophy and Technology.* [*Also* Philosophy and Technology Series, Vol. 1] 1983 ISBN 90-277-1576-9
81. M. Marković: *Dialectical Theory of Meaning.* Translated from Serbo-Croat. 1984
ISBN 90-277-1596-3
82. R.S. Cohen and M.W. Wartofsky (eds.): *Physical Sciences and History of Physics.* 1984.
ISBN 90-277-1615-3

Boston Studies in the Philosophy of Science

83. É. Meyerson: *The Relativistic Deduction.* Epistemological Implications of the Theory of Relativity. Translated from French. With a Review by Albert Einstein and an Introduction by Milič Čapek. 1985 ISBN 90-277-1699-4
84. R.S. Cohen and M.W. Wartofsky (eds.): *Methodology, Metaphysics and the History of Science.* In Memory of Benjamin Nelson. 1984 ISBN 90-277-1711-7
85. G. Tamás: *The Logic of Categories.* Translated from Hungarian. Edited by R.S. Cohen. 1986 ISBN 90-277-1742-7
86. S.L. de C. Fernandes: *Foundations of Objective Knowledge.* The Relations of Popper's Theory of Knowledge to That of Kant. 1985 ISBN 90-277-1809-1
87. R.S. Cohen and T. Schnelle (eds.): *Cognition and Fact.* Materials on Ludwik Fleck. 1986 ISBN 90-277-1902-0
88. G. Freudenthal: *Atom and Individual in the Age of Newton.* On the Genesis of the Mechanistic World View. Translated from German. 1986 ISBN 90-277-1905-5
89. A. Donagan, A.N. Perovich Jr and M.V. Wedin (eds.): *Human Nature and Natural Knowledge.* Essays presented to Marjorie Grene on the Occasion of Her 75th Birthday. 1986 ISBN 90-277-1974-8
90. C. Mitcham and A. Hunning (eds.): *Philosophy and Technology II.* Information Technology and Computers in Theory and Practice. [*Also* Philosophy and Technology Series, Vol. 2] 1986 ISBN 90-277-1975-6
91. M. Grene and D. Nails (eds.): *Spinoza and the Sciences.* 1986 ISBN 90-277-1976-4
92. S.P. Turner: *The Search for a Methodology of Social Science.* Durkheim, Weber, and the 19th-Century Problem of Cause, Probability, and Action. 1986. ISBN 90-277-2067-3
93. I.C. Jarvie: *Thinking about Society.* Theory and Practice. 1986 ISBN 90-277-2068-1
94. E. Ullmann-Margalit (ed.): *The Kaleidoscope of Science.* The Israel Colloquium: Studies in History, Philosophy, and Sociology of Science, Vol. 1. 1986 ISBN 90-277-2158-0; Pb 90-277-2159-9
95. E. Ullmann-Margalit (ed.): *The Prism of Science.* The Israel Colloquium: Studies in History, Philosophy, and Sociology of Science, Vol. 2. 1986 ISBN 90-277-2160-2; Pb 90-277-2161-0
96. G. Márkus: *Language and Production.* A Critique of the Paradigms. Translated from French. 1986 ISBN 90-277-2169-6
97. F. Amrine, F.J. Zucker and H. Wheeler (eds.): *Goethe and the Sciences: A Reappraisal.* 1987 ISBN 90-277-2265-X; Pb 90-277-2400-8
98. J.C. Pitt and M. Pera (eds.): *Rational Changes in Science.* Essays on Scientific Reasoning. Translated from Italian. 1987 ISBN 90-277-2417-2
99. O. Costa de Beauregard: *Time, the Physical Magnitude.* 1987 ISBN 90-277-2444-X
100. A. Shimony and D. Nails (eds.): *Naturalistic Epistemology.* A Symposium of Two Decades. 1987 ISBN 90-277-2337-0
101. N. Rotenstreich: *Time and Meaning in History.* 1987 ISBN 90-277-2467-9
102. D.B. Zilberman: *The Birth of Meaning in Hindu Thought.* Edited by R.S. Cohen. 1988 ISBN 90-277-2497-0
103. T.F. Glick (ed.): *The Comparative Reception of Relativity.* 1987 ISBN 90-277-2498-9
104. Z. Harris, M. Gottfried, T. Ryckman, P. Mattick Jr, A. Daladier, T.N. Harris and S. Harris: *The Form of Information in Science.* Analysis of an Immunology Sublanguage. With a Preface by Hilary Putnam. 1989 ISBN 90-277-2516-0
105. F. Burwick (ed.): *Approaches to Organic Form.* Permutations in Science and Culture. 1987 ISBN 90-277-2541-1

Boston Studies in the Philosophy of Science

106. M. Almási: *The Philosophy of Appearances.* Translated from Hungarian. 1989
ISBN 90-277-2150-5
107. S. Hook, W.L. O'Neill and R. O'Toole (eds.): *Philosophy, History and Social Action.* Essays in Honor of Lewis Feuer. With an Autobiographical Essay by L. Feuer. 1988
ISBN 90-277-2644-2
108. I. Hronszky, M. Fehér and B. Dajka: *Scientific Knowledge Socialized.* Selected Proceedings of the 5th Joint International Conference on the History and Philosophy of Science organized by the IUHPS (Veszprém, Hungary, 1984). 1988 ISBN 90-277-2284-6
109. P. Tillers and E.D. Green (eds.): *Probability and Inference in the Law of Evidence.* The Uses and Limits of Bayesianism. 1988 ISBN 90-277-2689-2
110. E. Ullmann-Margalit (ed.): *Science in Reflection.* The Israel Colloquium: Studies in History, Philosophy, and Sociology of Science, Vol. 3. 1988
ISBN 90-277-2712-0; Pb 90-277-2713-9
111. K. Gavroglu, Y. Goudaroulis and P. Nicolacopoulos (eds.): *Imre Lakatos and Theories of Scientific Change.* 1989 ISBN 90-277-2766-X
112. B. Glassner and J.D. Moreno (eds.): *The Qualitative-Quantitative Distinction in the Social Sciences.* 1989 ISBN 90-277-2829-1
113. K. Arens: *Structures of Knowing.* Psychologies of the 19th Century. 1989
ISBN 0-7923-0009-2
114. A. Janik: *Style, Politics and the Future of Philosophy.* 1989 ISBN 0-7923-0056-4
115. F. Amrine (ed.): *Literature and Science as Modes of Expression.* With an Introduction by S. Weininger. 1989 ISBN 0-7923-0133-1
116. J.R. Brown and J. Mittelstrass (eds.): *An Intimate Relation.* Studies in the History and Philosophy of Science. Presented to Robert E. Butts on His 60th Birthday. 1989
ISBN 0-7923-0169-2
117. F. D'Agostino and I.C. Jarvie (eds.): *Freedom and Rationality.* Essays in Honor of John Watkins. 1989 ISBN 0-7923-0264-8
118. D. Zolo: *Reflexive Epistemology.* The Philosophical Legacy of Otto Neurath. 1989
ISBN 0-7923-0320-2
119. M. Kearn, B.S. Philips and R.S. Cohen (eds.): *Georg Simmel and Contemporary Sociology.* 1989 ISBN 0-7923-0407-1
120. T.H. Levere and W.R. Shea (eds.): *Nature, Experiment and the Science.* Essays on Galileo and the Nature of Science. In Honour of Stillman Drake. 1989 ISBN 0-7923-0420-9
121. P. Nicolacopoulos (ed.): *Greek Studies in the Philosophy and History of Science.* 1990
ISBN 0-7923-0717-8
122. R. Cooke and D. Costantini (eds.): *Statistics in Science.* The Foundations of Statistical Methods in Biology, Physics and Economics. 1990 ISBN 0-7923-0797-6
123. P. Duhem: *The Origins of Statics.* Translated from French by G.F. Leneaux, V.N. Vagliente and G.H. Wagner. With an Introduction by S.L. Jaki. 1991 ISBN 0-7923-0898-0
124. H. Kamerlingh Onnes: *Through Measurement to Knowledge.* The Selected Papers, 1853-1926. Edited and with an Introduction by K. Gavroglu and Y. Goudaroulis. 1991
ISBN 0-7923-0825-5
125. M. Čapek: *The New Aspects of Time: Its Continuity and Novelties.* Selected Papers in the Philosophy of Science. 1991 ISBN 0-7923-0911-1
126. S. Unguru (ed.): *Physics, Cosmology and Astronomy, 1300–1700.* Tension and Accommodation. 1991 ISBN 0-7923-1022-5

Boston Studies in the Philosophy of Science

127. Z. Bechler: *Newton's Physics on the Conceptual Structure of the Scientific Revolution.* 1991 ISBN 0-7923-1054-3
128. É. Meyerson: *Explanation in the Sciences.* Translated from French by M-A. Siple and D.A. Siple. 1991 ISBN 0-7923-1129-9
129. A.I. Tauber (ed.): *Organism and the Origins of Self.* 1991 ISBN 0-7923-1185-X
130. F.J. Varela and J-P. Dupuy (eds.): *Understanding Origins.* Contemporary Views on the Origin of Life, Mind and Society. 1992 ISBN 0-7923-1251-1
131. G.L. Pandit: *Methodological Variance.* Essays in Epistemological Ontology and the Methodology of Science. 1991 ISBN 0-7923-1263-5
132. G. Munévar (ed.): *Beyond Reason.* Essays on the Philosophy of Paul Feyerabend. 1991 ISBN 0-7923-1272-4
133. T.E. Uebel (ed.): *Rediscovering the Forgotten Vienna Circle.* Austrian Studies on Otto Neurath and the Vienna Circle. Partly translated from German. 1991 ISBN 0-7923-1276-7
134. W.R. Woodward and R.S. Cohen (eds.): *World Views and Scientific Discipline Formation.* Science Studies in the [former] German Democratic Republic. Partly translated from German by W.R. Woodward. 1991 ISBN 0-7923-1286-4
135. P. Zambelli: *The Speculum Astronomiae and Its Enigma.* Astrology, Theology and Science in Albertus Magnus and His Contemporaries. 1992 ISBN 0-7923-1380-1
136. P. Petitjean, C. Jami and A.M. Moulin (eds.): *Science and Empires.* Historical Studies about Scientific Development and European Expansion. ISBN 0-7923-1518-9
137. W.A. Wallace: *Galileo's Logic of Discovery and Proof.* The Background, Content, and Use of His Appropriated Treatises on Aristotle's *Posterior Analytics.* 1992 ISBN 0-7923-1577-4
138. W.A. Wallace: *Galileo's Logical Treatises.* A Translation, with Notes and Commentary, of His Appropriated Latin Questions on Aristotle's *Posterior Analytics.* 1992 ISBN 0-7923-1578-2
     Set (137 + 138) ISBN 0-7923-1579-0
139. M.J. Nye, J.L. Richards and R.H. Stuewer (eds.): *The Invention of Physical Science.* Intersections of Mathematics, Theology and Natural Philosophy since the Seventeenth Century. Essays in Honor of Erwin N. Hiebert. 1992 ISBN 0-7923-1753-X
140. G. Corsi, M.L. dalla Chiara and G.C. Ghirardi (eds.): *Bridging the Gap: Philosophy, Mathematics and Physics.* Lectures on the Foundations of Science. 1992 ISBN 0-7923-1761-0
141. C.-H. Lin and D. Fu (eds.): *Philosophy and Conceptual History of Science in Taiwan.* 1992 ISBN 0-7923-1766-1
142. S. Sarkar (ed.): *The Founders of Evolutionary Genetics.* A Centenary Reappraisal. 1992 ISBN 0-7923-1777-7
143. J. Blackmore (ed.): *Ernst Mach – A Deeper Look.* Documents and New Perspectives. 1992 ISBN 0-7923-1853-6
144. P. Kroes and M. Bakker (eds.): *Technological Development and Science in the Industrial Age.* New Perspectives on the Science–Technology Relationship. 1992 ISBN 0-7923-1898-6
145. S. Amsterdamski: *Between History and Method.* Disputes about the Rationality of Science. 1992 ISBN 0-7923-1941-9
146. E. Ullmann-Margalit (ed.): *The Scientific Enterprise.* The Bar-Hillel Colloquium: Studies in History, Philosophy, and Sociology of Science, Volume 4. 1992 ISBN 0-7923-1992-3
147. L. Embree (ed.): *Metaarchaeology.* Reflections by Archaeologists and Philosophers. 1992 ISBN 0-7923-2023-9
148. S. French and H. Kamminga (eds.): *Correspondence, Invariance and Heuristics.* Essays in Honour of Heinz Post. 1993 ISBN 0-7923-2085-9
149. M. Bunzl: *The Context of Explanation.* 1993 ISBN 0-7923-2153-7

Boston Studies in the Philosophy of Science

150. I.B. Cohen (ed.): *The Natural Sciences and the Social Sciences.* Some Critical and Historical Perspectives. 1994 ISBN 0-7923-2223-1
151. K. Gavroglu, Y. Christianidis and E. Nicolaidis (eds.): *Trends in the Historiography of Science.* 1994 ISBN 0-7923-2255-X
152. S. Poggi and M. Bossi (eds.): *Romanticism in Science.* Science in Europe, 1790–1840. 1994 ISBN 0-7923-2336-X
153. J. Faye and H.J. Folse (eds.): *Niels Bohr and Contemporary Philosophy.* 1994 ISBN 0-7923-2378-5
154. C.C. Gould and R.S. Cohen (eds.): *Artifacts, Representations, and Social Practice.* Essays for Marx W. Wartofsky. 1994 ISBN 0-7923-2481-1
155. R.E. Butts: *Historical Pragmatics.* Philosophical Essays. 1993 ISBN 0-7923-2498-6
156. R. Rashed: *The Development of Arabic Mathematics: Between Arithmetic and Algebra.* Translated from French by A.F.W. Armstrong. 1994 ISBN 0-7923-2565-6
157. I. Szumilewicz-Lachman (ed.): *Zygmunt Zawirski: His Life and Work.* With Selected Writings on Time, Logic and the Methodology of Science. Translations by Feliks Lachman. Ed. by R.S. Cohen, with the assistance of B. Bergo. 1994 ISBN 0-7923-2566-4
158. S.N. Haq: *Names, Natures and Things.* The Alchemist Jābir ibn Hayyān and His *Kitāb al-Ahjār* (Book of Stones). 1994 ISBN 0-7923-2587-7
159. P. Plaass: *Kant's Theory of Natural Science.* Translation, Analytic Introduction and Commentary by Alfred E. and Maria G. Miller. 1994 ISBN 0-7923-2750-0
160. J. Misiek (ed.): *The Problem of Rationality in Science and its Philosophy.* On Popper vs. Polanyi. The Polish Conferences 1988–89. 1995 ISBN 0-7923-2925-2
161. I.C. Jarvie and N. Laor (eds.): *Critical Rationalism, Metaphysics and Science.* Essays for Joseph Agassi, Volume I. 1995 ISBN 0-7923-2960-0
162. I.C. Jarvie and N. Laor (eds.): *Critical Rationalism, the Social Sciences and the Humanities.* Essays for Joseph Agassi, Volume II. 1995 ISBN 0-7923-2961-9
Set (161–162) ISBN 0-7923-2962-7
163. K. Gavroglu, J. Stachel and M.W. Wartofsky (eds.): *Physics, Philosophy, and the Scientific Community.* Essays in the Philosophy and History of the Natural Sciences and Mathematics. In Honor of Robert S. Cohen. 1995 ISBN 0-7923-2988-0
164. K. Gavroglu, J. Stachel and M.W. Wartofsky (eds.): *Science, Politics and Social Practice.* Essays on Marxism and Science, Philosophy of Culture and the Social Sciences. In Honor of Robert S. Cohen. 1995 ISBN 0-7923-2989-9
165. K. Gavroglu, J. Stachel and M.W. Wartofsky (eds.): *Science, Mind and Art.* Essays on Science and the Humanistic Understanding in Art, Epistemology, Religion and Ethics. Essays in Honor of Robert S. Cohen. 1995 ISBN 0-7923-2990-2
Set (163–165) ISBN 0-7923-2991-0
166. K.H. Wolff: *Transformation in the Writing.* A Case of Surrender-and-Catch. 1995 ISBN 0-7923-3178-8
167. A.J. Kox and D.M. Siegel (eds.): *No Truth Except in the Details.* Essays in Honor of Martin J. Klein. 1995 ISBN 0-7923-3195-8
168. J. Blackmore: *Ludwig Boltzmann, His Later Life and Philosophy, 1900–1906.* Book One: A Documentary History. 1995 ISBN 0-7923-3231-8
169. R.S. Cohen, R. Hilpinen and R. Qiu (eds.): *Realism and Anti-Realism in the Philosophy of Science.* Beijing International Conference, 1992. 1996 ISBN 0-7923-3233-4
170. I. Kuçuradi and R.S. Cohen (eds.): *The Concept of Knowledge.* The Ankara Seminar. 1995 ISBN 0-7923-3241-5

Boston Studies in the Philosophy of Science

171. M.A. Grodin (ed.): *Meta Medical Ethics*: The Philosophical Foundations of Bioethics. 1995 ISBN 0-7923-3344-6
172. S. Ramirez and R.S. Cohen (eds.): *Mexican Studies in the History and Philosophy of Science.* 1995 ISBN 0-7923-3462-0
173. C. Dilworth: *The Metaphysics of Science*. An Account of Modern Science in Terms of Principles, Laws and Theories. 1995 ISBN 0-7923-3693-3
174. J. Blackmore: *Ludwig Boltzmann, His Later Life and Philosophy, 1900–1906* Book Two: The Philosopher. 1995 ISBN 0-7923-3464-7
175. P. Damerow: *Abstraction and Representation.* Essays on the Cultural Evolution of Thinking. 1996 ISBN 0-7923-3816-2
176. M.S. Macrakis: *Scarcity's Ways: The Origins of Capital.* A Critical Essay on Thermodynamics, Statistical Mechanics and Economics. 1997 ISBN 0-7923-4760-9
177. M. Marion and R.S. Cohen (eds.): *Québec Studies in the Philosophy of Science.* Part I: Logic, Mathematics, Physics and History of Science. Essays in Honor of Hugues Leblanc. 1995 ISBN 0-7923-3559-7
178. M. Marion and R.S. Cohen (eds.): *Québec Studies in the Philosophy of Science.* Part II: Biology, Psychology, Cognitive Science and Economics. Essays in Honor of Hugues Leblanc. 1996 ISBN 0-7923-3560-0
Set (177–178) ISBN 0-7923-3561-9
179. Fan Dainian and R.S. Cohen (eds.): *Chinese Studies in the History and Philosophy of Science and Technology.* 1996 ISBN 0-7923-3463-9
180. P. Forman and J.M. Sánchez-Ron (eds.): *National Military Establishments and the Advancement of Science and Technology.* Studies in 20th Century History. 1996 ISBN 0-7923-3541-4
181. E.J. Post: *Quantum Reprogramming*. Ensembles and Single Systems: A Two-Tier Approach to Quantum Mechanics. 1995 ISBN 0-7923-3565-1
182. A.I. Tauber (ed.): *The Elusive Synthesis: Aesthetics and Science.* 1996 ISBN 0-7923-3904-5
183. S. Sarkar (ed.): *The Philosophy and History of Molecular Biology: New Perspectives.* 1996 ISBN 0-7923-3947-9
184. J.T. Cushing, A. Fine and S. Goldstein (eds.): *Bohmian Mechanics and Quantum Theory: An Appraisal.* 1996 ISBN 0-7923-4028-0
185. K. Michalski: *Logic and Time.* An Essay on Husserl's Theory of Meaning. 1996 ISBN 0-7923-4082-5
186. G. Munévar (ed.): *Spanish Studies in the Philosophy of Science.* 1996 ISBN 0-7923-4147-3
187. G. Schubring (ed.): *Hermann Günther Graßmann (1809–1877): Visionary Mathematician, Scientist and Neohumanist Scholar.* Papers from a Sesquicentennial Conference. 1996 ISBN 0-7923-4261-5
188. M. Bitbol: *Schrödinger's Philosophy of Quantum Mechanics.* 1996 ISBN 0-7923-4266-6
189. J. Faye, U. Scheffler and M. Urchs (eds.): *Perspectives on Time.* 1997 ISBN 0-7923-4330-1
190. K. Lehrer and J.C. Marek (eds.): *Austrian Philosophy Past and Present.* Essays in Honor of Rudolf Haller. 1996 ISBN 0-7923-4347-6
191. J.L. Lagrange: *Analytical Mechanics.* Translated and edited by Auguste Boissonade and Victor N. Vagliente. Translated from the *Mécanique Analytique, novelle édition* of 1811. 1997 ISBN 0-7923-4349-2
192. D. Ginev and R.S. Cohen (eds.): *Issues and Images in the Philosophy of Science.* Scientific and Philosophical Essays in Honour of Azarya Polikarov. 1997 ISBN 0-7923-4444-8

Boston Studies in the Philosophy of Science

193. R.S. Cohen, M. Horne and J. Stachel (eds.): *Experimental Metaphysics*. Quantum Mechanical Studies for Abner Shimony, Volume One. 1997 ISBN 0-7923-4452-9
194. R.S. Cohen, M. Horne and J. Stachel (eds.): *Potentiality, Entanglement and Passion-at-a-Distance*. Quantum Mechanical Studies for Abner Shimony, Volume Two. 1997 ISBN 0-7923-4453-7; Set 0-7923-4454-5
195. R.S. Cohen and A.I. Tauber (eds.): *Philosophies of Nature: The Human Dimension*. 1997 ISBN 0-7923-4579-7
196. M. Otte and M. Panza (eds.): *Analysis and Synthesis in Mathematics*. History and Philosophy. 1997 ISBN 0-7923-4570-3
197. A. Denkel: *The Natural Background of Meaning*. 1999 ISBN 0-7923-5331-5
198. D. Baird, R.I.G. Hughes and A. Nordmann (eds.): *Heinrich Hertz: Classical Physicist, Modern Philosopher*. 1999 ISBN 0-7923-4653-X
199. A. Franklin: *Can That be Right?* Essays on Experiment, Evidence, and Science. 1999 ISBN 0-7923-5464-8
200. D. Raven, W. Krohn and R.S. Cohen (eds.): *The Social Origins of Modern Science*. 2000 ISBN 0-7923-6457-0
201. Reserved
202. Reserved
203. B. Babich and R.S. Cohen (eds.): *Nietzsche, Theories of Knowledge, and Critical Theory*. Nietzsche and the Sciences I. 1999 ISBN 0-7923-5742-6
204. B. Babich and R.S. Cohen (eds.): *Nietzsche, Epistemology, and Philosophy of Science*. Nietzsche and the Science II. 1999 ISBN 0-7923-5743-4
205. R. Hooykaas: *Fact, Faith and Fiction in the Development of Science*. The Gifford Lectures given in the University of St Andrews 1976. 1999 ISBN 0-7923-5774-4
206. M. Fehér, O. Kiss and L. Ropolyi (eds.): *Hermeneutics and Science*. 1999 ISBN 0-7923-5798-1
207. R.M. MacLeod (ed.): *Science and the Pacific War*. Science and Survival in the Pacific, 1939-1945. 1999 ISBN 0-7923-5851-1
208. I. Hanzel: *The Concept of Scientific Law in the Philosophy of Science and Epistemology*. A Study of Theoretical Reason. 1999 ISBN 0-7923-5852-X
209. G. Helm; R.J. Deltete (ed./transl.): *The Historical Development of Energetics*. 1999 ISBN 0-7923-5874-0
210. A. Orenstein and P. Kotatko (eds.): *Knowledge, Language and Logic*. Questions for Quine. 1999 ISBN 0-7923-5986-0
211. R.S. Cohen and H. Levine (eds.): *Maimonides and the Sciences*. 2000 ISBN 0-7923-6053-2
212. H. Gourko, D.I. Williamson and A.I. Tauber (eds.): *The Evolutionary Biology Papers of Elie Metchnikoff*. 2000 ISBN 0-7923-6067-2
213. S. D'Agostino: *A History of the Ideas of Theoretical Physics*. Essays on the Nineteenth and Twentieth Century Physics. 2000 ISBN 0-7923-6094-X
214. S. Lelas: *Science and Modernity*. Toward An Integral Theory of Science. 2000 ISBN 0-7923-6303-5
215. E. Agazzi and M. Pauri (eds.): *The Reality of the Unobservable*. Observability, Unobservability and Their Impact on the Issue of Scientific Realism. 2000 ISBN 0-7923-6311-6
216. P. Hoyningen-Huene and H. Sankey (eds.): *Incommensurability and Related Matters*. 2001 ISBN 0-7923-6989-0
217. A. Nieto-Galan: *Colouring Textiles*. A History of Natural Dyestuffs in Industrial Europe. 2001 ISBN 0-7923-7022-8

Boston Studies in the Philosophy of Science

218. J. Blackmore, R. Itagaki and S. Tanaka (eds.): *Ernst Mach's Vienna 1895–1930*. Or Phenomenalism as Philosophy of Science. 2001 ISBN 0-7923-7122-4
219. R. Vihalemm (ed.): *Estonian Studies in the History and Philosophy of Science*. 2001 ISBN 0-7923-7189-5
220. W. Lefèvre (ed.): *Between Leibniz, Newton, and Kant*. Philosophy and Science in the Eighteenth Century. 2001 ISBN 0-7923-7198-4
221. T.F. Glick, M.Á. Puig-Samper and R. Ruiz (eds.): *The Reception of Darwinism in the Iberian World*. Spain, Spanish America and Brazil. 2001 ISBN 1-4020-0082-0
222. U. Klein (ed.): *Tools and Modes of Representation in the Laboratory Sciences*. 2001 ISBN 1-4020-0100-2

Also of interest:

R.S. Cohen and M.W. Wartofsky (eds.): *A Portrait of Twenty-Five Years Boston Colloquia for the Philosophy of Science, 1960-1985*. 1985 ISBN Pb 90-277-1971-3

Previous volumes are still available.

KLUWER ACADEMIC PUBLISHERS – DORDRECHT / BOSTON / LONDON

MIX
Papier aus verantwortungsvollen Quellen
Paper from responsible sources
FSC® C105338

If you have any concerns about our products,
you can contact us on
ProductSafety@springernature.com

In case Publisher is established outside the EU,
the EU authorized representative is:
Springer Nature Customer Service Center GmbH
Europaplatz 3, 69115 Heidelberg, Germany

Printed by Libri Plureos GmbH
in Hamburg, Germany